Geometrical methods of mathematical physics

Geometrical methods of mathematical physics

BERNARD F. SCHUTZ

Reader in General Relativity, University College, Cardiff

CAMBRIDGE
UNIVERSITY PRESS

PUBLISHED BY THE PRESS SYNDICATE OF THE UNIVERSITY OF CAMBRIDGE
The Pitt Building, Trumpington Street, Cambridge, United Kingdom

CAMBRIDGE UNIVERSITY PRESS
The Edinburgh Building, Cambridge CB2 2RU, UK http://www.cup.cam.ac.uk
40 West 20th Street, New York, NY 10011–4211, USA http://www.cup.org
10 Stamford Road, Oakleigh, Melbourne 3166, Australia

First published 1980
Reprinted 1985, 1987, 1988, 1990, 1993, 1999

A catalogue record for this book is available from the British Library

ISBN 0 521 23271 6 hardback
ISBN 0 521 29887 3 paperback

Transferred to digital printing 1999

CONTENTS

	Preface	ix
1	**Some basic mathematics**	1
1.1	The space R^n and its topology	1
1.2	Mappings	5
1.3	Real analysis	9
1.4	Group theory	11
1.5	Linear algebra	13
1.6	The algebra of square matrices	16
1.7	Bibliography	20
2	**Differentiable manifolds and tensors**	23
2.1	Definition of a manifold	23
2.2	The sphere as a manifold	26
2.3	Other examples of manifolds	28
2.4	Global considerations	29
2.5	Curves	30
2.6	Functions on M	30
2.7	Vectors and vector fields	31
2.8	Basis vectors and basis vector fields	34
2.9	Fiber bundles	35
2.10	Examples of fiber bundles	37
2.11	A deeper look at fiber bundles	38
2.12	Vector fields and integral curves	42
2.13	Exponentiation of the operator $\mathrm{d}/\mathrm{d}\lambda$	43
2.14	Lie brackets and noncoordinate bases	43
2.15	When is a basis a coordinate basis?	47
2.16	One-forms	49
2.17	Examples of one-forms	50
2.18	The Dirac delta function	51
2.19	The gradient and the pictorial representation of a one-form	52
2.20	Basis one-forms and components of one-forms	55
2.21	Index notation	56

2.22 Tensors and tensor fields 57
2.23 Examples of tensors 58
2.24 Components of tensors and the outer product 59
2.25 Contraction 59
2.26 Basis transformations 60
2.27 Tensor operations on components 63
2.28 Functions and scalars 64
2.29 The metric tensor on a vector space 64
2.30 The metric tensor field on a manifold 68
2.31 Special relativity 70
2.32 Bibliography 71

3 Lie derivatives and Lie groups 73
3.1 Introduction: how a vector field maps a manifold into itself 73
3.2 Lie dragging a function 74
3.3 Lie dragging a vector field 74
3.4 Lie derivatives 76
3.5 Lie derivative of a one-form 78
3.6 Submanifolds 79
3.7 Frobenius' theorem (vector field version) 81
3.8 Proof of Frobenius' theorem 83
3.9 An example: the generators of S^2 85
3.10 Invariance 86
3.11 Killing vector fields 88
3.12 Killing vectors and conserved quantities in particle dynamics 89
3.13 Axial symmetry 89
3.14 Abstract Lie groups 92
3.15 Examples of Lie groups 95
3.16 Lie algebras and their groups 101
3.17 Realizations and representations 105
3.18 Spherical symmetry, spherical harmonics and representations of the rotation group 108
3.19 Bibliography 112

4 Differential forms 113
A The algebra and integral calculus of forms 113
4.1 Definition of volume – the geometrical role of differential forms 113
4.2 Notation and definitions for antisymmetric tensors 115
4.3 Differential forms 117
4.4 Manipulating differential forms 119
4.5 Restriction of forms 120
4.6 Fields of forms 120

4.7 Handedness and orientability 121
4.8 Volumes and integration on oriented manifolds 121
4.9 N-vectors, duals, and the symbol $\epsilon_{ij\ldots k}$ 125
4.10 Tensor densities 128
4.11 Generalized Kronecker deltas 130
4.12 Determinants and $\epsilon_{ij\ldots k}$ 131
4.13 Metric volume elements 132
B The differential calculus of forms and its applications 134
4.14 The exterior derivative 134
4.15 Notation for derivatives 135
4.16 Familiar examples of exterior differentiation 136
4.17 Integrability conditions for partial differential equations 137
4.18 Exact forms 138
4.19 Proof of the local exactness of closed forms 140
4.20 Lie derivatives of forms 142
4.21 Lie derivatives and exterior derivatives commute 143
4.22 Stokes' theorem 144
4.23 Gauss' theorem and the definition of divergence 147
4.24 A glance at cohomology theory 150
4.25 Differential forms and differential equations 152
4.26 Frobenius' theorem (differential forms version) 154
4.27 Proof of the equivalence of the two versions of Frobenius' theorem 157
4.28 Conservation laws 158
4.29 Vector spherical harmonics 160
4.30 Bibliography 161

5 Applications in physics 163
A Thermodynamics 163
5.1 Simple systems 163
5.2 Maxwell and other mathematical identities 164
5.3 Composite thermodynamic systems: Caratheodory's theorem 165
B Hamiltonian mechanics 167
5.4 Hamiltonian vector fields 167
5.5 Canonical transformations 168
5.6 Map between vectors and one-forms provided by $\tilde{\omega}$ 169
5.7 Poisson bracket 170
5.8 Many-particle systems: symplectic forms 170
5.9 Linear dynamical systems: the symplectic inner product and conserved quantities 171
5.10 Fiber bundle structure of the Hamiltonian equations 174
C Electromagnetism 175
5.11 Rewriting Maxwell's equations using differential forms 175

5.12 Charge and topology 179
5.13 The vector potential 180
5.14 Plane waves: a simple example 181
D Dynamics of a perfect fluid 181
5.15 Role of Lie derivatives 181
5.16 The comoving time-derivative 182
5.17 Equation of motion 183
5.18 Conservation of vorticity 184
E Cosmology 186
5.19 The cosmological principle 186
5.20 Lie algebra of maximal symmetry 190
5.21 The metric of a spherically symmetric three-space 192
5.22 Construction of the six Killing vectors 195
5.23 Open, closed, and flat universes 197
5.24 Bibliography 199

6 Connections for Riemannian manifolds and gauge theories 201
6.1 Introduction 201
6.2 Parallelism on curved surfaces 201
6.3 The covariant derivative 203
6.4 Components: covariant derivatives of the basis 205
6.5 Torsion 207
6.6 Geodesics 208
6.7 Normal coordinates 210
6.8 Riemann tensor 210
6.9 Geometric interpretation of the Riemann tensor 212
6.10 Flat spaces 214
6.11 Compatibility of the connection with volume-measure or the metric 215
6.12 Metric connections 216
6.13 The affine connection and the equivalence principle 218
6.14 Connections and gauge theories: the example of electromagnetism 219
6.15 Bibliography 222

Appendix: solutions and hints for selected exercises 224

Notation 244

Index 246

PREFACE

Why study geometry?

This book aims to introduce the beginning or working physicist to a wide range of analytic tools which have their origin in differential geometry and which have recently found increasing use in theoretical physics. It is not uncommon today for a physicist's mathematical education to ignore all but the simplest geometrical ideas, despite the fact that young physicists are encouraged to develop mental 'pictures' and 'intuition' appropriate to physical phenomena. This curious neglect of 'pictures' of one's mathematical tools may be seen as the outcome of a gradual evolution over many centuries. Geometry was certainly extremely important to ancient and medieval natural philosophers; it was in geometrical terms that Ptolemy, Copernicus, Kepler, and Galileo all expressed their thinking. But when Descartes introduced coordinates into Euclidean geometry, he showed that the study of geometry could be regarded as an application of algrebra. Since then, the importance of the study of geometry in the education of scientists has steadily declined, so that at present a university undergraduate physicist or applied mathematician is not likely to encounter much geometry at all.

One reason for this suggests itself immediately: the relatively simple geometry of the three-dimensional Euclidean world that the nineteenth-century physicist believed he lived in can be mastered quickly, while learning the great diversity of analytic techniques that must be used to solve the differential equations of physics makes very heavy demands on the student's time. Another reason must surely be that these analytic techniques were developed at least partly in response to the profound realization by physicists that the laws of nature could be expressed as differential equations, and this led most mathematical physicists genuinely to neglect geometry until relatively recently.

However, two developments in this century have markedly altered the balance between geometry and analysis in the twentieth-century physicist's outloook. The first is the development of the theory of relativity, according to which the Euclidean three-space of the nineteenth-century physicist is only an approximation to the correct description of the physical world. The second development, which is only beginning to have an impact, is the realization by twentieth-century mathematicians, led by Cartan, that the relation between geometry and analysis is a two-way street: on the one hand analysis may be the foundation of the study of geometry, but on the other hand the study of geometry leads naturally to the development of certain analytic tools (such as the Lie derivative and the exterior calculus) and certain concepts (such as the manifold, the fiber bundle, and the identification of vectors with derivatives) that have great power in applications of analysis. In the modern view, geometry remains subsidiary to analysis. For example, the basic concept of differential geometry, the differentiable manifold, is defined in terms of real numbers and differentiable functions. But this is no disadvantage: it means that concepts from analysis can be expressed geometrically, and this has considerable heuristic power.

Because it has developed this intimate connection between geometrical and analytic ideas, modern differential geometry has become more and more important in theoretical physics, where it has led to a greater simplicity in the mathematics and a more fundamental understanding of the physics. This revolution has affected not only special and general relativity, the two theories whose content is most obviously geometrical, but other fields where the geometry involved is not always that of physical space but rather of a more abstract space of variables: electromagnetism, thermodynamics, Hamiltonian theory, fluid dynamics, and elementary particle physics.

Aims of this book

In this book I want to introduce the reader to some of the more important notions of twentieth-century differential geometry, trying always to use that geometrical or 'pictorial' way of thinking that is usually so helpful in developing a physicist's intuition. The book attempts to teach mathematics, not physics. I have tried to include a wide range of applications of this mathematics to branches of physics which are familiar to most advanced undergraduates. I hope these examples will do more than illustrate the mathematics: the new mathematical formulation of familiar ideas will, if I have been successful, give the reader a deeper understanding of the physics.

I will discuss the background I have assumed of the reader in more detail below, but here it may be helpful to give a brief list of some of the 'familiar' ideas which are seen in a new light in this book: vectors, tensors, inner products, special relativity, spherical harmonics and the rotation group (and angular-momentum operators), conservation laws, volumes, theory of integration, curl and cross-product, determinants of matrices, partial differential equations and their integrability conditions, Gauss' and Stokes' integral theorems of vector calculus, thermodynamics of simple systems, Caratheodory's theorem (and the second law of thermodynamics), Hamiltonian systems in phase space, Maxwell's equations, fluid dynamics (including the laws governing the conservation of circulation), vector calculus in curvilinear coordinate systems, and the quantum theory of a charged scalar field. Besides these more or less familiar subjects, there are a few others which are not usually taught at undergraduate level but which most readers would certainly have heard of: the theory of Lie groups and symmetry, open and closed cosmologies, Riemannian geometry, and gauge theories of physics. That all of these subjects can be studied by the methods of differential geometry is an indication of the importance differential geometry is likely to have in theoretical physics in the future.

I believe it is important for the reader to develop a pictorial way of thinking and a feeling for the 'naturalness' of certain geometrical tools in certain situations. To this end I emphasize repeatedly the idea that tensors are geometrical objects, defined independently of any coordinate system. The role played by components and coordinate transformations is submerged into a secondary position: whenever possible I write equations without indices, to emphasize the coordinate-independence of the operations. I have made no attempt to present the material in a strictly rigorous or axiomatic way, and I have had to ignore many aspects of our subject which a mathematician would regard as fundamental. I do, of course, give proofs of all but a handful of the most important results (references for the exceptions are provided), but I have tried wherever possible to make the main geometrical ideas in the proof stand out clearly from the background of manipulation. I want to show the beauty, elegance, and naturalness of the mathematics with the minimum of obscuration.

How to use this book

The first chapter contains a review of the sort of elementary mathematics assumed of the reader plus a short introduction to some concepts, particularly in topology, which undergraduates may not be familiar with. The next chapters are the core of the book: they introduce tensors, Lie derivatives, and differential forms. Scattered through these chapters are some applications, but most of the physical applications are left for systematic treatment in chapter 5. The final chapter, on Riemannian geometry, is more advanced and makes contact with areas of particle physics and general relativity in which differential geometry is an everyday tool.

The material in this book should be suitable for a one-term course, provided the lecturer exercises some selection in the most difficult areas. It should also be possible to teach the most important points as a unit of, say, ten lectures in an advanced course on mathematical methods. I have taught such a unit to graduate students, concentrating mainly on §§ 2.1–2.3, 2.5–2.8, 2.12–2.14, 2.16, 2.17, 2.19–2.28, 3.1–3.13, 4.1–4.6, 4.8, 4.14–4.18, 4.20–4.23, 4.25, 4.26, 5.1, 5.2, 5.4–5.7, and 5.15–5.18. I hope lecturers will experiment with their own choices of material, especially because there are many people for whom geometrical reasoning is easier and more natural than purely analytic reasoning, and for them an early exposure to geometrical ideas can only be helpful. As a general guide to selecting material, section headings within chapters are printed in two different styles. Fundamental material is marked by **boldface** headings, while more advanced or supplementary topics are marked by ***boldface italics***. All of the last chapter falls into this category. The same convention of type-face distinguishes those exercises which are central to the development of the mathematics from those which are peripheral.

The exercises form an integral part of the book. They are inserted in the middle of the text, and they are designed to be worked when they are first encountered. Usually the text after an exercise will assume that the reader has worked and understood the exercise. The reader who does not have the time to work an exercise should nevertheless read it and try to understand its result. Hints and some solutions will be found at the end of the book.

Background assumed of the reader

Most of this book should be understandable to an advanced undergraduate or beginning graduate student in theoretical physics or applied mathematics. It presupposes reasonable facility with vector calculus, calculus of many variables, matrix algebra (including eigenvectors and determinants), and a little operator theory of the sort one learns in elementary quantum mechanics. The physical applications are drawn from a variety of fields, and not everyone will feel at home with them all. It should be possible to skip many sections on physics without undue loss of continuity, but it would probably be unrealistic to attempt this book without some familiarity with classical mechanics, special relativity, and electromagnetism. The bibliography at the end of chapter 1 lists some books which provide suitable background.

I want to acknowledge my debt to the many people, both colleagues and teachers, who have helped me to appreciate the beauty of differential geometry and understand its usefulness in physics. I am especially indebted to Kip Thorne, Rafael Sorkin, John Friedman, and Frank Estabrook. I also want to thank the first two and many patient students at University College, Cardiff, for their comments on earlier versions of this book. Two of my students, Neil Comins and Brian Wade, deserve special mention for their careful and constructive suggestions. It is also a pleasure to thank Suzanne Ball, Jane Owen, and Margaret Wilkinson for their fast and accurate typing of the manuscript through all its revisions. Finally, I thank my wife for her patience and encouragement, particularly during the last few hectic months.

Cardiff, 30 June 1979

Bernard Schutz

1 SOME BASIC MATHEMATICS

This chapter reviews the elementary mathematics upon which the geometrical development of later chapters relies. Most of it should be familiar to most readers, but we begin with two topics, topology and mappings, which many readers may find unfamiliar. The principal reason for including them is to enable us to define precisely what is meant by a manifold, which we do early in chapter 2. Readers to whom topology is unfamiliar may wish to skip the first two sections initially and refer back to them only after chapter 2 has given them sufficient motivation.

1.1 The space R^n and its topology

The space R^n is the usual n-dimensional space of vector algebra: a point in R^n is a sequence of n real numbers $(x_1, x_2, \ldots, x_n)$, also called an *n-tuple* of real numbers. Intuitively we have the idea that this is a *continuous* space, that there are points of R^n arbitrarily close to any given point, that a line joining any two points can be subdivided into arbitrarily many pieces that also join points of R^n. These notions are in contrast to properties we would ascribe to, say, a lattice, such as the set of all n-tuples of integers $(i_1, i_2, \ldots, i_n)$. The concept of continuity in R^n is made precise in the study of its *topology*. The word 'topology' has two distinct meanings in mathematics. The one we are discussing now may be called *local topology*. The other is *global topology*, which is the study of large-scale features of the space, such as those which distinguish the sphere from the torus. We shall have something to say about global topology later, particularly in the chapter on differential forms. But first we must take a brief look at local topology.

The fundamental concept is that of a neighborhood of a point in R^n, which we can define after introducing a *distance function* between any two points $\mathbf{x} = (x_1, \ldots, x_n)$ and $\mathbf{y} = (y_1, \ldots, y_n)$ of R^n:

$$d(\mathbf{x}, \mathbf{y}) = [(x_1 - y_1)^2 + (x_2 - y_2)^2 + \ldots + (x_n - y_n)^2]^{1/2}. \qquad (1.1)$$

A neighborhood of radius r of the point $\mathbf{x}$ in R^n is the set of points $N_r(\mathbf{x})$ whose distance from $\mathbf{x}$ is less than r. For R^2 this is illustrated in figure 1.1. The continuity of the space can now be more precisely defined by considering very small neighborhoods. A set of points of R^n is *discrete* if each point has a neighborhood which contains no other points of the set. Clearly R^n itself is not discrete. A set of points S of R^n is said to be *open* if every point $\mathbf{x}$ in S has a neighborhood entirely contained in S. Clearly, discrete sets are not open, and from now on we will have no use for discrete sets. A simple example of an open set in R^1 (also known simply as R) is all points x for which $a < x < b$ for two real numbers a and b. An important thing to understand is that the set of points for which $a \leqslant x < b$ is *not* open, because the point $x = a$ does not have a neighborhood entirely contained in the set: some points of *any* neighborhood of $x = a$ must be less than a and therefore outside the set. This is illustrated in figure 1.2. This is, of course, a very general property: any reasonable 'chunk' of R^n will be open if we do not include the boundary of the chunk in the set.

Fig. 1.1. The distance function $d(\mathbf{x}, \mathbf{y})$ defines a neighborhood in R^2 which is the interior of the disc bounded by the circle of radius r. The circle itself is not part of this neighborhood.

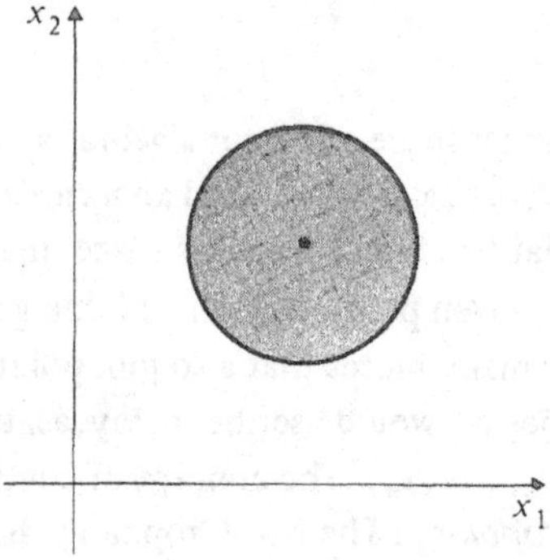

Fig. 1.2. (a) Any neighborhood of the point $x = a$ must include points to the left of a, while (b) any point to the right of a has a neighborhood entirely to the right of a.

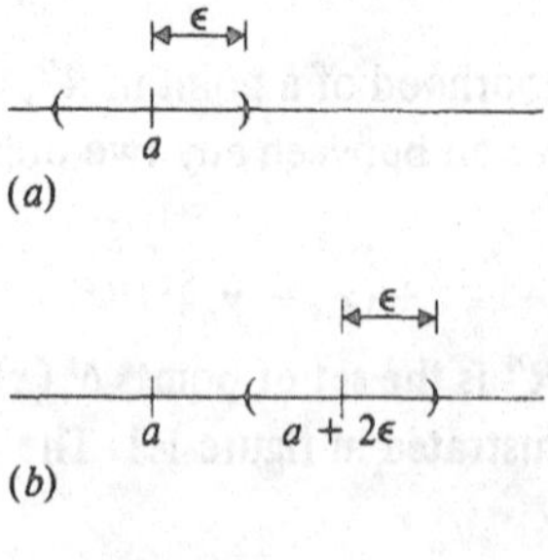

The idea that a line joining any two points of R^n can be infinitely subdivided can be made more precise by saying that any two points of R^n have neighborhoods which do not intersect. (They will also have some neighborhoods which *do* intersect, but if we choose small enough neighborhoods we can make them disjoint.) This is called the *Hausdorff property* of R^n. It is possible to construct non-Hausdorff spaces, but for our purposes they are artificial and we shall ignore them.

Notice that we have used the distance function $d(\mathbf{x}, \mathbf{y})$ to define neighborhoods and thereby open sets. We say that $d(\mathbf{x}, \mathbf{y})$ has *induced a topology* on R^n. By this we mean that it has enabled us to define open sets of R^n which have the properties:

(Ti) if O_1 and O_2 are open, so is their intersection, $O_1 \cap O_2$; and

(Tii) the union of any collection (possibly infinite in number) of open sets is open.

In order to make (Ti) apply to all open sets of R^n, we *define* the empty set (or null set) to be open, and in order to make (Tii) work we likewise define R^n itself to be open. (In more advanced treatments one defines a *topological space* to be a collection of points with a definition of open sets satisfying (Ti) and (Tii). In this sense the distance function $d(\mathbf{x}, \mathbf{y})$ has enabled us to make R^n into a topological space.)

At this point we must ask whether the induced topology depends very much on the precise form of $d(\mathbf{x}, \mathbf{y})$. Suppose, for example, that we use a different distance function

$$d'(\mathbf{x}, \mathbf{y}) = [4(x_1 - y_1)^2 + (x_2 - y_2)^2 + \ldots + (x_n - y_n)^2]^{1/2}. \qquad (1.2)$$

This also defines neighborhoods and open sets, as shown in figure 1.3 for R^2.

Fig. 1.3. The distance function $d'(\mathbf{x}, \mathbf{y}) = [4(x_1 - y_1)^2 + (x_2 - y_2)^2]^{1/2}$ defines a neighborhood in R^2 which is the interior of the disc bounded by the ellipse $4(x_1 - y_1)^2 + (x_2 - y_2)^2 = r^2$. As in figure 1.1, the ellipse itself is not in the neighborhood.

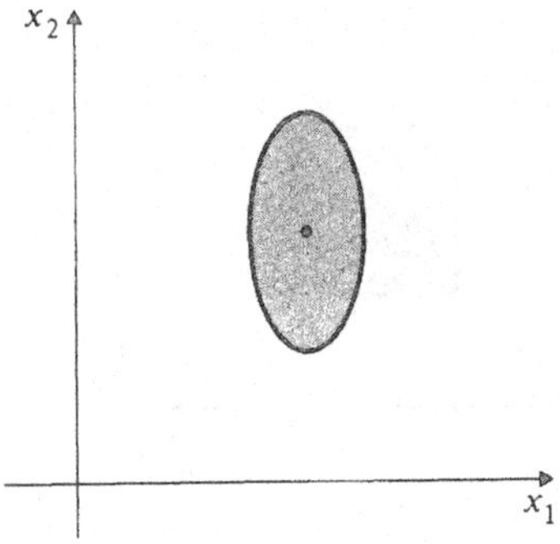

The key point is that any set which is open according to $d'(\mathbf{x}, \mathbf{y})$ is also open according to $d(\mathbf{x}, \mathbf{y})$, and vice versa. The proof of this is not hard, and it rests on the fact that any given d-type neighborhood of $\mathbf{x}$ contains a d'-type neighborhood entirely within it, and vice versa. That is, given a d-type neighborhood of radius ϵ about $\mathbf{x}$, one can choose a number δ so small that a d'-type neighborhood of $\mathbf{x}$ of radius δ is entirely within the original (see figure 1.4). So we can conclude that if a set is open as defined by $d(\mathbf{x}, \mathbf{y})$ it is also open as defined by $d'(\mathbf{x}, \mathbf{y})$, and vice versa. We therefore say that both d and d' induce the same topology on R^n. The reader may wish to show that the distance functions

Fig. 1.4. In R^2 a d-neighborhood of radius ϵ (bounded by the circle) entirely contains a d'-neighborhood of radius δ (bounded by the ellipse defined in figure 1.3) if $\delta < \epsilon$. If $\delta > 2\epsilon$ the inclusion is reversed.

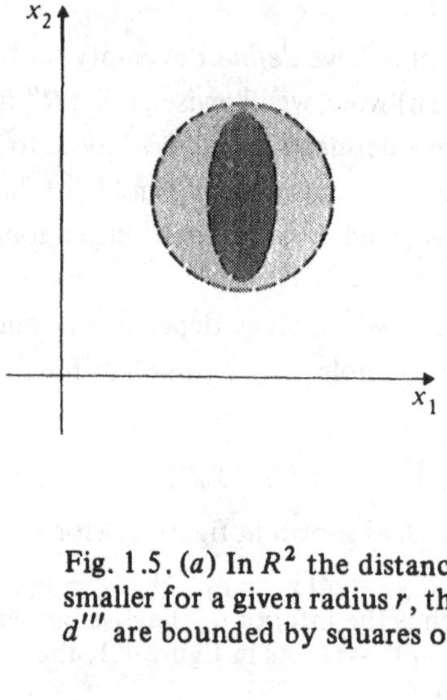

Fig. 1.5. (a) In R^2 the distance function d'' has circular neighborhoods smaller for a given radius r, than those of d. (b) The neighborhoods of d''' are bounded by squares of side $2r$.

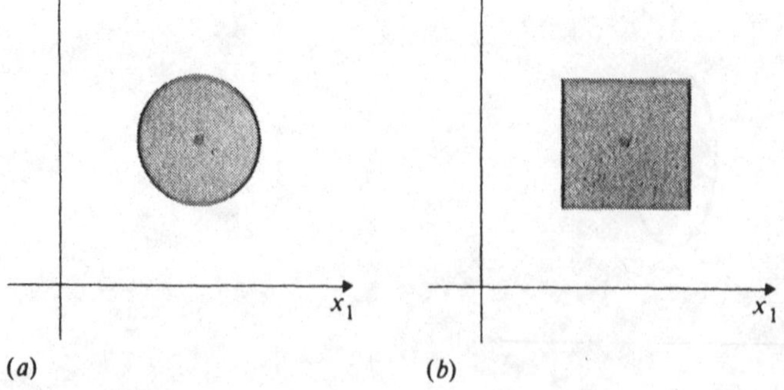

$$d''(\mathbf{x}, \mathbf{y}) = \exp[d(\mathbf{x}, \mathbf{y})] - 1, \tag{1.3}$$

$$d'''(\mathbf{x}, \mathbf{y}) = \text{maximum}\,(|x_1 - y_1|, |x_2 - y_2|, \ldots, |x_n - y_n|) \tag{1.4}$$

also induce the same topology. Their neighborhoods in R^2 are illustrated in figure 1.5. So although we began with the usual Euclidean distance function $d(\mathbf{x}, \mathbf{y})$, the topology we have defined is not very dependent on the form of d. This is called the 'natural' topology of R^n. Topology is a more 'primitive' concept than distance. We do not need to know the actual distance between points, since many different distance definitions will do. What we need is only a notion that the distance between points can be made arbitrarily small and that no two distinct points have zero distance between them.

Our definition of a neighborhood was tied to a particular distance function, but because the topology of a manifold is more general than any particular distance function the word 'neighborhood' is often used in a different sense. We will often find it convenient to let a neighborhood of a point $\mathbf{x}$ be any set containing an open set containing $\mathbf{x}$. It should always be clear from the context which sense of 'neighborhood' is intended.

1.2 Mappings

The concept of a mapping, simple though it is, will be so useful later that it is well to spend some time discussing it. A map f from a space M to a space N is a rule which associates with an element x of M a unique element y of N. It is useful to keep in one's mind a general picture of a map, such as figure 1.6. The simplest example of a map is an ordinary real-valued function on R. The function f associates a point x in R with a point $f(x)$ also in R. (This illustrates the fact that the spaces M and N need not be distinct.) Such a map is shown in the usual way in figure 1.7. Notice that the map gives a unique $f(x)$ for every x, but not necessarily a unique x for every $f(x)$. In the figure, both x_0 and

Fig. 1.6. A pictorial representation of the mapping f: $M \to N$ showing $x \mapsto f(x)$.

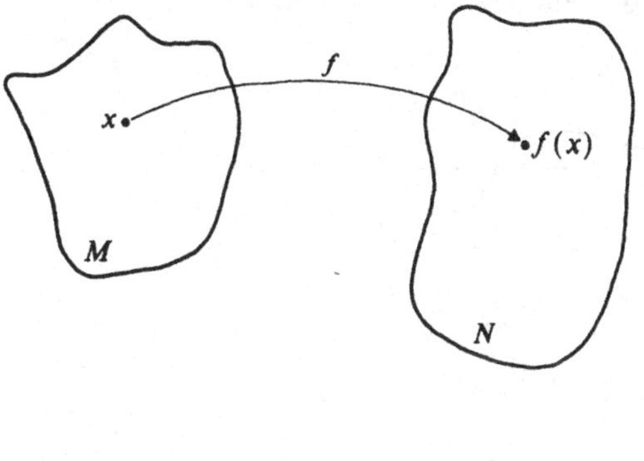

x_1 map into the same value. Such a map is called *many-to-one*. More generally, if f maps M to N then for any set S in M the elements in N mapped from points of S form a set T called the *image* of S under f, denoted by $f(S)$. Conversely, the set S is called the *inverse image* of T, denoted by $f^{-1}(T)$. If the map is many-to-one then the inverse image of a single point of N is not a single point of M, so there is no *map* f^{-1} from N to M, since every map must have a unique image. So in general the symbol $f^{-1}(T)$ must be read as a single symbol: it is not the image of T under a map f^{-1} but simply a set called $f^{-1}(T)$. On the other hand, if every point in $f(S)$ has a unique inverse image point in S, then f is said to be *one-to-one* (abbreviated 1–1) and there does exist another 1–1 map f^{-1}, called the *inverse* of f, which maps the image of M to M. These concepts, if not the words used to describe them, are familiar from elementary calculus. The function $f(x) = \sin x$ is many-to-one, since $f(x) = f(x + 2n\pi) = f((2n + 1)\pi - x)$ for any integer n. Therefore, a true inverse function does not exist. The usual inverse function, arcsin y or $\sin^{-1} y$, is obtained by restricting the original sine function to the 'principal' values, $-\pi/2 < x \leqslant \pi/2$, on which it is indeed 1–1 and invertible.

Another example of a 1–1 map is a geographical map of part of the Earth's surface: this maps a point of the Earth's surface to a point of a piece of paper. Yet another map is a rotation of a sphere about some diameter: this maps a point of the sphere to another one a fixed angular distance away as measured about the axis of rotation.

We shall now introduce some standard notation and terminology regarding maps. The statement that f maps M to N is abbreviated $f: M \rightarrow N$. The statement that f maps a particular element x of M to y of N has its own special notation, $f: x \mapsto y$. If the name of a map is f, the image of a point x is $f(x)$. When the map is a real-valued function of, say, n variables (so $f: R^n \rightarrow R$), it is conventional among physicists to use the symbol $f(x)$ to denote both the value of f on x and the function itself. When there is no chance of confusion we will follow that convention. If we have two maps, f and g, $f: M \rightarrow N$ and $g: N \rightarrow P$, then there is a

Fig. 1.7. A many-to-one map (function) of R to R.

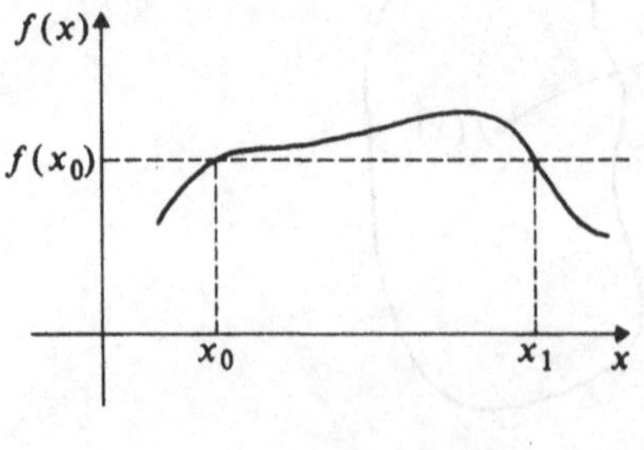

map called the *composition* of f and g, denoted by $g \circ f$, which maps M to P ($g \circ f$: $M \to P$). This is defined in the obvious way: take a point x of M, find the point $f(x)$ of N, and use g to map it to P: $(g \circ f)(x) = g(f(x))$. It is conventional to write the composition $g \circ f$ in such a way that the map acting first is the one on the right.

If a map is defined for all points of M, then we say it is a mapping from M *into* N. If, in addition, every point of N has an inverse image (not necessarily a unique one), we say it is a mapping from M *onto* N. As mentioned above, if the inverse image is unique, the map is one-to-one. (A map which is both 1–1 and onto is called a bijection.) As an example, let N be the unit open disc in R^2, the set of all points x for which $d(\mathbf{x}, \mathbf{0}) < 1$ (where $\mathbf{0}$ is the origin of R^2). Let M be the surface of the hemisphere $\theta < \pi/2$ of the unit sphere (see figure 1.8). Clearly there is a map f which is 1–1 from M onto N.

The terminology of mapping theory, combined with what we have learned of topology, enables us to give a useful and compact definition of a continuous function, or in fact of any continuous map. A map $f: M \to N$ is *continuous at x* in M if any open set of N containing $f(x)$ contains the image of an open set of M containing x. (This presupposes, of course, that M and N are topological spaces. Otherwise continuity has no meaning.) More generally, f is *continuous on M* (or, simply, continuous) if it is continuous at all points of M. Let us see how this is related to the usual elementary calculus definition of a continuous function.

Suppose f is a real-valued function of one real variable. That is, f is a map of R to R, taking a number x to a number $f(x)$. (In the notation above, $f: R \to R$.) Then in the elementary calculus view f is continuous at a point x_0 if for every $\epsilon > 0$ there exists a $\delta > 0$ such that $|f(x) - f(x_0)| < \epsilon$ for all x for which $|x - x_0| < \delta$ (see figure 1.9). To re-express this in terms of open sets, notice that for R the distance function $d'''(\mathbf{x}, \mathbf{x}_0)$ defined in §1.1 above just reduces to

Fig. 1.8. By imagining the disc to be the equatorial section of the ball bounded by the sphere, one constructs a simple map by projecting perpendicular to the disc.

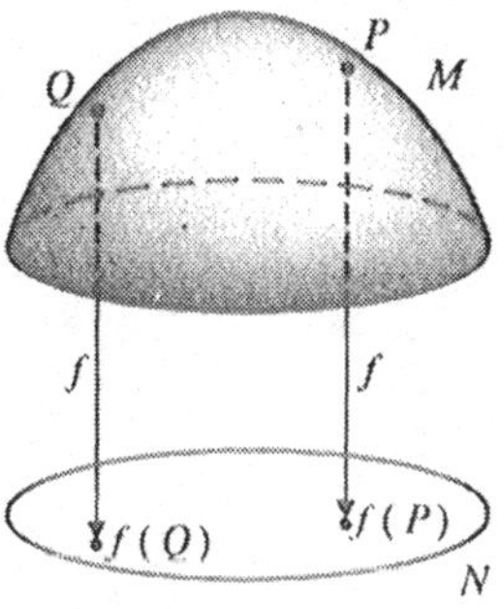

$d'''(x, x_0) = |x - x_0|$. Therefore this definition says that f is continuous at x_0 if every d'''-neighborhood of $f(x_0)$ contains the image of a d'''-neighborhood of x_0. Since these neighborhoods are open sets, the new definition of continuity given in the previous paragraph contains the elementary-calculus definition as a special case. Conversely, the elementary-calculus definition implies the other because any open set in R containing $f(x_0)$ contains a d'''-neighborhood of $f(x_0)$, which in turn contains the image of an open set containing x_0 (namely, that of a d'''-neighborhood of x_0). The two definitions are equivalent.

The condition that a map be continuous on all of M is even easier to phrase, because it is a theorem that $f: M \to N$ is continuous if and only if the inverse image of every open set of N is open in M. The proof of this is not difficult. If f is continuous at all x, then the inverse image of any open set is open because it contains an open set containing each point in the inverse image. Conversely, if the inverse image of every open set of N is open, then it contains an open set about any of its points, so f is continuous at each of these points.

The open-set definition of continuity is much easier to use and to understand than the ϵ–δ one, particularly for functions of more than one variable, and it is of course the only possible definition applicable to general maps between topological spaces.

Having defined continuity, we can go on to define differentiation of functions in the usual way. If $f(x_1, \ldots, x_n)$ is a function defined on some open region S of R^n, then it is said to be *differentiable of class* C^k if all its partial derivatives of order less than or equal to k exist and are continuous functions on S. As a shorthand, such a function f is said to be a C^k function. Special cases are C^0 (a continuous function) and C^∞ (a function, all of whose derivatives exist:

Fig. 1.9. A continuous function, as defined in the text. Notice that f maps the neighborhood of x_0 of radius δ into that of $f(x_0)$ of radius ϵ, while the inverse image of the latter neighborhood includes the former but is not necessarily identical to it. It can contain other regions of the x-axis, such as the one on the right. If f is continuous in this second neighborhood as well then the inverse image will be an open set.

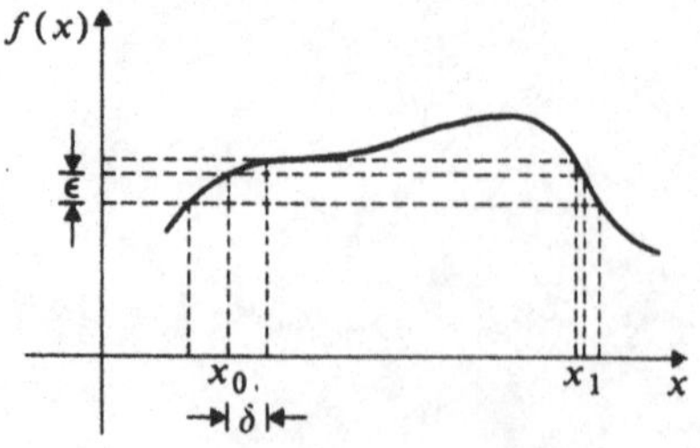

usually called an infinitely differentiable function). Obviously, a function of class C^k is also of class C^j for all $0 \leqslant j < k$. It is also possible to define derivatives of more general continuous maps. These are usually called the differentials of the map. The interested reader may consult Choquet-Bruhat, DeWitt-Morette & Dillard-Bleick (1977), or Warner (1971) in the bibliography.

If f is a 1–1 map of an open set M of R^n onto another open set N of R^n, it can be expressed concretely as

$$y_i = f_i(x_1, x_2, \ldots, x_n), \quad \text{or } \mathbf{y} = \mathbf{f}(\mathbf{x}),$$

where $\{x_i, i = 1, \ldots, n\}$ define a point $\mathbf{x}$ of M and $\{y_i, i = 1, \ldots, n\}$ likewise define a point $\mathbf{y}$ of N. If the functions $\{f_i, i = 1, \ldots, n\}$ are all C^k-differentiable, then the map is said to be C^k-differentiable. The *Jacobian matrix* of a C^1 map is the matrix of partial derivatives $\partial f_i/\partial x_j$. The determinant of this matrix is simply called the *Jacobian*, J, and is often denoted by

$$J = \partial(f_1, \ldots, f_n)/\partial(x_1, \ldots, x_n). \tag{1.5}$$

If the Jacobian at a point $\mathbf{x}$ is nonzero, then the *inverse function theorem* assures us that the map f is 1–1 and onto in some neighborhood of $\mathbf{x}$ (see Choquet-Bruhat *et al.*, 1977, for a proof).

If a function $g(x_1, \ldots, x_n)$ is mapped into a function $g_*(y_1, \ldots, y_n)$ by the rule

$$g_*(f_1(x_1, \ldots, x_n), \ldots, f_n(x_1, \ldots, x_n)) = g(x_1, \ldots, x_n)$$

(that is, g_* has the same value at $\mathbf{f}(\mathbf{x})$ as g has at $\mathbf{x}$), then the integral of g over M equals the integral of $g_* J$ over N:

$$\int_M g(x_1, \ldots, x_n)\mathrm{d}x_1 \ldots \mathrm{d}x_n = \int_N g_*(y_1, \ldots, y_n) J \,\mathrm{d}y_1 \ldots \mathrm{d}y_n. \tag{1.6}$$

Since g and g_* have the same value at appropriate points, it is often said that the volume-element $\mathrm{d}x_1 \ldots \mathrm{d}x_n$ has changed to $J\,\mathrm{d}y_1 \ldots \mathrm{d}y_n$. This is a particularly useful point of view if the map f is viewed as a coordinate change. While this should be familiar to readers from the calculus of many variables, we will examine it in more detail in §2.25 and §4.8.

1.3 Real analysis

As just mentioned, it is assumed that the reader is familiar with the calculus of many variables. In this section we will cover just a few important points.

A real function of a single real variable, $f(x)$, is said to be *analytic* at $x = x_0$ if it has a *Taylor expansion* about x_0 which converges to $f(x)$ in some neighborhood of x_0:

$$f(x) = f(x_0) + (x - x_0)\left(\frac{\mathrm{d}f}{\mathrm{d}x}\right)_{x_0} + \frac{1}{2}(x - x_0)^2\left(\frac{\mathrm{d}^2f}{\mathrm{d}x^2}\right)_{x_0} + \frac{1}{3!}(x - x_0)^3\left(\frac{\mathrm{d}^3f}{\mathrm{d}x^3}\right)_{x_0} + \ldots \tag{1.7}$$

Naturally, functions which are not infinitely differentiable at x_0 (i.e. for which $(\mathrm{d}^nf/\mathrm{d}x^n)_{x_0}$ does not exist for some n) are not analytic. But there are infinitely-differentiable functions which are not analytic. A famous example is $\exp(-1/x^2)$, whose value and all of whose derivatives are zero at $x = 0$, but which is not identically zero in any neighborhood of $x = 0$. (This is explained by the fact that the analytic extension of this function into the complex plane has an essential singularity at $z = 0$; nevertheless, it is perfectly well-behaved on the real line.) However, it is reassuring to know that analytic functions are good approximations to many nonanalytic functions in the following sense. A real-valued function $g(x_1, \ldots, x_n)$ defined on an open region S of R^n is said to be *square-integrable* if the multiple integral

$$\int_S [g(x_1, \ldots, x_n)]^2 \,\mathrm{d}x_1\,\mathrm{d}x_2 \ldots \mathrm{d}x_n \tag{1.8}$$

exists. It is a theorem of functional analysis that any square-integrable function g may be approximated by an analytic function g' in such a way that the integral of $(g - g')^2$ over S may be made as small as one wishes. For this reason physicists typically do not hesitate to assume that a given function is analytic if this helps to establish a result, and we will follow this practice on occasion. Since a C^∞ function need not be analytic, there is a special notation for analytic functions: C^ω. Naturally, a C^ω function is C^∞.

An *operator* A on functions defined on R^n is a map which takes one function f into another one, $A(f)$. If $A(f)$ is just gf, where g is another function, then the operator is simply multiplicative. Other operators on functions on R might be simple differentiation,

$$D(f) = \partial f/\partial x,$$

for example, or integration using a fixed kernel function g,

$$(G(f))(x) = \int_0^x f(y)g(x,y)\,\mathrm{d}y,$$

or a more complicated operation like

$$E(f) = f^2 + \partial^3 f/\partial x^3.$$

In each case the operator may or may not be defined on all functions f. For example, D may not be defined on a function which is not C^1, while G is undefined on functions which give unbounded integrals. Specifying the set of functions on which an operator is allowed to act in fact forms part of the definition of the operator; this set is called its *domain*.

The *commutator* of two operators A and B, called $[A,B]$, is another operator defined by

$$[A,B](f) = (AB-BA)(f) = A(B(f))-B(A(f)). \tag{1.9}$$

If two operators have vanishing commutator, they are said to *commute*. Here one has to be careful about the domains of the operators: the domain of $[A,B]$ may not be as large as that of either A or B. For example, if $A = \mathrm{d}/\mathrm{d}x$ and $B = x\,\mathrm{d}/\mathrm{d}x$, then we we may take both their domains to be all C^1 functions. But for not all C^1 functions will the successive operator $A(B(f))$ be defined, since it involves second derivatives. The operators AB and BA can be given all C^2 functions as domains, and this is a smaller set than C^1 functions. Then, at least at first, the commutator $[A,B]$ also has only C^2 functions in its domain. We can enlarge the domain (also called extending the operator) in this case, though not always, by the following observation. It is easy to work out that on any C^2 function f

$$[A,B](f) = [\mathrm{d}/\mathrm{d}x, x\,\mathrm{d}/\mathrm{d}x]f = \mathrm{d}f/\mathrm{d}x,$$

because the second derivatives in AB and BA cancel out. So we can identify $[A, B]$ simply with $\mathrm{d}/\mathrm{d}x$ (i.e. with A itself) and thereby extend its domain to all C^1 functions. The lesson is that the commutator may be defined even on functions on which the products in the commutator are not. When dealing with differential operators it is often best, at least initially, to define their domain to be C^∞ functions. This will be our approach later, as it eliminates the need to worry about domains.

1.4 Group theory

A collection of elements G together with a binary operation (called $\cdot$) is a group if it satisfies the axioms:

(Gi) Associativity: if x, y, and z are in G, then

$$x\cdot(y\cdot z) = (x\cdot y)\cdot z.$$

(Gii) Right identity: G contains an element e such that, for any x in G,

$$x\cdot e = x.$$

(Giii) Right inverse: for every x in G there is an element called x^{-1}, also in G, for which

$$x\cdot x^{-1} = e.$$

A group is Abelian (commutative) if in addition

(Giv) $x\cdot y = y\cdot x$ for all x, y in G.

A familiar example of a group with a finite set of elements is the group of all permutations of n objects; the binary composition of two permutations is simply the permutation obtained by following one permutation by the other. This group has $n!$ elements. Its identity element is the 'permutation' which leaves all objects fixed.

We should note that a few simple conclusions can be deduced from (Gi)–(Giii): the identity element e is unique; it is also a left-identity ($e \cdot x = x$); the inverse element x^{-1} is unique for any x; and it is also a left-inverse ($x^{-1} \cdot x = e$). It is common to omit the symbol $\cdot$ when there is no risk of confusion: $x \cdot y$ is simply xy.

The most important kind of group in modern physics is the Lie group, about which we will have much to say later. We will give a precise definition in chapter 2, but here it is enough to say that it is a *continuous* group: any open set of elements of a Lie group has a 1–1 map onto an open set of R^n for some n. An example of a Lie group is the translation group of R^n ($\mathbf{x} \rightarrow \mathbf{x} + \mathbf{a}$, $\mathbf{a} = \text{const}$). Each point $\mathbf{a}$ of R^n corresponds to an element of the group, so the group has in fact a 1–1 map onto all of R^n. The group composition law is simply addition: two elements $\mathbf{a} = (a_1, \ldots, a_n)$ and $\mathbf{b} = (b_1, \ldots, b_n)$ compose to form $\mathbf{c} = (a_1 + b_1, \ldots, a_n + b_n)$, denoted symbolically as $\mathbf{c} = \mathbf{a} + \mathbf{b}$. This example illustrates the fact that one need not always use the symbol $\cdot$ to represent the group operation. With Abelian groups, as this one is, it is more common to use the symbol $+$.

A *subgroup* S of a group G is a collection of elements of G which themselves form a group with the same binary operation. (The prefix 'sub' always denotes a subset having the same properties as the larger set. We shall encounter many 'subs': vector subspaces, submanifolds, Lie subalgebras, and Lie subgroups.) As a group, a subgroup must have an identity element. Since the group's identity e is unique, any subgroup must also contain e. In the example of the permutation group, one can invent many subgroups. The permutations of n objects which do not change the position of the first object form a subgroup of the permutation group of n objects because, (i) the identity e leaves the first object fixed; (ii) the inverse of such a permutation also leaves the first object fixed; and (iii) the composition of any two such permutations still leaves the first object fixed. In fact this subgroup is identical to the group of permutations of $n - 1$ objects. The reader should try to prove that the set of all *even* permutations is also a subgroup of the permutation group, and that the set of all odd permutations is *not* a subgroup.

Our statement that a certain subgroup of the permutation group of n objects 'is identical to' the group of permutations of $n - 1$ objects is an example of a *group isomorphism*. Two groups, G_1 and G_2, with binary operations $\cdot$ and $*$

respectively, are isomorphic (which just means identical in their group properties) if there is a 1–1 map f of G_1 onto G_2 which respects the group operations:

$$f(x \cdot y) = f(x) * f(y). \tag{1.10}$$

The isomorphism f for our example is trivial: an element of the subgroup of the n-permutation group which permutes only the last $n-1$ objects is mapped to the same permutation in the $(n-1)$-permutation group. But an isomorphism is not always so trivial. Let G_1 be the group of positive real numbers with the operation of multiplication, and let G_2 be the group of all real numbers with the operation of addition. (Why are these groups?) Then if x is a number in G_1, $f(x) = \log x$ defines a map $f\colon G_1 \to G_2$ which satisfies (1.10):

$$\log(xy) = \log x + \log y.$$

The two groups are isomorphic and f is an isomorphism.

Another useful relation between groups is called a *group homomorphism*. This is like an isomorphism except that the map can be many-to-one and may only be into. (See §1.2 for terminology.) Equation (1.10) must still be satisfied. A trivial homomorphism of a group into itself is a map which maps every element of the group to the identity e. Less trivial is the homomorphism from the permutation group onto the multiplicative group whose only elements are $\{1, -1\}$. This homomorphism maps any even permutation to 1 and any odd one to -1. The reader should verify (1.10) for this example, i.e. that the composition of two odd permutations is even, that of an odd and even is odd, and that of two evens is even.

1.5 Linear algebra

A set V is a *vector space* (over the real numbers) if it has a binary operation called $+$ with which it is an Abelian group (see above) and if multiplication $(\cdot)$ by real numbers is defined to satisfy the following axioms (in which $\bar{x}$ and $\bar{y}$ are vectors and a, b real numbers):

(Vi) $a \cdot (\bar{x} + \bar{y}) = (a \cdot \bar{x}) + (a \cdot \bar{y})$,
(Vii) $(a + b) \cdot \bar{x} = (a \cdot \bar{x}) + (b \cdot \bar{x})$,
(Viii) $(ab) \cdot \bar{x} = a \cdot (b \cdot \bar{x})$,
(Viv) $1 \cdot \bar{x} = \bar{x}$.

The identity element of V is called $\bar{0}$, or simply 0. Apart from the usual examples of vector spaces, note that the following are vector spaces:

(i) The set of all $n \times n$ matrices, where '+' means adding corresponding entries and '·' means multiplying each entry by the real number.

(ii) The set of all real continuous functions $f(x)$ defined on the interval $a \leqslant x \leqslant b$.

It is usual to drop the multiplication dot and parentheses used in these axioms. An expression like

$$a\bar{x} + b\bar{y} + c\bar{z} \tag{1.11}$$

is called a *linear combination* of the vectors $\bar{x}, \bar{y}$, and $\bar{z}$. A set of elements $\{\bar{x}_1, \bar{x}_2, \ldots, \bar{x}_m\}$ of V is *linearly independent* if it is impossible to find real numbers $\{a_1, a_2, \ldots, a_m\}$ not all zero for which

$$a_1\bar{x}_1 + a_2\bar{x}_2 + \ldots + a_m\bar{x}_m = 0. \tag{1.12}$$

The set is a *maximal* linearly independent set if including *any* other vector of V in it would make it linearly dependent. By definition this means that any other vector in V can be expressed as a linear combination of elements of a maximal set, and so a maximal set forms a *basis* for V. For example, if V is the set of $n \times n$ real matrices, then one basis is the collection of the n^2 different matrices that each have zeroes everywhere except for a one in a single entry. In general, the number of vectors in a basis is the *dimension* of V. (All bases have the same number of elements, if that number is finite.) Let the vectors $\{\bar{x}_i, i = 1, \ldots, n\}$ be a basis. Then an arbitrary vector y is expressible as

$$\bar{y} = \sum_{i=1}^{n} a_i\bar{x}_i. \tag{1.13}$$

The numbers $\{a_i, i = 1, \ldots, n\}$ are the *components* of $\bar{y}$ on this basis.

A *subspace* of a vector space V is a subset of V that is itself a vector space. (Compare this with the definition of a subgroup in §1.4.) In particular, it must include the zero vector and all linear combinations of any of its elements. Any set of vectors $\{\bar{y}_1, \ldots, \bar{y}_m\}$ is said to *generate* the subspace of V which is formed by all possible linear combinations

$$a_1\bar{y}_1 + a_2\bar{y}_2 + \ldots + a_m\bar{y}_m.$$

If $m < n$ this is necessarily a proper subspace, i.e. one not identical with V. In any case, the dimension of the subspace is the maximum number of linearly independent vectors among the generators.

So far nothing has been said about inner products or magnitudes of vectors. These are additional concepts which may or may not be useful in particular applications involving vectors: there is no necessity to impose them on a vector space. One way to introduce them is to define a *norm* on a vector space. A normed vector space V is a vector space with a mapping from V into the real numbers (i.e. a function that assigns to every vector a real number called its norm), where the map satisfies the axioms

(Ni) $n(\bar{x}) \geqslant 0$ for all $\bar{x}$ in V, and $n(\bar{x}) = 0$ if and only if $\bar{x} = \bar{0}$;

(Nii) $n(a\bar{x}) = |a| n(\bar{x})$ for all a in R and $\bar{x}$ in V;

(Niii) $n(\bar{x} + \bar{y}) \leqslant n(\bar{x}) + n(\bar{y})$ for all $\bar{x}, \bar{y}$ in V.

There are many functions that can satisfy these axioms. Consider, for example, R^n itself as a vector space, where vector addition is defined by

$$\mathbf{x}+\mathbf{y} = (x_1+y_1, \ldots, x_n+y_n), \tag{1.14}$$

and multiplication by real numbers by

$$a\mathbf{x} = (ax_1, \ldots, ax_n). \tag{1.15}$$

Then, corresponding to three of the four distance functions defined in §1.1, we can define a norm, the 'distance' of a vector from the origin:

$$n(\mathbf{x}) = [(x_1)^2 + (x_2)^2 + \ldots + (x_n)^2]^{1/2}, \tag{1.16}$$

$$n'(\mathbf{x}) = [4(x_1)^2 + (x_2)^2 + \ldots + (x_n)^2]^{1/2}, \tag{1.17}$$

$$n'''(\mathbf{x}) = \text{maximum}\ (|x_1|, |x_2|, \ldots, |x_n|). \tag{1.18}$$

The reader should verify that each norm satisfies axioms (Ni)–(Niii). In addition, the reader should verify that $d''(\mathbf{x}, \mathbf{y})$ does *not* define a norm.

The first two norms are distinguished from the third by satisfying an additional axiom which one may wish to impose, the *parallelogram rule*:

(Niv) $[n(\bar{x}+\bar{y})]^2 + [n(\bar{x}-\bar{y})]^2 = 2[n(\bar{x})]^2 + 2[n(\bar{y})]^2$.

Such a norm permits one to define a bilinear symmetric *inner product* between two vectors

$$\bar{x}\cdot\bar{y} = \tfrac{1}{4}[n(\bar{x}+\bar{y})]^2 - \tfrac{1}{4}[n(\bar{x}-\bar{y})]^2. \tag{1.19}$$

Bilinearity means:

$$(a\bar{x}+b\bar{y})\cdot\bar{z} = a(\bar{x}\cdot\bar{z}) + b(\bar{y}\cdot\bar{z}), \tag{1.20}$$

and

$$\bar{z}\cdot(a\bar{x}+b\bar{y}) = a(\bar{z}\cdot\bar{x}) + b(\bar{z}\cdot\bar{y}). \tag{1.21}$$

Symmetry means:

$$\bar{x}\cdot\bar{y} = \bar{y}\cdot\bar{x}. \tag{1.22}$$

In addition, the inner product is positive-definite, i.e.

$$\bar{x}\cdot\bar{x} \geqslant 0 \quad \text{and} \quad \bar{x}\cdot\bar{x} = 0 \quad \text{only if} \quad \bar{x} = \bar{0}. \tag{1.23}$$

This follows trivially since $\bar{x}\cdot\bar{x} = [n(\bar{x})]^2$.

The norm $n(\mathbf{x})$ on R^n defined above is called the *Euclidean norm*. When we regard R^n as a vector space with this norm we denote it by E^n and call it n-dimensional Euclidean space. It is important to bear in mind the distinction between R^n and E^n: R^n is simply the set of all n-tuples $(x_1, \ldots, x_n)$, without any implication of distance, vector properties, or norms. The purpose of making this distinction will become clear in chapter 2.

To define the inner product and show it was bilinear and symmetric, only axioms (Nii) and (Niv) of norms are in fact necessary. A *pseudo-norm* is one which violates (Ni) and (Niii): the inner product of a vector with itself is not necessarily positive. Special relativity is an example of a physical theory using a pseudo-norm, and we will look in some detail at it later.

While we have defined only a vector space over the real numbers, we can just as easily define one over the complex numbers by allowing the numbers a and b in (Vi)–(Viv) to be complex. Then a vector will have complex components. Such vector spaces are commonly used in quantum mechanics.

1.6 The algebra of square matrices

A *linear transformation* T on a vector space V is a map from V onto itself which obeys the rule of linearity (cf. equations (1.20) and (1.21))

$$T(a\bar{x} + b\bar{y}) = aT(\bar{x}) + bT(\bar{y}). \tag{1.24}$$

If we have a basis $\{\bar{e}_i, i = 1, \ldots, n\}$ for V, then

$$\bar{x} = \sum_{i=1}^{n} a_i \bar{e}_i, \tag{1.25}$$

$$T(\bar{x}) = T\left(\sum_{i=1}^{n} a_i \bar{e}_i\right) = \sum_{i=1}^{n} a_i T(\bar{e}_i)$$

$$= \sum_{i=1}^{n} a_i \sum_{j=1}^{n} T_{ij} \bar{e}_j, \tag{1.26}$$

where we have replaced each vector $T(\bar{e}_i)$ by its component form $\Sigma_{j=1}^{n} T_{ij} \bar{e}_j$. The numbers T_{ij} are called the components of T, and can be represented as a square $n \times n$ matrix.

A very important algebraic result with which the reader should be comfortable is the following:

$$\sum_{i=1}^{n} A_i \left(\sum_{j=1}^{m} B_{ij} C_j\right) = \sum_{j=1}^{m} C_j \left(\sum_{i=1}^{n} B_{ij} A_i\right). \tag{1.27}$$

That is, the order in which the sums are performed makes no difference. Consequently, it is customary to write the above expression as

$$\sum_{i=1}^{n} \sum_{j=1}^{m} A_i B_{ij} C_j, \quad \text{or simply} \quad \sum_{i,j} A_i B_{ij} C_j, \tag{1.28}$$

emphasizing that the sum is simply the sum of various products over all possible combinations of indices.

Two successive linear transformations T and U acting on the space V produce the transformation UT:

$$UT(\bar{x}) = U(T(\bar{x}))$$

$$= U\left(\sum_{i,j} a_i T_{ij} \bar{e}_j\right)$$

$$= \sum_{ijk} a_i T_{ij} U_{jk} \bar{e}_k$$

$$= \sum_{ik} a_i \left(\sum_j T_{ij} U_{jk} \right) \bar{e}_k. \tag{1.29}$$

From this it follows that the components of UT are

$$\sum_j T_{ij} U_{jk}. \tag{1.30}$$

It is important to realize that if we represent T_{ij} as a matrix (i being the row index and j the column index), and similarly for U_{jk}, then the sum (1.30) is just the *matrix* product of their respective matrices. Generally speaking, if A_{ij} and B_{ij} are matrices, then their matrix products are

$$(AB)_{ik} = \sum_j A_{ij} B_{jk} = \sum_j B_{jk} A_{ij}, \tag{1.31}$$

$$(BA)_{ik} = \sum_j B_{ij} A_{jk} = \sum_j A_{jk} B_{ij}. \tag{1.32}$$

Notice that the third expression in equation (1.31) equals the second simply because each A_{ij} and B_{ij} is a number, and multiplication of numbers is commutative. By comparing the third expression of equation (1.31) with the second of (1.32), we see that what is important is not the order of the factors but the positions of the summation index and of the free indices. The inequivalence of these two expressions means that matrix multiplication is generally not commutative.

The *transpose* A^{T} of a matrix A has elements

$$(A^{\mathrm{T}})_{ij} = A_{ji}. \tag{1.33}$$

(If A is complex then we define the adjoint A^* of A by $(A^*)_{ij} = \bar{A}_{ji}$, where a bar denotes complex conjugation.) The *unit* matrix, I, has ones on the main diagonal and zeroes elsewhere; this is symbolized by

$$(I)_{ij} = \delta_{ij}, \tag{1.34}$$

where δ_{ij} is the Kronecker delta symbol, which is 1 if $i = j$, and 0 if $i \neq j$. The *identity transformation* is the one which maps any vector $\bar{x}$ into itself. It has components δ_{ij} on *any* basis. The *inverse* A^{-1} of a matrix A is a matrix such that

$$A^{-1}A = AA^{-1} = I. \tag{1.35}$$

Not every matrix has an inverse, the zero matrix being an obvious one. When an inverse exists it is unique. Clearly A is the inverse of A^{-1}. If A^{-1} exists, A is said to be *nonsingular*. (Otherwise it is singular.) The set of all nonsingular $n \times n$ matrices forms a group with the operation of matrix multiplication. The group identity is the matrix I. This group is an extremely important Lie group called $GL(n, R)$ and we will study it carefully in chapter 3.

The *determinant* of a 2×2 matrix

$$A = \begin{pmatrix} a & b \\ c & d \end{pmatrix}$$

is called det (A), and is defined as

$$\det(A) = ad - bc. \tag{1.36}$$

The determinant of an $n \times n$ matrix is defined by induction on $(n-1) \times (n-1)$ matrices by the following *rule of cofactors*. The *cofactor* of an element a_{ij} of A is called a^{ij} and is defined as $(-1)^{i+j}$ times the determinant of the $(n-1) \times (n-1)$ matrix formed by eliminating from A the row and column that a_{ij} belongs to. Thus, in the matrix

$$A = \begin{pmatrix} a & b & c \\ d & e & f \\ g & h & k \end{pmatrix} \tag{1.37}$$

the cofactor of a is $ek - fh$, while that of f is $bg - ah$. Then the determinant of A is defined to be

$$\det(A) = \left(\sum_{j=1}^{n} a_{ij}a^{ij}\right), \text{ for any fixed } i. \tag{1.38}$$

For the matrix (1.37), taking $i = 1$ gives

$$\det(A) = a(ek - fh) + b(fg - dk) + c(dh - eg),$$

while taking $i = 2$ gives

$$\det(A) = d(hc - bk) + e(ak - cg) + f(bg - ah),$$

both of which are the same. This rule always looks very mysterious when presented this way. It will make much more sense when we look at it again in §4.12.

The rows (or columns) of an $n \times n$ matrix may each be thought of as giving the components of a vector in some n-dimensional vector space. The determinant of a matrix vanishes if and only if the n vectors defined by its rows (or columns) are linearly dependent. This follows from some other properties of determinants: if a single row is multiplied by a constant λ, the determinant is multiplied by λ; if one row is replaced element-by-element by the sum of itself and any multiple of another row, the determinant is unchanged; and if any two rows are exchanged, the determinant changes sign. These properties are equally true if 'row' is replaced by 'column'. Again, they will make more sense after studying §4.12.

Suppose we construct a matrix B from a matrix A in such a way that

$$b_{ij} = a^{ji}/\det(A). \tag{1.39}$$

Then (1.38) shows that

$$\sum_{j=1}^{n} a_{ij}b_{ji} = 1, \quad \text{for any fixed } i.$$

It turns out (as experimentation can convince you) that

$$\sum_{j=1}^{n} a_{ij}b_{jk} = \delta_{ik},$$

or, in other words, that *the inverse of A* is the matrix B whose elements are given by (1.39). If follows that A is nonsingular if and only if $\det(A) \neq 0$.

The *trace* of a matrix A is the sum of its diagonal elements:

$$\operatorname{tr}(A) = \sum_i a_{ii}. \tag{1.40}$$

A *similarity transformation* of A by a nonsingular matrix B is a map $A \mapsto B^{-1}AB$. The following list of useful formulae may be proved from the definitions we have given:

$$(AB)^{\mathrm{T}} = B^{\mathrm{T}}A^{\mathrm{T}}, \tag{1.41}$$

$$(AB)^{-1} = B^{-1}A^{-1}, \tag{1.42}$$

$$\det(AB) = \det(A)\det(B), \tag{1.43}$$

$$\operatorname{tr}(B^{-1}AB) = \operatorname{tr}(A) \tag{1.44}$$

$$\det(B^{-1}AB) = \det(A), \tag{1.45}$$

$$\det(A^{\mathrm{T}}) = \det(A). \tag{1.46}$$

The $n \times n$ matrix A has an *eigenvalue* λ and an *eigenvector* $\bar{V} \neq \bar{0}$ if the following equation holds:

$$A(\bar{V}) = \lambda \bar{V}, \tag{1.47}$$

where on the left A acts as a linear transformation on $\bar{V}$. In component form we can write this as the n equations

$$\sum_j (a_{ij} - \lambda\delta_{ij})V_j = 0. \tag{1.48}$$

This has a nonzero solution for $\bar{V}$ if and only if

$$\det(A - \lambda I) = 0. \tag{1.49}$$

The eigenvalues of A are solutions to equation (1.49), which is clearly a polynomial equation of nth order. To any real solution λ there corresponds an eigenvector. If a solution is complex there is no real eigenvector, so if $\bar{V}$ is in a real vector space there is a fundamental difference between real and complex eigenvalues. If the vector space (and the matrix A) are complex, there is no particular distinction that need be made. Suppose $\{\lambda_1, \ldots, \lambda_n\}$ are the n roots of (1.49), repeated as often as appropriate if they happen to be multiple roots. There are three important results which we shall quote:

$$\{\text{eigenvalues of } A^T\} = \{\text{eigenvalues of } A\}, \quad (1.50)$$

$$\det(A) = \lambda_1 \lambda_2 \ldots \lambda_n, \quad (1.51)$$

$$\text{tr}(A) = \lambda_1 + \lambda_2 + \ldots + \lambda_n. \quad (1.52)$$

The last two can be proved by inspecting the polynomial in (1.49) closely. The first follows from (1.46).

1.7 Bibliography

The following textbooks treat their subjects at roughly the level I have assumed for this book. The list is intended to be a guide to the level of the material; it is certainly not a comprehensive list for background reading.

Elementary calculus: G. B. Thomas, *Calculus and Analytic Geometry* (Addison-Wesley, Reading, Mass., 1960).

Mechanics: K. R. Symon, *Mechanics* (Addison-Wesley, Reading, Mass., 1953); or H. Goldstein, *Classical Mechanics* (Addison-Wesley, Reading, Mass., 1950); or L. Landau & E. M. Lifshitz, *Mechanics* (Pergamon, New York, 1960).

Thermodynamics: M. W. Zemansky, *Heat and Thermodynamics* (McGraw-Hill, New York, 1957); or E. Fermi, *Thermodynamics* (Dover, New York, 1956).

Electromagnetism: J. R. Reitz & F. J. Milford, *Foundations of Electromagnetic Theory* (Addison-Wesley, Reading, Mass., 1960); or J. B. Marion, *Classical Electromagnetic Radiation* (Academic Press, New York, 1965).

Special relativity: A. P. French, *Special Relativity* (Nelson, London, 1968); or E. F. Taylor & J. A. Wheeler, *Spacetime Physics* (Freeman, San Francisco, 1965).

Quantum mechanics: L. I. Schiff, *Quantum Mechanics* (McGraw-Hill, New York, 1955).

The material in this chapter can be studied in greater depth in the following references:

R. A. Dean, *Elements of Abstract Algebra* (Wiley, New York, 1967);

W. Rudin, *Principles of Mathematical Analysis* (McGraw-Hill, New York, 1964);

E. W. Packel, *Functional Analysis, a Short Course* (International Textbook Co., Glasgow, 1974);

F. Riesz & B. Sz.-Nagy, *Functional Analysis* (Ungar, New York, 1955);

A. Wallace, *Differential Topology: First Steps* (Benjamin, Reading, Mass., 1968).

There are some basic references for most of the material in later chapters. These books generally treat them in greater depth and more rigor. They will be listed here and referred to later by their authors' names:

Y. Choquet-Bruhat, C. DeWitt-Morette & M. Dillard-Bleick, *Analysis, Manifolds, and Physics* (North-Holland, Amsterdam, 1977). A comprehensive book aimed at mathematically literate physicists. Especially strong on differential equation theory.

R. Abraham & J. E. Marsden, *Foundations of Mechanics*, revised 2nd edn (Benjamin/Cummings, Reading, Mass., 1978). As its title indicates, a book with narrower scope than Choquet-Bruhat *et al.*, but with consequently greater depth. Notable for its attention to global problems and its large bibliography. Considerably larger than the first edition.

W. Thirring, *A Course in Mathematical Physics I: Classical Dynamical Systems* (Springer Verlag, Vienna, 1978). A more elementary introduction to the modern way of doing mechanics than Abraham & Marsden.

V. I. Arnold, *Mathematical Methods of Classical Mechanics* (Springer Verlag, Berlin, 1978) is another good introduction to this method, with a good range of topics of varying difficulty.

F. W. Warner, *Foundations of Differentiable Manifolds and Lie Groups* (Scott, Foresman, Glenview, Ill., 1971). A readable introduction for mathematics undergraduates. Particularly strong on Lie groups and cohomology theory.

M. Spivak, *A Comprehensive Introduction to Differential Geometry*, four volumes (Publish or Perish, Boston, 1970). Just what the title says, this work is aimed at undergraduate mathematicians. It has a relaxed, often humorous style, many good exercises, and lots of detail.

There are many other books on differential geometry that are worth consulting. At an introductory level the reader may like N. J. Hicks, *Notes on Differential Geometry* (D. Van Nostrand, New York, 1965); or R. L. Bishop & S. I. Goldberg, *Tensor Analysis on Manifolds* (Macmillan, London, 1968). In the same 'modern' spirit as the present book is C. T. J. Dodson & T. Poston, *Tensor Geometry* (Pitman, London, 1977). This treats in a more leisurely manner many subjects we only touch on, but it does not deal with general Lie derivatives or the calculus of differential forms. Readers who find the leap from the present book to, say, Choquet-Bruhat *et al.* too great may find the first half of Dodson & Poston helpful. Authoritative works include S. Kobayashi & K. Nomizu, *Foundations of Differential Geometry*, two volumes (Interscience, New York, 1963 and 1969), which concentrates on positive-definite metric geometry; and J. A. Schouten, *Ricci Calculus* (Springer, Berlin, 1954), which is written in a rather old-fashioned notation.

Not surprisingly, some of the best introductions to differential geometry for physicists are in textbooks on general relativity. I recommend C. W. Misner, K. S. Thorne & J. A. Wheeler, *Gravitation* (Freeman, San Francisco, 1973), which devotes several chapters to a development in the same spirit and similar notation as I use here. More advanced and compact is the chapter on differential geometry in S. W. Hawking & G. F. R. Ellis, *The Large Scale Structure of Space–Time* (Cambridge University Press, 1973). An introduction with considerable range is the article Differential geometry, by C. W. Misner in *Relativity, Groups, and Topology*, ed. C. DeWitt & B. DeWitt (Gordon & Breach, New York, 1964). Another valuable introduction is the article Differential geometry from a modern standpoint, by B. Schmidt in *Relativity, Astrophysics and Cosmology*, ed. W. Israel (Reidel, Dordrecht, 1973). Also aimed at physicists are R. Hermann, *Differential Geometry and the Calculus of Variations* (Academic Press, New York, 1968); D. Lovelock & H. Rund, *Tensors, Differential Forms and Variational Principles* (Wiley, New York, 1975); H. Flanders, *Differential Forms* (Academic Press, New York, 1963); and Von Westenholz, *Differential Forms in Mathematical Physics* (North-Holland, Amsterdam, 1979).

2 DIFFERENTIABLE MANIFOLDS AND TENSORS

It is hard to imagine a physical problem which does not involve some sort of continuous space. It might be physical three-dimensional space, four-dimensional spacetime, phase space for a problem in classical or quantum mechanics, the space of all thermodynamic equilibrium states, or some still more abstract space. All these spaces have different geometrical properties, but they all share something in common, something which has to do with their being continuous spaces rather than, say, lattices of discrete points. The key to differential geometry's importance to modern physics is that it studies precisely those properties common to all such spaces. The most basic of these properties go into the definition of the differentiable manifold, which is the mathematically precise substitute for the word 'space'.

2.1 Definition of a manifold

As in §1.1, we denote by R^n the set of all n-tuples of real numbers $(x_1, x_2, \ldots, x_n)$. A set (of 'points') M is defined to be a *manifold* if each point of M has an open neighborhood which has a continuous 1–1 map onto an open set of R^n for some n. (The reader unsure of what a 1–1 map *onto* something means should look at §1.2.) This simply means that M is locally 'like' R^n. The *dimension* of M is, of course, n. It is important that the definition involves only open sets and not the whole of M and R^n, because we do not want to restrict the global topology of M. This will be clear in the example of the sphere in §2.2. Notice that the map is only required to be 1–1, not to preserve lengths or angles or any other geometrical notion. Length is not even defined at this level of geometry, and we shall encounter physical applications in which we will not want to introduce a notion of distance between points of our manifolds. At this elementary ('primitive') geometrical level we are only trying to ensure that the local topology of our space (as described in §1.1) is the same as that of R^n. A manifold is a space with this topology.

By definition, the map associates with a point P of M an n-tuple $(x_1(P), \ldots, x_n(P))$. These numbers $x_1(P), \ldots, x_n(P)$ are called the *coordinates* of P under this map, as illustrated in figure 2.1. One way of thinking about an

n-dimensional manifold is that it is simply any set which can be given n independent coordinates in some neighborhood of any point, since these coordinates actually *define* the required map to R^n. We shall adopt the standard notation of writing the index of the coordinate as a superscript: $x^1(P), x^2(P), \ldots, x^n(P)$ are the n coordinates of P (*not* powers of $x(P)$!) under the map.

From the discussion so far, we ought now to have a general idea of what a manifold is, but to do any better than this we must examine the nature of these coordinate maps. Suppose f is a 1–1 map from a neighborhood U of a point P of M onto an open set $f(U)$ of R^n. As stressed above, the neighborhood U does not necessarily include *all* of M (we shall see in §2.2 that on the sphere it *cannot* include the whole sphere), so there will be other neighborhoods with their own maps, and each point of M must lie in at least one such neighborhood. The pair consisting of a neighborhood and its map is called a *chart*. It is easy to see that these open neighborhoods must have overlaps if all points of M are to be included in at least one, and it is these overlaps which enable us to give a further characterization of the manifold (refer to figure 2.2). Suppose V is a neighborhood overlapping U, and that V has a map g onto an open region of R^n. This open region may be completely distinct from the one that f maps U onto. The intersection of V and U is open (by axiom (Ti) of §1.1) and is given two different coordinate systems by the two maps. There is thus some equation relating these coordinate systems. To find it, pick a point in the image of the overlap under f (i.e. a point in R^n), say the point $(x^1, x^2, \ldots, x^n)$ in figure 2.3. The map f has an inverse f^{-1}, so there is a unique point S in the overlap which has these coordinates under f. Now let g take us from S to another point in R^n, say

Fig. 2.1. A region U of M has a 1–1 map f onto a region $f(U)$ of R^n. This map associates any point, say P, with a unique n-tuple of numbers $(x_1, x_2, \ldots, x_n)$. In this way U acquires a coordinate system, illustrated by drawing the dashed lines that are the images under f^{-1} of the usual coordinate lines of R^n.

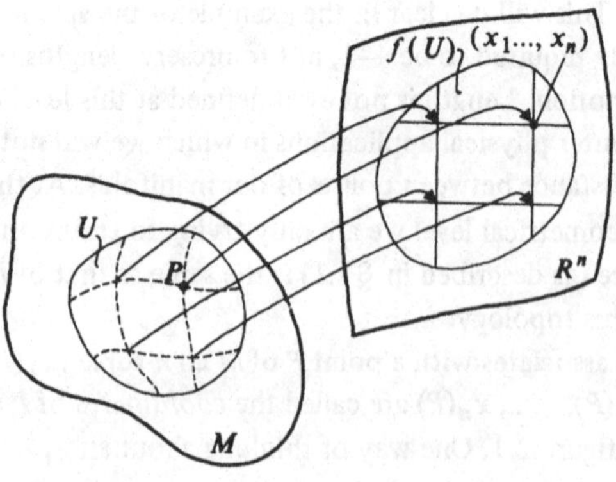

$(y^1, y^2, \ldots, y^n)$. (What we have done is constructed the composite map of $R^n \to R^n$, called $g \circ f^{-1}$.) In this way we obtain a functional relationship (a *coordinate transformation*)

$$\begin{aligned} y^1 &= y^1(x^1, x^2, \ldots, x^n) \\ y^2 &= y^2(x^1, x^2, \ldots, x^n) \\ &\vdots \\ y^n &= y^n(x^1, x^2, \ldots, x^n). \end{aligned}$$

If the partial derivatives of order k or less of all these functions $\{y^i\}$ with respect to all the $\{x^i\}$ exist and are continuous, then the maps f and g (strictly, the charts (U, f) and (V, g)) are said to be C^k-related. (This is the notation introduced in §1.2 for differentiability.) If it is possible to construct a whole system of charts (called, appropriately enough, an *atlas*) in such a way that every point

Fig. 2.2. The neighborhoods U and V in M overlap (shaded area). Their respective maps to R^n, f and g, give two different maps (hence two coordinate systems) to the overlap region. The relation between these coordinates characterizes the differentiability class of the manifold.

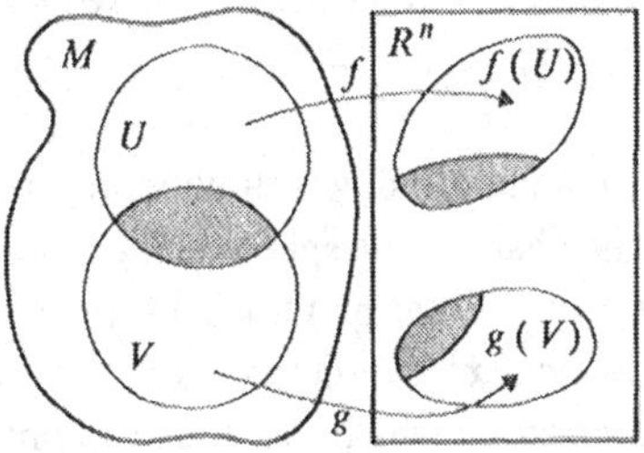

Fig. 2.3. A magnification of figure 2.2, which shows how the overlap makes a map from R^n to R^n, which is f^{-1} followed by g (called $g \circ f^{-1}$).

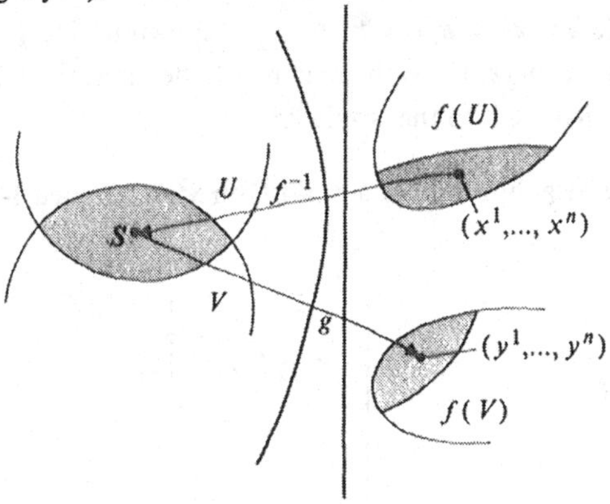

of M is in at least one neighborhood and every chart is C^k-related to every other one it overlaps with, then the manifold M is said to be a C^k *manifold*. A manifold of class C^1 (which includes C^k for $k > 1$) is called a *differentiable manifold.*

The differentiability of a manifold endows it with an enormous amount of structure: the possibility of defining tensors, differential forms, and Lie derivatives. This differential structure is our main subject. Remember, we have not introduced the concept of distance on M, and we have no notion of the 'shape' or 'curvature' of M. We only know that locally it is smooth, and that is all we need for what follows.

In most applications we will assume a C^∞ manifold, but usually this is not strictly necessary. There will be times when we shall find it convenient to assume an analytic manifold (C^ω: the functions $\{y^i\}$ are analytic functions of $\{x^i\}$), but this will be in the physicist's spirit of invoking analyticity where convenient, as mentioned in §1.3. We will take the view that in learning this subject for the first time it is better to make rather strong assumptions about the manifold in order to see what is going on in the differential geometry. After the student is more comfortable with the subject he can worry about relaxing his assumptions. Accordingly, the reader should assume throughout the book that any manifold is sufficiently differentiable for whatever argument we happen to be using.

2.2 The sphere as a manifold

One of the simplest examples of a manifold, which illustrates the importance of allowing for more than one chart, is the sphere. (The word 'sphere' *always* means the *surface* of the sphere, not its interior.) Consider the two-sphere (called S^2), the set of points in R^3 for which $(x^1)^2 + (x^2)^2 + (x^3)^2 = \text{const.}$ Any point has a sufficiently small neighborhood which as a 1–1 map onto a disc in R^2 (see figure 2.4). This shows that the map involved certainly will not preserve lengths or angles. As a specific example of a map, consider the usual spherical coordinates, with $\theta \equiv x^1$ and $\phi \equiv x^2$. Then the sphere appears to be mapped onto the rectangle $0 \leqslant x^1 \leqslant \pi, 0 \leqslant x^2 \leqslant 2\pi$, as shown in figure 2.5. But there are some funny features here. First, the map breaks down at the pole $\theta = 0$, where one point is 'mapped' to the whole line $x^1 = 0, 0 \leqslant x^2 \leqslant 2\pi$. So

Fig. 2.4. A small neighborhood of a point P on S^2 is mapped 1–1 onto a disc in R^2.

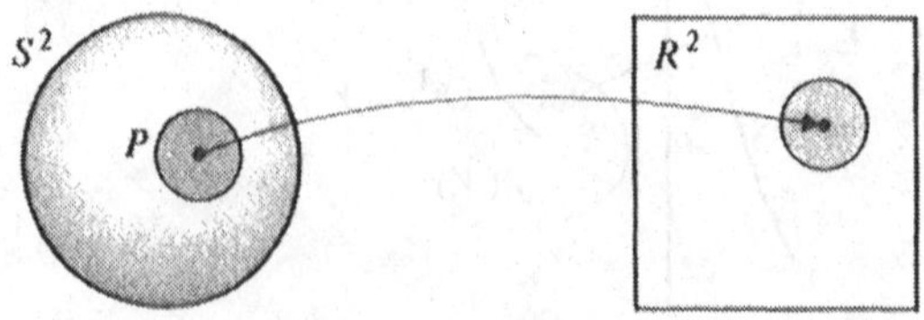

at the pole there is not even a map. The second difficulty is that the points having $\phi = 0$ are 'mapped' to two places, $x^2 = 0$ and $x^2 = 2\pi$: again, there is no map. To get around these problems we must restrict the map to the open region $0 < x^1 < \pi, 0 < x^2 < 2\pi$. Then the two poles and the semicircle $\phi = 0$ joining them are left out of the map. So here at least two maps are needed to cover the sphere completely. The second one could be another spherical coordinate system, this time with its line $\phi = 0$ in the equator of the first system, say from $\phi = \pi/2$ to $\phi = 3\pi/2$. Then every point on the sphere is in at least one of these two charts. The overlap functions, expressing the second system's coordinates in terms of the first one's, will be complicated, but it should be clear that they will be analytic. So the sphere is an analytic manifold.

A better map of S^2 onto a region of R^2, which fails at only one point, is the so-called stereographic map of the sphere onto the plane, shown in figure 2.6 in a vertical cross-sectional view. The sphere is tangent to the plane, and a line is drawn from the point N on the sphere diametrically opposite the point of tangency. This line intersects the sphere at P, and R^2 at Q. This defines the map:

Fig. 2.5. Ordinary spherical coordinates appear to give a map from S^2 to R^2, which is good for ordinary points like P. But where is the image of the north pole? And which of two points is the image of Q on the line $\phi = 0$?

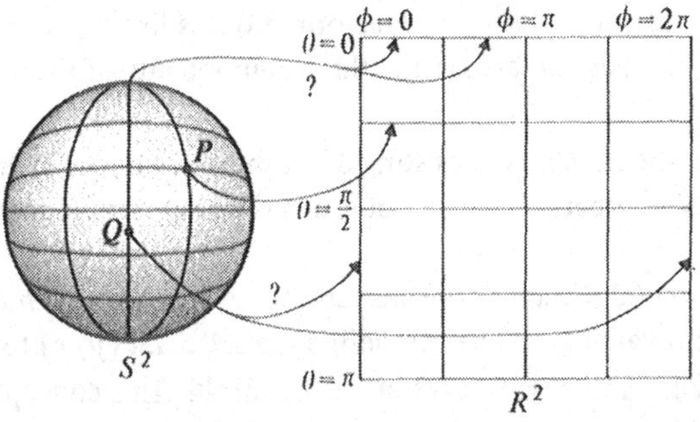

Fig. 2.6. The stereographic map of S^2 to R^2. The set S^2 with the single point N removed is open, and this set is mapped onto all of R^2. The map fails at N itself.

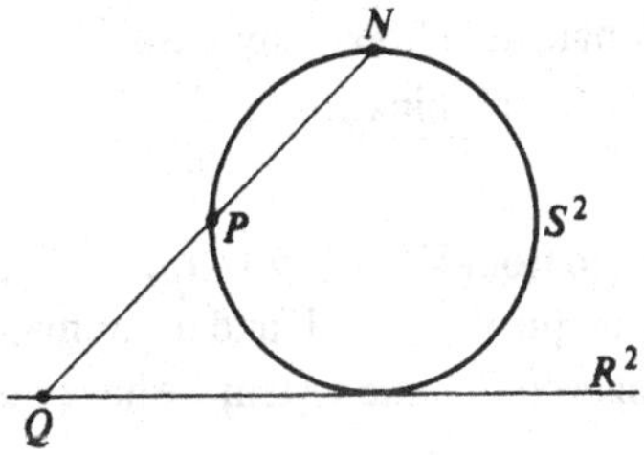

P is mapped to Q, or in other words the coordinates of P in S^2 are just the coordinates of Q in R^2. This map is 1–1 except at N, for as the line from N becomes horizontal (P approaching N), the point Q goes to infinity. But no matter in what direction in R^2 the point Q goes to infinity, the point P always approaches N. So N is mapped into all of 'infinity' and another coordinate patch must be used near N. There is *no* mapping which is good on all of S^2. Notice that this whole discussion really depends only on the global topology of S^2: exactly the same remarks apply to the surface of, say a bowl or a wine glass, which are simply deformations of S^2. On the other hand, the two-dimensional interior of the annulus bounded by two concentric circles in R^2 *can* be covered by a single coordinate patch. Try to find it!

2.3 Other examples of manifolds

The usefulness of the concept of a manifold really comes from its generality, the fact that it embraces sets which one might not ordinarily regard as spaces. By definition, *any* set M that can be parameterized continuously is a manifold whose dimension is the number of independent parameters. For example:

(i) The set of all rotations of a rigid object in three dimensions is a manifold, since it can be continuously parameterized by the three 'Euler angles' (cf. Goldstein, 1950).

(ii) The set of all (pure boost) Lorentz transformations is likewise a three-dimensional manifold; the parameters are the three components of the velocity of the boost.

(iii) For N particles, the numbers consisting of all their positions ($3N$ numbers) and velocities ($3N$ numbers) define a point in a $6N$-dimensional manifold, called phase space.

(iv) Given an equation (algebraic or differential) for a dependent variable y in terms of an independent variable x, one can define the set of *all* (y, x) to be a manifold; any particular solution is a curve in this manifold. This concept is easily extended to arbitrary numbers of dependent and independent variables.

(v) A particularly common manifold is a vector space, whose definition is given in § 1.5. (Here we are dealing with vector spaces over the real numbers.) To see that such a space is a manifold we will construct a map from it to some R^n. Suppose the vector space V is n-dimensional, and choose any basis $\{\bar{e}_1, \ldots, \bar{e}_n\}$. *Any* vector $\bar{y}$ is then representable as a linear combination

$$\bar{y} = a_1\bar{e}_1 + \ldots + a_n\bar{e}_n. \tag{2.1}$$

But $\bar{y}$ is a point in V, so this establishes a map from V to R^n, $\bar{y} \mapsto (a_1, \ldots, a_n)$. In fact every point of R^n corresponds to a unique vector in V under this map, so not only is V covered entirely by the single coordinate system we have just constructed, but V is identical, as a manifold, with R^n. In the language of group theory (§1.4), V and R^n are isomorphic. This is an important result. It means that every vector space may be thought of, when convenient, simply as R^n.

(vi) Example (i) above is an example of a *Lie group*, which we are now in a position to define. A Lie group G is a group which is also a C^∞ manifold, with the restriction that the group operation induces a C^∞ map of the manifold into itself. What this means is the following. Pick out any element a of the group. This element induces a map of G into itself, taking any element b of G into ba, $b \mapsto ba$. This map must be C^∞; in concrete terms in whatever coordinates are used on G, the coordinates of ba must be C^∞ functions of those of b. The demand for such a map is really a compatibility requirement, to ensure that the manifold property is compatible with the group property. In example (i) above, then, the set of all rotations forms a group, and it is not hard to show that this group structure is indeed compatible with the three-dimensional manifold structure. (This Lie group is called $SO(3)$.) This definition of Lie groups may seem abstract and perhaps rather arid at first, but we shall become much more familiar with them in chapter 3. A simple example of a Lie group is R^n. It is a vector space (see (v) above) and therefore a group, and it is also a manifold: R^n is in fact the simplest Lie group.

2.4 Global considerations

Because every manifold is locally the same as some R^n, any two manifolds of the same dimension (and same differentiability class) are locally indistinguishable at this level of differential geometry. But this is certainly not the case when we consider their global structure, as the comparison of S^2 with R^2 in §2.2 showed. Manifolds therefore divide up into classes according to their global properties. As an example, the sphere S^2 and the surface of a crayon have the same global structure. Although neither has a single map onto R^2, each has a single perfectly good 1–1 map onto the other, as illustrated in figure 2.7.

Fig. 2.7. A smooth (C^∞) crayon can be mapped 1–1 onto a sphere S^2. The map is global, not restricted to patches. It is a diffeomorphism, and so is its inverse.

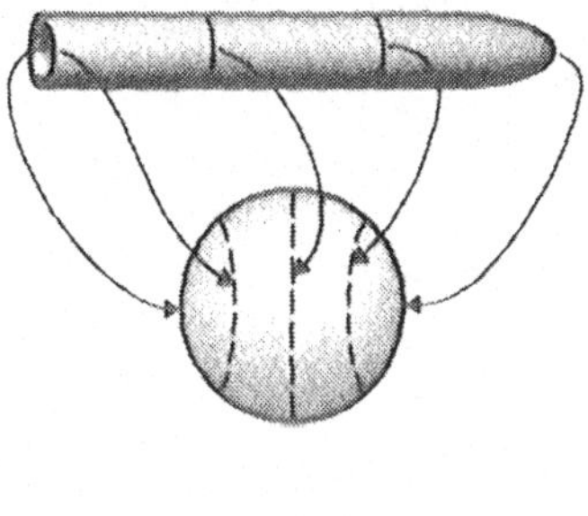

(Strictly speaking, the crayon should have very smooth edges to be identical to S^2 as a C^∞ manifold.) Such a map directly from one C^∞ manifold M to another N, which is 1–1 and C^∞ (a map is C^∞ if the coordinates of a point in N are infinitely differentiable functions of the coordinates of the inverse image of the point in M) and whose inverse is also C^∞, is called a *diffeomorphism* of M onto N. The manifolds M and N are said to be diffeomorphic if such a map exists. The surface of a teacup is diffeomorphic to the torus (doughnut) because each has just one hole: one can smoothly deform one into the other.

Most of the geometry we will study in this book will be local, depending only on the differential structure. But there will be occasions, such as our studies of fiber bundles and of integration of functions, when the global properties of our manifolds will become very important.

2.5 Curves

Curves in the manifold will be of great importance to us. One's ordinary idea of a curve is that it is a continuous series of points in M. It is convenient here to make a somewhat different definition: a curve is a (differentiable) mapping from an open set of R^1 into M (see figure 2.8). Thus, one associates with each point of R^1 (which is a real number, say λ) a point in M, which is called the *image point* of λ. The set of all image points is the ordinary notion of the curve, but our definition gives each point a value of λ. Clearly we have a *parameterized* curve, with parameter λ. Thus, two curves are different even if they have the same image in M, provided they assign a different parameter value to the image points. Again, by a 'differentiable' mapping we simply mean that the coordinates of the image point, $\{x^i(\lambda), i = 1, \ldots, n\}$ are differentiable functions of λ.

2.6 Functions on M

A function on M is a rule that assigns a real number (the *value* of the function) to each point of M. When a region of M is mapped differentiably onto

Fig. 2.8. A curve in M is a map from R^1 into M. The point λ in R^1 is mapped to P in M. The image of the open interval from a to b in R^1 is the line shown in M.

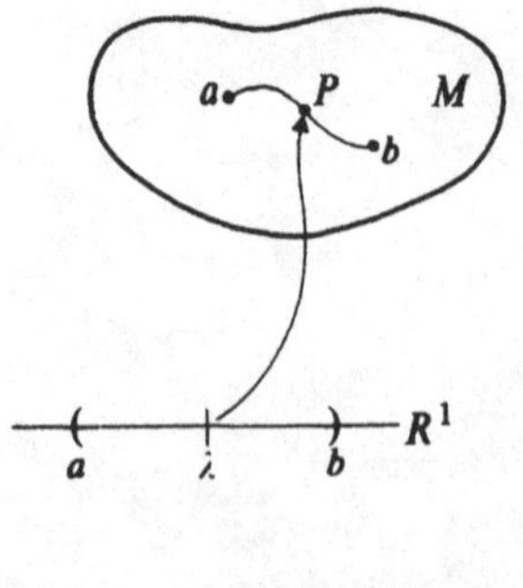

a region of R^n, the function becomes a function on R^n (see figure 2.9). If this function is *differentiable* in R^n, then it is said to be a differentiable function on M. We can say the same thing in another way: abstractly, the function may be written as $f(P)$, where P is a point of M. But P has coordinates, so one can express the value of the function by some algebraic expression $f(x^1, x^2, \ldots, x^n)$. Then if this expression is differentiable in its arguments, the function is differentiable. The coordinates themselves, of course, are continuous and infinitely-differentiable functions. For example, x^3 is the function such that $x^3(P)$ is the value of the third coordinate of the point P.

From now on we shall avoid referring to the mapping from M to R^n directly, although we shall occasionally refer to the coordinates (which describe the mapping). The purpose of discussing mappings up till now has been to establish the fundamental concepts in as precise a way as possible. From now on we shall be more interested in using these concepts to develop the differential structure of the manifold, so we will always assume that we can place coordinates $\{x^i, i = 1, \ldots, n\}$ on the manifold, and that any sufficiently-differentiable set of equations $y^i = y^i(x^j)$ which is locally invertible (i.e. whose Jacobian is nonzero – see §1.2) constitutes an acceptable coordinate transformation to new coordinates $\{y^i, i = 1, \ldots, n\}$.

2.7 Vectors and vector fields

Consider a curve passing through the point P of M, described by the equations $x^i = x^i(\lambda)$, $i = 1, \ldots, n$. Consider also a differentiable function $f(x^1, \ldots, x^n)$ (abbreviated $f(x^i)$) on M. At each point of the curve, f has a value. Therefore, along the curve there is a differentiable function $g(\lambda)$ which gives the value of f at the point whose parameter value is λ:

Fig. 2.9. The function f on M is a map from M to R^1. The coordinate map g from a region U of M containing P onto a region $g(U)$ of R^n has an inverse. The composite map $f \circ g^{-1}$ gives a map from R^n to R^1, which is a function on R^n. This is just the expression of $f(P)$ in terms of the coordinates of P.

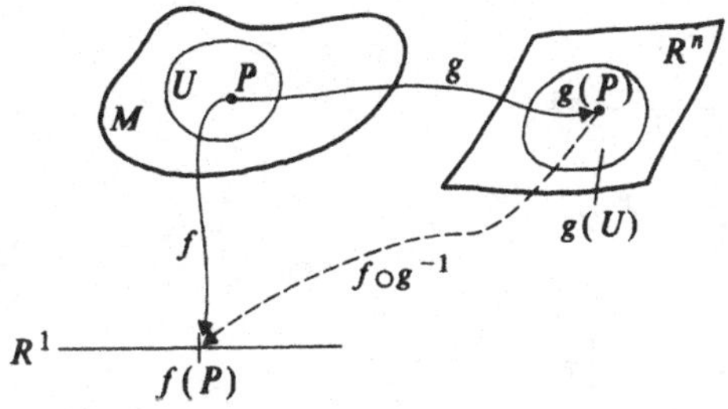

$$g(\lambda) = f(x^1(\lambda), \ldots, x^n(\lambda)) = f(x^i(\lambda)).$$

Differentiating and using the chain rule gives

$$\frac{dg}{d\lambda} = \sum_i \frac{dx^i}{d\lambda} \frac{\partial f}{\partial x^i}. \tag{2.2}$$

This is true for any function g, so we can write

◆
$$\frac{d}{d\lambda} = \sum_i \frac{dx^i}{d\lambda} \frac{\partial}{\partial x^i}. \tag{2.3}$$

Now, in the ordinary view of vectors in Euclidean space, one would say that the set of numbers $\{dx^i/d\lambda\}$ are components of a vector tangent to the curve $x^i(\lambda)$; one can see this by realizing that $\{dx^i\}$ are infinitesimal displacements along the curve, and that dividing them by $d\lambda$ only change the scale, not the direction, of this displacement. In fact, since a curve has a unique parameter, to every curve there is a unique set $\{dx^i/d\lambda\}$, which are then said to be components of *the* tangent to the curve. Thus, with our definition of a curve, every curve has a unique tangent vector.

Of course, every vector is the tangent to an infinite number of different curves through P, for two different reasons. The first is that there are many curves which are tangent to one another and have the same tangent vector at P, and the second is that the same path may be re-parameterized in such a way as to give the same tangent at P. These are illustrated in figure 2.10. As an example of this, consider the simple curve $x^i(\lambda) = \lambda a^i$, where the numbers $\{a^i\}$ are constants. Then if P is the point $\lambda = 0$, the tangent there is $dx^i/d\lambda = a^i$. Another curve,

Fig. 2.10. (*a*) Two curves having the same tangent vector. (*b*) Two curves having the same path but different parameterizations. If the maps are called h_1 and h_2, then the map $h_2^{-1} \circ h_1$ gives a relation between the two parameters, $\lambda_2 = \lambda_2(\lambda_1)$. If $d\lambda_2/d\lambda_1 = 1$ at P the two tangent vectors will be the same at P.

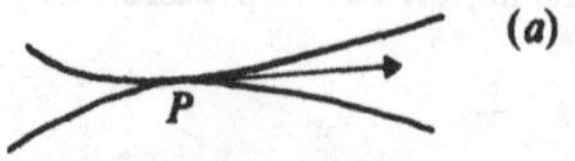

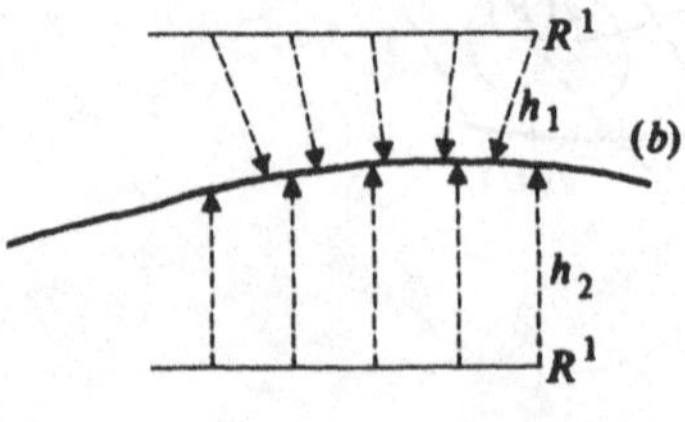

$x^i(\mu) = \mu^2 b^i + \mu a^i$, also passes through P at $\mu = 0$ and has the same tangent vector there, $dx^i/d\mu = a^i$. A re-parameterization of the first curve, $x^i = (\mu^3 + \mu)a^i$, passes through all the same points and at P ($\mu = 0$) has the same tangent, $dx^i/d\mu = a^i$. So each vector really characterizes a whole equivalence class of curves at that point.

This use of the term 'vector' relies on familiar concepts from Euclidean space, where vectors are defined by analogy with displacements Δx^i. However, since manifolds need have *no* distance relation between points, we shall need a definition of vector which relies only on infinitesimal neighborhoods of points of M. Suppose a and b are two numbers, and $x^i = x^i(\mu)$ is another curve through P. Then at P we have

$$\frac{d}{d\mu} = \sum_i \frac{dx^i}{d\mu}\frac{\partial}{\partial x^i},$$

and

$$a\frac{d}{d\lambda} + b\frac{d}{d\mu} = \sum_i \left(a\frac{dx^i}{d\lambda} + b\frac{dx^i}{d\mu}\right)\frac{\partial}{\partial x^i}.$$

Now, the numbers $\{a dx^i/d\lambda + b dx^i/d\mu\}$ are components of a new vector, which is certainly the tangent to *some* curve through P. So there must exist a curve with parameter, say, ϕ such that at P

$$\frac{d}{d\phi} = \sum_i \left(a\frac{dx^i}{d\lambda} + b\frac{dx^i}{d\mu}\right)\frac{\partial}{\partial x^i}.$$

Collecting these results, we get, at P,

$$a\frac{d}{d\lambda} + b\frac{d}{d\mu} = \frac{d}{d\phi}.$$

Therefore, the directional derivatives along curves, like $d/d\lambda$, form a *vector space* at P.† There are in any coordinate system special curves, the coordinate lines themselves. The derivations along them are clearly $\partial/\partial x^i$, and equation (2.3) shows that any $d/d\lambda$ can be written as a linear combination of the particular derivatives $\partial/\partial x^i$. It follows that $\{\partial/\partial x^i\}$ are a *basis* for this vector space. Then (2.3) shows that $d/d\lambda$ has components $\{dx^i/d\lambda\}$ on this basis. We therefore have the remarkable result that *the space of all tangent vectors at P and the space of all derivatives along curves at P are in 1–1 correspondence*. For this reason the mathematician says that $d/d\lambda$ *is* the tangent vector to the curve $x^i(\lambda)$. We shall adopt this point of view, since it has three advantages. First, it is precise, since it does not involve displacements over finite separations. Second, it makes no

† The derivatives must, of course, obey the other axioms of § 1.5 if they are to form a vector space, but closure under linear combinations is the only nontrivial one.

mention of coordinates; in particular, it does not rely on notions like 'transforms the same way as . . .'. Third, a derivative is a kind of 'motion' along the curve, which is what, conceptually, a tangent vector generates; this association of a concept from analysis – the derivative – with one from geometry – the vector – has very powerful consequences.

One can still maintain the same 'picture' of a vector as an arrow tangent to the curve, since the components are just the same. Now, however, one must realize that only vectors at the same point P can be added together. Vectors at two different points have no relation with one another. The vectors lie, not in M, but in the *tangent space* to M at P, which is called T_P. For ordinary manifolds, like the surface of a sphere, this tangent space is easy enough to visualize as a plane tangent to the sphere at that point. For more abstract manifolds it may be harder.

We shall use the term *vector* to refer to a vector at a given point P of M. The term *vector field* refers to a rule for defining a vector at each point of M.

2.8 Basis vectors and basis vector fields

At any point P, the space T_P is a vector space with the same dimension n as the manifold. Any collection of n linearly independent vectors in T_P is a *basis* for T_P. By choosing a basis in each T_P for all points P of M, we arrive at a basis for vector fields. If we have a coordinate system $\{x^i\}$ in a neighborhood U of P, then the coordinates define the *coordinate basis* $\{\partial/\partial x^i\}$ at all points in U.

But one need not use the coordinate basis; one could refer vectors to some arbitrary basis $\{\bar{e}_i\}$. Here the subscript i is used as a label to distinguish one basis vector from another. It does *not* denote the component of anything. At a point P, an arbitrary vector $\bar{V}$ can be written as

$$\bar{V} = \sum V^i \frac{\partial}{\partial x^i} = \sum_j V^{j'} \bar{e}_j.$$

The numbers $\{V^i\}$ are the components of $\bar{V}$ on $\{\partial/\partial x^i\}$. The numbers $\{V^{j'}\}$ are the components of $\bar{V}$ on $\{\bar{e}_j\}$, and are related to V^i by the usual vector transformation laws, which we will deal with later. If $\bar{V}$ and the bases $\{\partial/\partial x^i\}$ and $\{\bar{e}_j\}$ are regarded as vector fields, then the components $\{V^i\}$ and $\{V^{j'}\}$ of the field V are *functions* on M. A vector field is said to be differentiable if these functions are differentiable.

We have implicitly assumed above that the vectors $\{\partial/\partial x^i\}$ of an arbitrary coordinate system are in fact all linearly independent at any point P of U. What justification do we have for this? We shall show that this is just the condition for the coordinates to be *good* coordinates at P, i.e. for them to provide a 1–1 map of some neighborhood U of P onto a region V of R^n. Consider a set of coordinates on U which *are* good, say $\{y^i, i = 1, \ldots, n\}$. Then the map from $(x^1, \ldots, x^n)$ to U can be expressed by the equations

$$y^j = y^j(x^1, \ldots, x^n), \quad j = 1, \ldots, n.$$

By the inverse function theorem (§1.2) this map is 1–1 (has an inverse) in U if and only if the Jacobian matrix $\partial y^j/\partial x^i$ has a nonvanishing determinant. This means that at any point of U the vectors whose components are $(\partial y^1/\partial x^1, \partial y^2/\partial x^1, \ldots, \partial y^n/\partial x^1)$, $(\partial y^1/\partial x^2, \partial y^2/\partial x^2, \ldots, \partial y^n/\partial x^2), \ldots, (\partial y^1/\partial x^n, \partial y^2/\partial x^n, \ldots, \partial y^n/\partial x^n)$ are linearly independent. But these are just the components of the vectors $\{\partial/\partial x^i, i = 1, \ldots, n\}$ on the coordinate basis of the $\{y^i\}$ system, because by the chain rule

$$\frac{\partial}{\partial x^1} = \frac{\partial y^1}{\partial x^1}\frac{\partial}{\partial y^1} + \frac{\partial y^2}{\partial x^1}\frac{\partial}{\partial y^2} + \ldots + \frac{\partial y^n}{\partial x^1}\frac{\partial}{\partial y^n},$$

and similarly for the other x^is. So $\{x^i\}$ is in fact a good coordinate system in U if and only if $\{\partial/\partial x^i\}$ are a basis for vectors at each point of U. The reader may wish to look at the basis vectors of the spherical coordinates on the sphere to see how they go bad at the poles.

2.9 Fiber bundles

A particularly interesting manifold is formed by combining a manifold M with all its tangent spaces T_P. This is illustrated in figure 2.11 for the simplest case: a one-dimensional manifold M (a curve) and its tangent spaces (lines tangent to it at each point). In part (*a*) of figure 2.11 we draw the curve and a few tangent spaces; these are lines drawn tangent to the curve, and each must be thought of as extending infinitely far in both directions in order to allow for vectors of arbitrary length at each point. Now, when drawn this way the picture

Fig. 2.11. (*a*) A one-dimensional manifold and some of its tangent spaces. (*b*) The same, with the tangent spaces drawn parallel to one another to avoid spurious intersections.

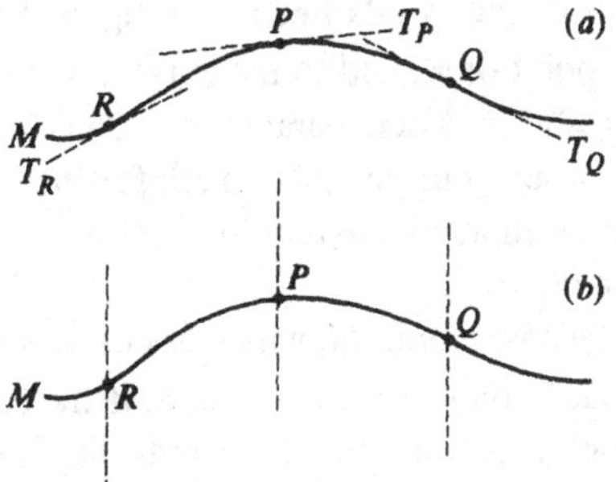

can be messy, since the various tangent spaces intersect one another and the curve M haphazardly: these spurious intersections have no meaning. A better way of drawing the picture is as in figure 2.11(b), where the tangent spaces are drawn parallel: they do not intersect each other, and they cross M only at the point where they are defined. This picture unfortunately does not show the fact that each T_P is 'tangent' to the curve, but that is the price to be paid for clarity. Each point on the vertical line T_P represents a vector, having that 'length' and being tangent to M at P. Figure 2.11(b) also shows something else: every point in the figure (a two-dimensional manifold) is a point of one and only one tangent space for M, say T_R for point R of M. To each point in that figure there is one and only one vector at one and only one point at M. So one is led to define a new manifold TM, consisting of all vectors at all points, which is thus two-dimensional. It is called a *fiber bundle*, and the fibers are the spaces T_P for each P. The term 'fiber' comes from drawing pictures like figure 2.11(b) above. To see that TM is indeed a two-dimensional manifold, let us construct a coordinate system for a portion of it. Let the one-dimensional manifold M have coordinate x, and let us find coordinates for the tangent spaces to points of M in the region $a < x < b$ for some a and b, assuming that the coordinate x itself is a good coordinate in this interval. (The reason for this assumption will be evident in §2.11 below.) Any tangent vector $\bar{V}$ at any point P can be written as

$$\bar{V} = y\partial/\partial x, \tag{2.4}$$

so that the component y is a coordinate for T_P (cf. equation (2.1)). It clearly is a good coordinate over the whole fiber T_P. Since each fiber has a fixed value of x, the coordinates (x, y) locate a particular vector (y) tangent to a particular point (x). Since every point of the fiber bundle must by definition lie in a region of this sort, we have proved that this TM is a manifold. Clearly, the construction is easily generalized to tangent fiber bundles of higher-dimensional manifolds. A coordinate system of this sort, in which coordinates for T_P are determined by those on M at P by expressing a vector on the coordinate basis (2.4) is called a *natural* coordinate system for TM.

Now, the curve in the fiber bundle drawn as a dashed line in figure 2.12 identifies a particular vector at each point of M, and so the curve defines a vector field on M. Such a curve (i.e. one which is nowhere parallel to a fiber) is called a *cross-section* of TM. Clearly, it is not usually meaningful to ask for the 'length' of the curve, and so here we have an example of a manifold on which one usually would not bother to define a metric.

A general fiber bundle consists of a *base manifold*, which in our case is the curve M, and one *fiber* attached to each point of the base space. If the base space is n-dimensional and each fiber is m-dimensional, then the bundle has $m + n$

dimensions. It is a special kind of manifold, since it has the property of being decomposable into fibers: the points of a single fiber are related to one another while points on different fibers are not. This is formalized by defining a *projection* map π, which maps any point of a fiber to the point of the base manifold the fiber is attached to. A general manifold does not have such a projection defined on it. The following examples illustrate the wide variety of spaces describable as fiber bundles.

2.10 *Examples of fiber bundles*

(i) The fiber bundle *TM* we have illustrated consists of a manifold and its tangent spaces, and is called the *tangent bundle.* It is one of the most important abstract manifolds in physics. For an n-dimensional manifold, *TM* has $2n$ dimensions.

(ii) Later in this chapter we will generalize from vector fields to tensor fields. There are corresponding bundles over any differentiable manifold for every type of tensor.

(iii) The fibers need not be related to the differential structure of the base space. Consider the 'internal' variables describing the state of an elementary particle, such as isospin. A bundle whose fibers are isospin space and whose base space is spacetime is capable of describing both the position variables (x, y, z, t) of the particle and its internal (isospin) state.

(iv) The view of spacetime taken by Newtonian physics has a natural fiber-bundle structure. To Newton and Galileo, time was absolute: everyone can agree what events are simultaneous, no matter where they occur. We can therefore construct a bundle whose base space is R^1 (time) and whose fibers are R^3 (space). This is illustrated in figure 2.13. There is no natural relation between points on different fibers (points of space at different times), because Newtonian physics has no 'absolute space': two different observers moving with respect to each other disagree as to what constitutes a *fixed* point of space. So there is no natural fiber structure with R^3 as a base, while there is with R^1. One effect of

Fig. 2.12. A cross-section (dashed line) of the fiber bundle *TM* of a one-dimensional manifold *M* (heavy line).

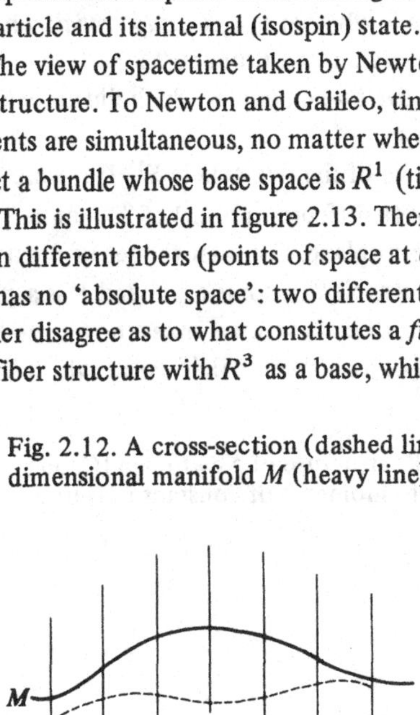

Einstein's relativity was to destroy this bundle structure and to substitute something else, a metric structure (see §2.31 below).

2.11 A deeper look at fiber bundles

There are two related aspects of fiber bundles which we should consider in order to appreciate the richness and usefulness of the bundle concept. These are their global properties and the importance of groups in their construction.

To understand the interesting global properties fiber bundles can have, we must first define a simpler concept, the *product space*. Two spaces M and N have an associated (Cartesian) product space $M \times N$ consisting of all ordered pairs (a, b) with a in M and b in N. For example, R^2 is defined as the product $R^1 \times R^1$. If M and N are manifolds, $M \times N$ is also a manifold in an obvious way: the set of coordinates $\{x^i, i = 1, \ldots, m\}$ of an open set U of M, taken together with $\{y^i, i = 1, \ldots, n\}$ of an open set V of N, form a set of $m + n$ coordinates for the open set (U, V) of $M \times N$. It is clear from our construction of fiber bundles above that they are, at least locally, product spaces, the product $U \times F$ of an open set U of the base manifold B with the space F representing a typical fiber (all fibers being identical to F). This in fact forms part of the definition of a fiber bundle: it is *locally trivial* (it is a product space when we look at a local region of B). The interesting question is whether it is *globally trivial*: whether the whole fiber bundle can be represented as the product $B \times F$.

The answer is usually no, and we give two examples which illustrate what both the question and the answer mean.

(i) Consider TS^2, the tangent bundle of the two-sphere S^2. If it were globally trivial, there would be a C^∞ 1–1 map (a diffeomorphism) of TS^2 onto $S^2 \times R^2$, since the typical fiber is R^2, the tangent plane. Consider the set of points in $S^2 \times R^2$ of the form $(P, \bar{V})$, where P is an arbitrary point of S^2 and $\bar{V}$ is a given fixed vector in R^2. Then the inverse of the above map gives a nowhere-zero *cross-section* of TS^2, i.e. a definition of a C^∞ vector field on S^2 which is

Fig. 2.13. The natural bundle structure of Newtonian (Galilean) spacetime, which is 'sliced up' into moments of constant universal time.

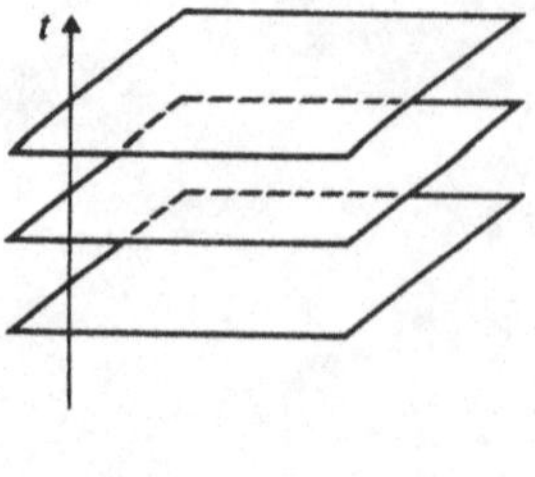

nowhere zero. But in fact there is *no* C^∞ vector field on S^2 which is nowhere zero. This is a consequence of the famous but difficult *fixed-point theorem* of the sphere, that every 1–1 map (diffeomorphism) of S^2 onto itself leaves at least one point of S^2 fixed, provided that the map is a member of a continuous family of maps containing the identity map. A nowhere-zero vector field would generate such a map with no fixed point, as we explain in §3.1 below. Therefore TS^2 does not have a global product structure. This is an example in which the bundle is nontrivial because of the topology of the base manifold, S^2.

(ii) The second example shows that one can actually make a bundle nontrivial even if the base space allows a trivial bundle. Consider TS^1, the tangent bundle of the circle S^1. Unlike S^2, the circle does allow a continuous nowhere-vanishing vector field, and TS^1 is identical to the product space $S^1 \times R$, as shown in figure 2.14. This is just the global version of the local picture shown in figure 2.11(b). But suppose we 'cut' the circle at P in figure 2.14 and unwrap the bundle, lying it flat, as in figure 2.15. To reconstruct figure 2.14 from 2.15 we simply identify point a with a', P with P', b with b', and so on. But we can reassemble the fiber bundle a different way by forming a Möbius band: identify a with b', P with P', b with a', and so on. This gives the strip a twist so that it looks like figure 2.16 when joined together. Locally it is still the same as figure 2.11(b); in fact the

Fig. 2.14. The trivial way of constructing TS^1 as the product space of the circle S^1 and the typical fiber R^1 (drawn vertically). Cf. figure 2.11(b).

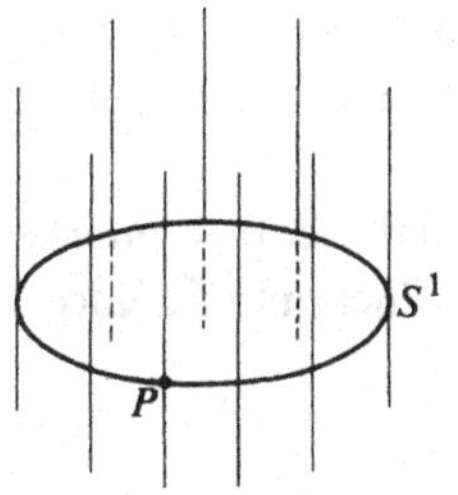

Fig. 2.15. TS^1 cut along one fiber and laid flat. The fibers extend infinitely far in the vertical direction.

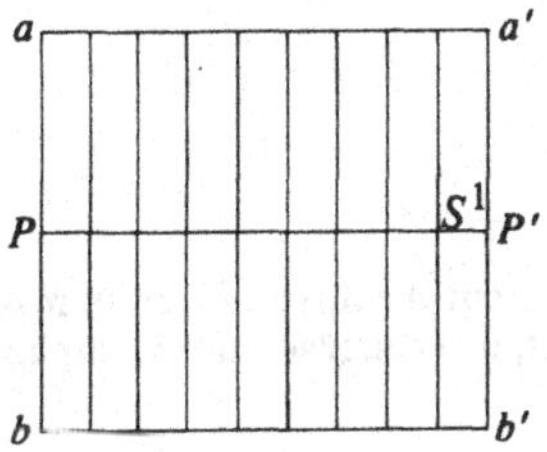

bundle over any connected open proper subset of S^1 ('proper' means not identical to S^1) has a 1–1 continuous map onto the same portion of figure 2.14. One has to go all the way around to see that there is no *continuous* 1–1 map of *all* of one bundle onto all of the other. Therefore, the Möbius band is not a product space, and the second bundle is nontrivial. Nontrivial constructions of bundles in analogous ways are used in modern particle physics to define the so-called 'instantons'.

The Möbius example has a lesson for us: it is not sufficient simply to say what the base and fiber of a bundle are, because there may be more than one way to construct such a bundle. We need a better definition of a fiber bundle, and this is where groups come in. The difference between the two bundles over S^1 is in what is called the bundles' *structure group*. To phrase the full definition of a fiber bundle more compactly, we need to define a *homeomorphism*, which is simply a 1–1 map from one space onto another, which is continuous and whose inverse is continuous.† (For an explanation of the terminology of maps, see §1.2.) We define a *fiber bundle* as a space E for which the following are given: a base manifold B, a projection $\pi: E \to B$, a typical fiber F, a structure group G of homeomorphisms of F onto itself, and a family $\{U_j\}$ of open sets covering B (i.e. open sets whose union is B), all of which satisfy the following restrictions.

(i) Locally the bundle is trivial, which means that the bundle over any set U_j, which is just $\pi^{-1}(U_j)$, has a homeomorphism onto the product space $U_j \times F$. We have noted this above. Part of this homeomorphism is a homeomorphism from each fiber, say $\pi^{-1}(x)$ where x is an element of B, onto F. Let us call this map $h_j(x)$, labelled not only by the point x which defines the fiber but also by the index j which denotes the set U_j containing x.

(ii) When two sets U_j and U_k overlap, a given point x in their intersection has two homeomorphisms $h_j(x)$ and $h_k(x)$ from its fiber onto F. Since a

Fig. 2.16. The Möbius-band version of this bundle: the fibers turn over once as one follows them around the circle. Locally it still has the same structure as figure 2.11(*b*).

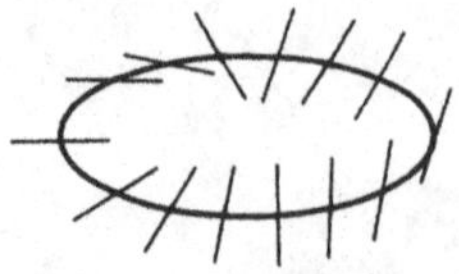

† A homeomorphism is a diffeomorphism without the differentiability requirement. For most of the bundles of physical interest, one can read 'diffeomorphisms' for 'homeomorphisms'.

homeomorphism is invertible, the map $h_j(x) \circ h_k^{-1}(x)$ is a homeomorphism of F onto F. This is required to be an element of the structure group G.

The second restriction contains the information about the global structure of the fiber bundle. To see how this works, we first give the complete definition of TS^1 (which has a straightforward generalization to TM for any M). The bundle $E = TS^1$ has base $B = S^1$, typical fiber $F = R^1$, and projection π: $(x, \bar{v}) \mapsto x$, where x is a point of S^1 and $\bar{v}$ is a vector in T_x. Let the covering $\{U_j\}$ be the open sets of any atlas of S^1. A typical family $\{U_j\}$ is illustrated in figure 2.17. Every U_j has a coordinate 'system', i.e. a parameterization of S^1, which we will call λ_j. The vector $\mathrm{d}/\mathrm{d}\lambda_j$ at x in U_j is a basis for T_x, so any vector $\bar{v}$ in T_x has the representation $\alpha_{(j)}\mathrm{d}/\mathrm{d}\lambda_j$ for any fixed j, where $\alpha_{(j)}$ is a real number. This is just equation (2.4) again. The homeomorphisms of T_x onto R which are part of the definition of TS^1 are *defined* to be $h_j(x)$: $\bar{v} \mapsto \alpha_{(j)}$. If x is in two neighborhoods U_j and U_k there are two such homeomorphisms from T_x onto R, and since λ_j and λ_k are unrelated, $\alpha_{(j)}$ and $\alpha_{(k)}$ can be any two nonzero real numbers. The homeomorphism $h_j(x) \circ h_k^{-1}(x)$: $F \to F$ maps $\alpha_{(k)} \mapsto \alpha_{(j)}$ and is therefore just multiplication by the number $r_{jk} = \alpha_{(j)}/\alpha_{(k)}$. Since r_{jk} is any real number other than zero, the structure group is $R^1 - \{0\}$, which is a group under multiplication. We note in passing that for an n-dimensional manifold M, the structure group of TM is the set of all $n \times n$ matrices with nonzero determinant, which is called $GL(n, \mathrm{R})$. We will study this group in chapter 3.

This defines TS^1. But what does it look like? It is possible to choose the coordinates λ_j in such a way that any two, say λ_j and λ_k, increase in the same direction in S^1 in the region where U_j and U_k overlap. (We say that S^1 is *orientable*; see §4.7.) With such a choice of coordinates it is not hard to see that all the 'overlap numbers' r_{jk} are positive, and the structure group reduces to R^+, multiplication by the positive real numbers. In fact we can do even better by scaling the coordinates in such a way that $\mathrm{d}\lambda_j/\mathrm{d}\lambda_k = 1$ in every overlap region. Then the group reduces to 1, the identity element. The structure group is trivial, and so is the bundle structure. This is the bundle represented in figure 2.14.

To characterize the structure of the Möbius band we must use different maps

Fig. 2.17. A set of neighborhoods of S^1 which cover S^1. The extent of each neighborhood is indicated by the parentheses. U_1 overlaps U_2, U_2 overlaps U_3, and so on until U_8 overlaps U_1.

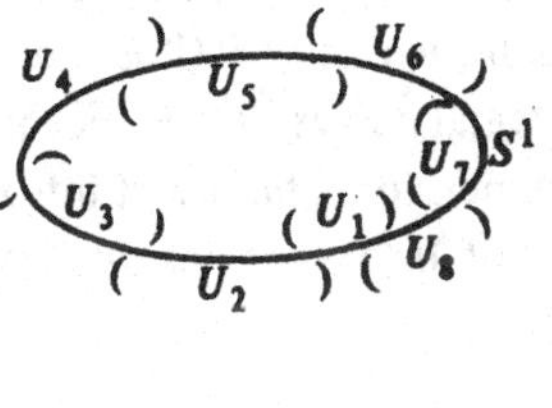

$h_j(x)$, and we must be careful *not* to try to interpret the bundle as a tangent bundle. The easiest procedure is to use the family $\{U_j, j = 1, \ldots, 8\}$ shown in figure 2.17 and to define $r_{12} = 1, r_{23} = 1, \ldots, r_{78} = 1$. But then the twist in the Möbius band forces us to use $r_{81} = -1$. The structure group consists of the elements $\{1, -1\}$ with multiplication as the group operation. We could have made other choices for the r_{jk}s, but we could not have found a smaller structure group.

The tangent bundle TS^1 had structure group $R^1 - \{0\}$, which is nearly the same as its typical fiber. The *frame bundle* of any manifold M has the same structure group as TM, but its fiber is the set of all *bases* for the tangent space (equivalently, for R^n). In the case of a one-dimensional manifold like S^1, this is the set of all nonzero vectors, which is identical to $R^1 - \{0\}$. So the frame bundle of S^1 has fibers homeomorphic to its structure group, and this is true of all frame bundles. Such a bundle is called a *principal fiber bundle.*

2.12 Vector fields and integral curves

As defined in §2.7, a *vector field* is a rule that gives a vector at every point of M. Each point has its own tangent vector space, so a vector field selects one vector from each space. Now, every curve has a tangent vector at every point, and the question arises of whether the converse is true: given an arbitrary vector field, is it possible to start at one point P and find a curve whose tangent vector is always the vector field at whatever point the curve passes through? The answer is yes, for C^1 vector fields, and such curves are called integral curves of the vector field. The proof is as follows. Let the components of the vector field be $V^i(P)$, functions of P. In some coordinate system $\{x^i\}$ we have $V^i(P) = v^i(x^j)$. The statement that this is a tangent vector to a curve with parameter λ is

$$\frac{\mathrm{d}x^i}{\mathrm{d}\lambda} = v^i(x^j). \tag{2.5}$$

This is just a set of first-order ordinary differential equations for $x^i(\lambda)$, and a unique solution always exists in some neighborhood of the initial point P. (This existence/uniqueness theorem for ordinary differential equations is proved in most textbooks on differential equations. A version may be found in Choquet-Bruhat, Dewitt-Morette & Dillard-Bleick (1977) in the bibliography.) Two particular vector fields are illustrated in figure 2.18.

Notice that the paths of different integral curves can never cross except possibly at a point where $V^i = 0$ for all i, because of the uniqueness of solutions to (2.5). Since some integral curve passes through each point P (it is found by solving (2.5) with initial conditions at P), the integral curves 'fill' M. For instance, if M is three-dimensional, then there is a two-dimensional family of integral curves for each vector field on M, and they cover all of M (except possibly isolated points where $V^i = 0$ for all i). Such a manifold-filling set of curves is called a *congruence*. The set of curves, incidentally, can usually be regarded as a manifold itself.

2.13 Exponentiation of the operator d/dλ

We now introduce an idea that will prove to be a useful tool in several subsequent calculations. Suppose we have an *analytic* manifold (C^ω), and the coordinate values $x^i(\lambda)$ of points along the integral curves of $\bar{Y} = \mathrm{d}/\mathrm{d}\lambda$ are analytic functions of λ. Then the coordinates of two points with parameters λ_0 and $\lambda_0 + \epsilon$ are related by the Taylor series

$$\begin{aligned} x^i(\lambda_0 + \epsilon) &= x^i(\lambda_0) + \epsilon \left(\frac{\mathrm{d}x^i}{\mathrm{d}\lambda}\right)_{\lambda_0} + \frac{1}{2!}\epsilon^2 \left(\frac{\mathrm{d}^2 x^i}{\mathrm{d}\lambda^2}\right)_{\lambda_0} + \dots \\ &= \left(1 + \epsilon \frac{\mathrm{d}}{\mathrm{d}\lambda} + \frac{1}{2}\epsilon^2 \frac{\mathrm{d}^2}{\mathrm{d}\lambda^2} + \dots\right) x^i \Big|_{\lambda_0} \\ &= \exp\left[\epsilon \frac{\mathrm{d}}{\mathrm{d}\lambda}\right] x^i \Big|_{\lambda_0}, \end{aligned} \tag{2.6}$$

where the 'exp' notation is an obvious and convenient shorthand for the differential operator which, when applied to $x^i(\lambda)$ and evaluated at λ_0, gives the Taylor series. It is called the exponentiation of the operator $\epsilon\mathrm{d}/\mathrm{d}\lambda$. Since $\epsilon\mathrm{d}/\mathrm{d}\lambda$ is an infinitesimal 'motion' along the integral curve, its exponentiation gives a finite motion. Other notations we will use include

$$\exp(\epsilon\mathrm{d}/\mathrm{d}\lambda) = \mathrm{e}^{\epsilon\mathrm{d}/\mathrm{d}\lambda} = \mathrm{e}^{\epsilon\bar{Y}}.$$

2.14 Lie brackets and noncoordinate bases

Given a coordinate system x^i, it is often convenient to adopt $\{\partial/\partial x^i\}$ as a basis for vector fields. However, *any* linearly independent set of vector fields

Fig. 2.18. Integral curves of two vector fields on R^2. (*a*) $\bar{V} = x\partial/\partial y - y\partial/\partial x$; (*b*) $\bar{V} = (x + y/r)\partial/\partial y - (y - x/r)\partial/\partial x$, with $r = (x^2 + y^2)^{1/2}$.

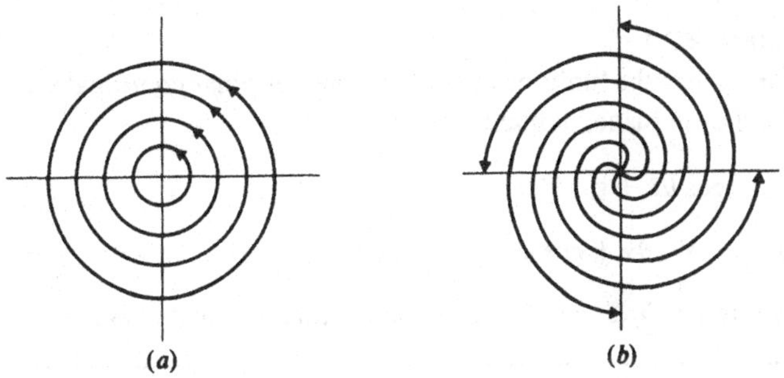

can serve as a basis, and one can easily show that not all of them are derivable from coordinate systems. This is because the operators $\partial/\partial x^i$ and $\partial/\partial x^j$ *commute* for all i, j. Two arbitrary vector fields do not commute: if $\bar{V} = \mathrm{d}/\mathrm{d}\lambda$ and $\bar{W} = \mathrm{d}/\mathrm{d}\mu$, then using equation (2.3)

$$\begin{aligned}\frac{\mathrm{d}}{\mathrm{d}\lambda}\frac{\mathrm{d}}{\mathrm{d}\mu} - \frac{\mathrm{d}}{\mathrm{d}\mu}\frac{\mathrm{d}}{\mathrm{d}\lambda} &= \sum_{i,j} V^i \frac{\partial}{\partial x^i} W^j \frac{\partial}{\partial x^j} - W^j \frac{\partial}{\partial x^j} V^i \frac{\partial}{\partial x^i} \\ &= \sum_{i,j} V^i W^j \left(\frac{\partial}{\partial x^i}\frac{\partial}{\partial x^j} - \frac{\partial}{\partial x^j}\frac{\partial}{\partial x^i}\right) \\ &\quad + \sum_{i,j} V^i \frac{\partial W^j}{\partial x^i}\frac{\partial}{\partial x^j} - \sum_{i,j} W^j \frac{\partial V^i}{\partial x^j}\frac{\partial}{\partial x^i} \\ &= \sum_{i,j}\left(V^i \frac{\partial W^j}{\partial x^i} - W^i \frac{\partial V^j}{\partial x^i}\right)\frac{\partial}{\partial x^j},\end{aligned} \tag{2.7}$$

where the last line follows from relabelling the summation indices in the final sum of the middle quantity. Therefore, the commutator

◆ $$\left[\frac{\mathrm{d}}{\mathrm{d}\lambda}, \frac{\mathrm{d}}{\mathrm{d}\mu}\right] \equiv \frac{\mathrm{d}}{\mathrm{d}\lambda}\frac{\mathrm{d}}{\mathrm{d}\mu} - \frac{\mathrm{d}}{\mathrm{d}\mu}\frac{\mathrm{d}}{\mathrm{d}\lambda} \tag{2.8}$$

is a *vector field* whose components do not vanish in general. If $\mathrm{d}/\mathrm{d}\lambda$ and $\mathrm{d}/\mathrm{d}\mu$ are two elements of a basis, then they will not be expressible as derivatives with respect to any coordinates. Such a basis is a noncoordinate basis.

It is important to realize that this distinction between coordinate and noncoordinate bases is one which can be made only over some region of the manifold, not at a single point. It depends on the *derivatives* of the components of the vectors, not just on their values at a point. The different properties of coordinate and noncoordinate bases therefore matter only over regions of a manifold, and are irrelevant in problems which involve only the tangent space T_P of a single point P.

Exercise 2.1

Show that the 'unit' basis vector fields for polar coordinates in the Euclidean plane, defined by

$$\hat{\mathbf{r}} = \cos\theta\,\hat{\mathbf{x}} + \sin\theta\,\hat{\mathbf{y}},$$

$$\hat{\boldsymbol{\theta}} = -\sin\theta\,\hat{\mathbf{x}} + \cos\theta\,\hat{\mathbf{y}},$$

where $\hat{\mathbf{x}} = \partial/\partial x$ and $\hat{\mathbf{y}} = \partial/\partial y$, are a noncoordinate basis.

The commutator [d/dλ, d/dμ] is called the *Lie bracket*† of $\bar{V}$ and $\bar{W}$, and we now look at its geometrical interpretation. In figure 2.19 we have drawn a coordinate grid on a two-dimensional manifold. Notice that by definition x^1 is constant along the lines of x^2, which are the integral curves of $\partial/\partial x^2$. That is why $\partial/\partial x^1$ and $\partial/\partial x^2$ commute: each is a derivative along a line on which the other is fixed. Now consider two arbitrary vector fields, $\bar{V} = \mathrm{d}/\mathrm{d}\lambda$ and $\bar{W} = \mathrm{d}/\mathrm{d}\mu$, whose integral curves are shown in figure 2.20. An integral curve of $\bar{W}$ is not necessarily a curve of constant λ, and vice versa. The derivative d/dμ is not a derivative holding λ fixed, so d/dλ and d/dμ do not commute. Although the $\bar{V}$ and $\bar{W}$ curves look like coordinate curves, their *parameterization* is not that of a coordinate system. Even the fact that they *look* like coordinate curves is an artefact of two dimensions: in three dimensions it may happen that curve (1) intersects curves (α) and (β) but (2) intersects only (α).

Fig. 2.19. Typical coordinate grid on a two-dimensional manifold.

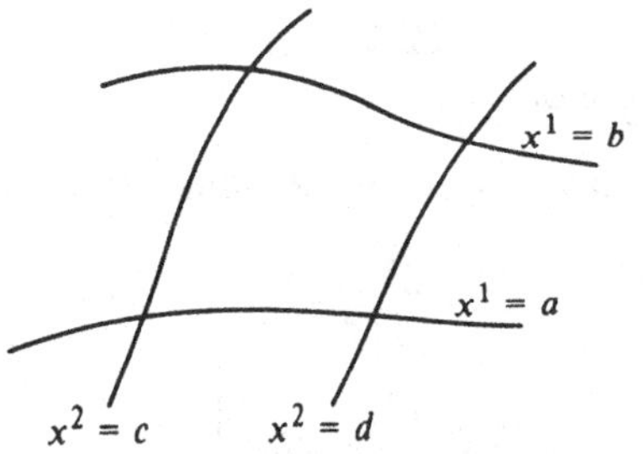

Fig. 2.20. Typical integral curves of two vector fields on a two-dimensional manifold.

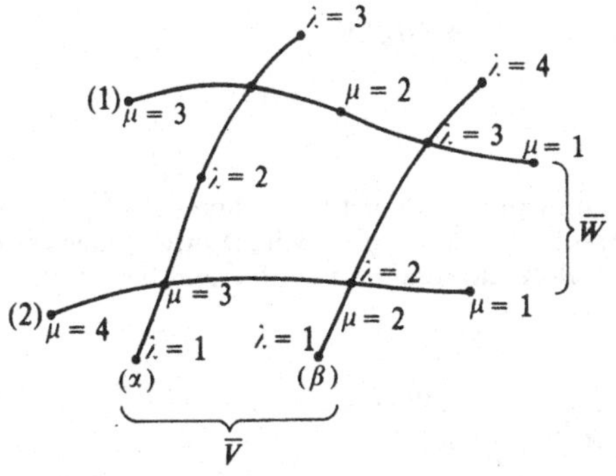

† The 'Lie' of Lie bracket is the same Lie as in Lie groups: Sophus Lie, the great mathematician of the late nineteenth century. The Lie bracket is, as we shall see, a special case of the Lie derivative. Readers who are familiar with Lie groups may recognize the Lie bracket as the commutator of the vector fields d/dλ and d/dμ which generate a Lie group of mappings. We discuss these mappings in chapter 3.

We can obtain a picture of the vector $[\bar{V}, \bar{W}]$ in the following manner. In figure 2.21, consider starting at P, moving $\Delta\lambda = \epsilon$ along the $\bar{V}$ curve through P, and then moving $\Delta\mu = \epsilon$ along a $\bar{W}$ curve. One winds up at A. Starting again at P and going first $\Delta\mu = \epsilon$ and then $\Delta\lambda = \epsilon$, takes one to $B \neq A$. We shall show that the vector stretching from A to B is $\epsilon^2\,[\bar{V}, \bar{W}]$, to lowest order in ϵ.

It is most convenient to use the exponentiation operator introduced earlier. It is clear that

$$x^i(R) = \exp\left[\epsilon\frac{\mathrm{d}}{\mathrm{d}\lambda}\right]x^i\Big|_P,$$

and

$$x^i(A) = \exp\left[\epsilon\frac{\mathrm{d}}{\mathrm{d}\mu}\right]\exp\left[\epsilon\frac{\mathrm{d}}{\mathrm{d}\lambda}\right]x^i\Big|_P. \tag{2.9}$$

Similarly, the path to point B from P gives us

$$x^i(B) = \exp\left[\epsilon\frac{\mathrm{d}}{\mathrm{d}\lambda}\right]\exp\left[\epsilon\frac{\mathrm{d}}{\mathrm{d}\mu}\right]x^i\Big|_P. \tag{2.10}$$

Then the difference in the coordinates of A and B is

$$x^i(B) - x^i(A) = [\mathrm{e}^{\epsilon\,\mathrm{d}/\mathrm{d}\lambda}, \mathrm{e}^{\epsilon\,\mathrm{d}/\mathrm{d}\mu}]\,x^i|_P, \tag{2.11}$$

just the commutator of the exponentiation operators. Returning to the Taylor series, we can write

$$\begin{aligned}[\mathrm{e}^{\epsilon\,\mathrm{d}/\mathrm{d}\lambda}, \mathrm{e}^{\epsilon\,\mathrm{d}/\mathrm{d}\mu}] &= \Big[1 + \epsilon\frac{\mathrm{d}}{\mathrm{d}\lambda} + \frac{1}{2}\epsilon^2\frac{\mathrm{d}^2}{\mathrm{d}\lambda^2} + O(\epsilon^3),\\ &\qquad 1 + \epsilon\frac{\mathrm{d}}{\mathrm{d}\mu} + \frac{1}{2}\epsilon^2\frac{\mathrm{d}^2}{\mathrm{d}\mu^2} + O(\epsilon^3)\Big]\\ &= \epsilon^2\left[\frac{\mathrm{d}}{\mathrm{d}\lambda}, \frac{\mathrm{d}}{\mathrm{d}\mu}\right] + O(\epsilon^3),\end{aligned}$$

$$x^i(B) - x^i(A) = \epsilon^2\,[\bar{V}, \bar{W}] + O(\epsilon^3). \tag{2.12}$$

Fig. 2.21. Geometric interpretation of the Lie bracket $[\bar{V}, \bar{W}]$ as the open part of an incomplete parallelogram whose other sides are equal parameter increments along integral curves of $\bar{V}$ and $\bar{W}$.

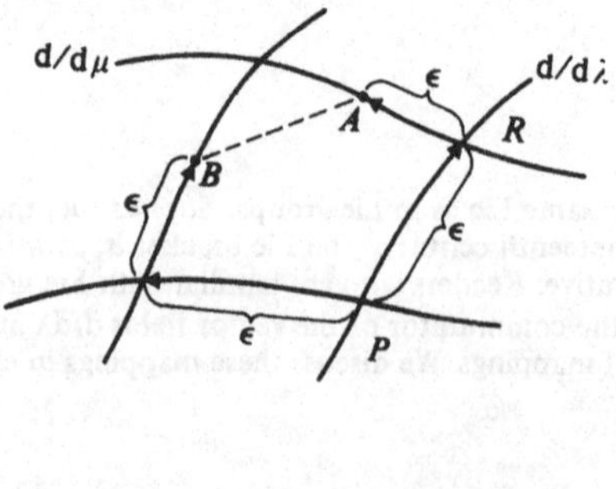

This is just the ith component of the Lie bracket, and (2.12) justifies the picture we have given for it.

Exercise 2.2

(a) Use (2.6) to prove (2.12).

(b) Prove that

$$\exp\,[a\mathrm{d}/\mathrm{d}\lambda + b\mathrm{d}/\mathrm{d}\mu] = \exp\,[a\mathrm{d}/\mathrm{d}\lambda]\,\exp\,[b\mathrm{d}/\mathrm{d}\mu] \qquad (2.13)$$

for all a and b if and only if $[\mathrm{d}/\mathrm{d}\lambda, \mathrm{d}/\mathrm{d}\mu] = 0$.

Exercise 2.3

Prove that any three twice-differentiable (i.e. C^2) vector fields $\bar{X}$, $\bar{Y}$ and $\bar{Z}$ satisfy the *Jacobi identity*

$$[[\bar{X}, \bar{Y}], \bar{Z}] + [[\bar{Y}, \bar{Z}], \bar{X}] + [[\bar{Z}, \bar{X}], \bar{Y}] = 0. \qquad (2.14)$$

A *Lie algebra* of vector fields on a region U of M is a set A of vector fields on U which is a vector space under addition (which means any linear combination *with constant coefficients* of fields in A is a field in A) and which is closed under the Lie-bracket operation (the Lie bracket of any two fields in A is another field in A). Clearly, the set of all C^∞ vector fields on U is a Lie algebra, but it is more interesting when a smaller set of vector fields singled out for some reason also forms a Lie algebra. These are closely related to the invariance properties of manifolds and to their associated invariance groups, which are usually Lie groups. We shall study this in greater detail in chapter 3, where we will also present a more general definition of a Lie algebra.

2.15 When is a basis a coordinate basis?

Suppose we are given two vector fields $\bar{A} = \mathrm{d}/\mathrm{d}\lambda$ and $\bar{B} = \mathrm{d}/\mathrm{d}\mu$ on a two-dimensional manifold M, and suppose that $\bar{A}$ and $\bar{B}$ are linearly independent at every point of some open neighborhood U of M, so that they form a basis for vector fields there. What condition would assure us that they are a coordinate basis, in other words that λ and μ are coordinates for U? It is clearly necessary that they commute

$$[\bar{A}, \bar{B}] = 0.$$

We shall show that this condition is sufficient as well. To do this we go right back to the basic definition of a manifold: we construct a 1–1 map from U onto a neighborhood in R^2. Beginning at some point P in U, and using arbitrary coordinates (x^1, x^2) in U, we move a parameter distance λ_1 from P along $\bar{A}$ to a point R whose coordinates are (by equation (2.6))

$$x^i(R) = e^{\lambda_1 d/d\lambda} x^i|_P.$$

If we go first a distance λ_1 along $\bar{A}$, then μ_1 along $\bar{B}$, we get to a point Q with coordinates

$$x^i(Q) = e^{\mu_1 d/d\mu} e^{\lambda_1 d/d\lambda} x^i|_P.$$

This equation defines an exponential-type map from some neighborhood V of the origin of R^2 into U: a given element of V, the pair (λ_1, μ_1), is mapped to the point Q. This map is illustrated in figure 2.22. In order for this map to define a coordinate system, it must be 1–1: it must have an inverse. We show below that it does have an inverse everywhere in U, but first we shall show that $\bar{A}$ and $\bar{B}$ are the coordinate basis vectors of this coordinate system if they commute in this neighborhood. Let us rewrite the map as the coordinate transformation from $\{\alpha, \beta\}$ to $\{x^1, x^2\}$:

$$x^i(\alpha, \beta) = e^{\beta d/d\mu} e^{\alpha d/d\lambda} x^i|_P.$$

The basis vectors $\partial/\partial\alpha$ and $\partial/\partial\beta$ have components (in the $\{x^i\}$ coordinate system) $\partial x^i/\partial\alpha$ and $\partial x^i/\partial\beta$, respectively. It is easy to show from (2.6) that

$$\frac{d}{d\alpha} e^{\alpha d/d\lambda} = e^{\alpha d/d\lambda} \frac{d}{d\lambda},$$

and since $d/d\mu$ and $d/d\lambda$ commute, we obtain

$$\frac{\partial x^i}{\partial\alpha} = e^{\beta d/d\mu} e^{\alpha d/d\lambda} \left.\frac{dx^i}{d\lambda}\right|_P,$$

$$\frac{\partial x^i}{\partial\beta} = e^{\beta d/d\mu} e^{\alpha d/d\lambda} \left.\frac{dx^i}{d\mu}\right|_P.$$

But $dx^i/d\lambda$ is just the component of $d/d\lambda$ in the $\{x^i\}$ coordinate system. Since this is an analytic function of M, operating on it with $\exp(\beta d/d\mu) \cdot \exp(\alpha d/d\lambda)$

Fig. 2.22. The map from R^2 to M described in the text. This provides a coordinate system in some neighborhood of P.

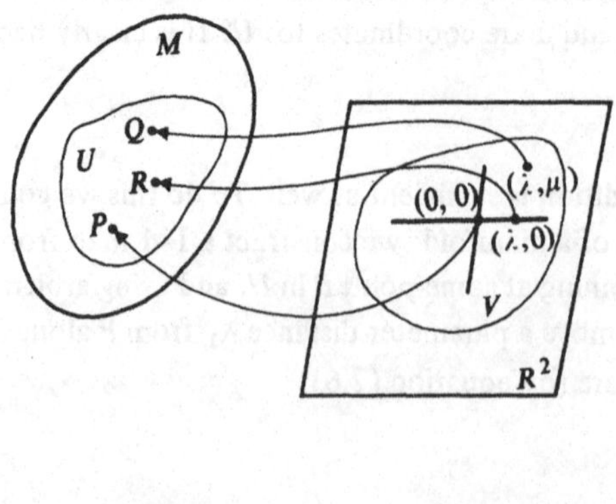

simply produces its value at the point whose coordinates are (α, β). Therefore we have everywhere in U

$$\partial/\partial\alpha = \mathrm{d}/\mathrm{d}\lambda \text{ and } \partial/\partial\beta = \mathrm{d}/\mathrm{d}\mu,$$

and we have proved the sufficiency of $[\bar{A}, \bar{B}] = 0$ as a condition that $\bar{A}$ and $\bar{B}$ be coordinate basis vectors.

We return now to the deferred proof that $\{\alpha, \beta\}$ do form a coordinate system in U. We must prove that the map $\{\alpha, \beta\} \to \{x^i\}$ has an inverse, and for this we use the inverse function theorem (see §1.2). This says that if the matrix

$$\begin{pmatrix} \dfrac{\partial x^1}{\partial\alpha} & \dfrac{\partial x^2}{\partial\alpha} \\ \dfrac{\partial x^1}{\partial\beta} & \dfrac{\partial x^2}{\partial\beta} \end{pmatrix}$$

has a nonzero determinant at some point $\{\alpha, \beta\}$, then the map has an inverse in some neighborhood of this point. The determinant will vanish if and only if the vectors $\partial x^i/\partial\alpha$, $\partial x^i/\partial\beta$ are linearly dependent, but from the above discussion it is clear that this will never happen because $\bar{A}$ and $\bar{B}$ are linearly independent in U. Therefore, everywhere in U the map is invertible and provides a coordinate system.

It is interesting to ask where this argument breaks down if $[\bar{A}, \bar{B}] \neq 0$. In this case the expression $\partial x^i/\partial\beta$ is more complicated. It is still true that, at least in some neighborhood of $\alpha = \beta = 0$, the map has an inverse. But because $\partial x^i/\partial\beta$ is no longer just $\mathrm{d}x^i/\mathrm{d}\mu$ at the point in question, the vectors $\bar{A}$ and $\bar{B}$ are not the basis vectors of the constructed coordinates.

The whole argument extends to n dimensions: if n vector fields $\{\bar{Y}_{(j)}, j = 1, \ldots, n\}$ on an n-dimensional manifold M are linearly independent and commute with one another in some open region U of M, then they are the coordinate basis vectors of the coordinate system $\{\alpha_j\}$, given in terms of an arbitrary system $\{x^i\}$ by

$$x^i(\alpha_1, \ldots, \alpha_n) = \exp\left[\sum_j \alpha_j \bar{Y}_{(j)}\right] x_i \Big|_P,$$

centred at an arbitrary point P in U.

2.16 One-forms

Let us go back to T_P, the space of all tangent vectors at P. As a first step towards tensors, we define a *one-form* as a linear, real-valued function of vectors. This means the following: a one-form $\tilde{\omega}$ at P associates with a vector $\bar{V}$ at P a real number, which we call $\tilde{\omega}(\bar{V})$. This notation expresses the idea that $\tilde{\omega}$ is a function on vectors. (A tilde (˜) over a letter always denotes a one-form, just as a bar (¯) denotes a vector.) The linearity of this function means

$$\tilde{\omega}(a\bar{V} + b\bar{W}) = a\tilde{\omega}(\bar{V}) + b\tilde{\omega}(\bar{W}), \tag{2.15}$$

where a and b are real numbers. We can define addition of one-forms and their multiplication by real numbers in a straightforward way: $a\tilde{\omega}$ is the one-form such that

$$(a\tilde{\omega})(\bar{V}) = a[\tilde{\omega}(\bar{V})] \tag{2.16a}$$

for all $\bar{V}$, and $\tilde{\omega} + \tilde{\sigma}$ is the one-form such that

$$(\tilde{\omega} + \tilde{\sigma})(\bar{V}) = \tilde{\omega}(\bar{V}) + \tilde{\sigma}(\bar{V}) \tag{2.16b}$$

for all $\bar{V}$. Thus one-forms at the point P satisfy the axioms of a vector space, which is called the dual vector space to T_P, and is denoted by $T^*{}_P$. The reason it is 'dual' is that vectors can also be regarded as linear, real-valued functions of one-forms, in the following manner. Given a vector $\bar{V}$, its value on any one-form $\tilde{\omega}$ is *defined* as $\tilde{\omega}(\bar{V})$. This is linear, since its value on $a\tilde{\omega} + b\tilde{\sigma}$ is, by (2.16) above,

$$\begin{aligned}(a\tilde{\omega} + b\tilde{\sigma})(\bar{V}) &= (a\tilde{\omega})(\bar{V}) + (b\tilde{\sigma})(\bar{V}) \\ &= a(\text{value of } \bar{V} \text{ on } \tilde{\omega}) + b(\text{value of } \bar{V} \text{ on } \tilde{\sigma}).\end{aligned} \tag{2.17}$$

It is thus the linearity property which enables us to regard each as a function taking the other as argument and producing a real number; vectors and one-forms are thus said to be *dual* to each other. Their value on one another is often represented in many ways:

$$\blacklozenge \quad \tilde{\omega}(\bar{V}) \equiv \bar{V}(\tilde{\omega}) \equiv \langle \tilde{\omega}, \bar{V} \rangle, \tag{2.18}$$

where the last expression emphasizes their equal status. The formation of the number $\tilde{\omega}(\bar{V})$ is often called the *contraction* of $\tilde{\omega}$ with $\bar{V}$. In older treatments of tensor algebra, vectors are often called 'contravariant vectors' and one-forms 'covariant vectors'. These names refer to the behavior of their components under a change of basis, which is something we will deal with in §2.26.

2.17 Examples of one-forms

Before going further with the mathematical development, let us look at some familiar examples of one-forms. One of the most common is the gradient of a function, which will be discussed in §2.19. Other examples include the following:

(i) In matrix algebra, if we call column vectors 'vectors', then row vectors are one-forms. This is because when multiplied (in the correct order) by the usual rules of matrix multiplication, they give a single real number. For example, in the two-dimensional case the row vector $(-1, 5)$ may be thought of as a function which takes an arbitrary column vector into a real number:

$$(-1, 5): \begin{pmatrix} x \\ y \end{pmatrix} \mapsto (-1, 5)\begin{pmatrix} x \\ y \end{pmatrix} = -x + 5y.$$

That this function is linear is easily checked.

(ii) In the Hilbert spaces used in quantum mechanics, the analogues of example (i) are Dirac kets $|\psi\rangle$ (vectors) and bras $\langle\phi|$ (one-forms), whose contraction is $\langle\phi|\psi\rangle$, a complex number. (The generalization of vector and tensor algebra to algebras over the complex numbers rather than the reals is trivial: one just replaces the word 'real' by 'complex'. In many ways, the generalization of our real manifolds to complex-analytic ones – where the maps are analytic maps to the space $(z^1, z^2, \ldots, z^n)$ for complex rather than real z^i – is also simple. But some features of complex manifolds, such as their global structure and curvature, present special problems, which we cannot treat in this book.) The notation $\langle\phi|\psi\rangle$ is, not accidentally, similar to (2.18).

In both examples (i) and (ii) one is used to switching between the vectors and one-forms, associating with a given vector its 'conjugate' or 'transpose', which is a one-form. We shall see in §2.29 below that this is equivalent to giving a metric or inner product to the vector space. This is a very important additional structure in a vector space, but the reader should bear in mind that there is no *a priori*, 'natural' way of associating a particular one-form with a particular vector.

2.18 *The Dirac delta function*

In quantum mechanics one often deals with *function spaces.* Consider the set $C[-1, 1]$ of all C^∞ real-valued functions defined on the interval $-1 \leqslant x \leqslant 1$ of R^1. This set is a group under addition (the sum of any two C^∞ functions is C^∞, etc.) and a vector space under multiplication by real constants (if f is a C^∞ function, so is cf for any constant c). Its dual space of one-forms is called the *distributions*. An example of a distribution is the *Dirac delta 'function'* $\delta(x)$, which is defined as that one-form whose value on a C^∞ function $f(x)$ is $f(0)$:

$$\langle\delta(x), f(x)\rangle = f(0). \tag{2.19}$$

In one sense $\delta(x)$ is a true function: it is a mapping $C[-1, 1] \to R$. It is customary to apply the word 'distribution' only to a continuous function of this sort. But any notion of continuity requires a topology, in this case a topology for $C[-1, 1]$. This is an infinite-dimensional vector space (there are an infinite number of linearly independent C^∞ functions), and a discussion of its topology is well outside the scope of this book. The interested reader can consult Choquet-Bruhat *et al.* (1977). What is important for one to understand at present is that this sense of function is not what Dirac and his contemporaries meant when they called $\delta(x)$ the delta function. To see what they had in mind we have to look again at a way of transforming a function in $C[-1, 1]$ into a one-form on $C[-1, 1]$.

For any function g in $C[-1, 1]$ it is possible to define a one-form $\tilde{g}$ whose value on a function f in $C[-1, 1]$ is

$$\langle \tilde{g}, f \rangle = \int_{-1}^{1} g(x) f(x)\, dx. \tag{2.20}$$

This is indeed a linear function mapping f to the integral's value. (Since g and f are continuous on $-1 \leqslant x \leqslant 1$, they are bounded there and the integral always exists.) The name 'delta function' was used as a loose way of turning this relation around: if $\delta(x)$ is a one-form, then one ought to be able to talk about it as a function of x in the ordinary sense whose integral with $f(x)$ produced $f(0)$:

$$\int_{-1}^{1} \delta(x) f(x)\, dx = f(0).$$

This idea caused great distress to mathematicians, some of whom even declared that Dirac was wrong despite the fact that he kept getting consistent and useful results. Wisely, the physicists rejected these extreme criticisms and followed their intuition. We can now see why they were 'wrong' and still succeeded. They were 'wrong' because they spoke of $\delta(x)$ as a function $R^1 \to R^1$, which it cannot be in any precise sense, and because they treated it as a function by integrating it and even differentiating it:

$$\int_{-1}^{1} \delta'(x) f(x)\, dx = -\int_{-1}^{1} \delta(x) f'(x)\, dx = -f'(0).$$

But they were 'right' because they never used $\delta(x)$ outside integrals with sufficiently-differentiable functions $f(x)$: they never used it except to map functions to real numbers. In this sense they employed the machinery but not the words of distribution theory, which was devised expressly in order to give delta functions a sound basis. Notice, however, that distribution theory has one big simplification over the older physicists' view: it can define the delta function without referring to any rule like (2.20) for turning a function into a one-form. As remarked in example (ii) above, such a rule is an extra structure on a vector space, which we now see is unnecessary for understanding delta functions.

We should remark in passing that we restricted the word 'distribution' to *continuous* one-forms, in keeping with the usual practice in defining the dual of any vector space. However, we did not include the word 'continuous' in our definition of one-forms in §2.16: have we been inconsistent? The answer is no, because on a finite-dimensional vector space a linear function is *always* continuous. (See Choquet-Bruhat *et al.* (1977) or Rudin (1964), in the bibliography of chapter 1.)

2.19 The gradient and the pictorial representation of a one-form

A *field* of one-forms, by analogy with vector fields, is a rule giving a one-form at every point. The rules in equation (2.16) extend to fields; in this case, a is a function on M, not necessarily constant. Differentiability of one-form fields can be defined in terms of that of vector fields and functions. For example, on a C^∞ manifold, a given one-form field $\tilde{\omega}$ defines, when supplied with a vector field $\bar{V}$, a function $\tilde{\omega}(\bar{V})$. If this function is C^∞ for any C^∞ $\bar{V}$ then $\tilde{\omega}$ is C^∞. (We will give an easier definition of differentiability after defining components of one-forms in §2.20.) As with vector fields, there is a fiber bundle called the *cotangent bundle* T^*M with M as base and T^*_P as the fiber over the point P. Cross-sections of T^*M are one-form fields.

A most useful and instructive one-form field is the gradient of a function f, which we denote by $\tilde{d}f$. Although elementary treatments of vector calculus call the gradient a vector, it is properly a one-form. Thus, the *gradient* $\tilde{d}f$ (*not* the 'infinitesimal' df, which we rarely use)† is defined by

$$\blacklozenge \qquad \tilde{d}f(d/d\lambda) = df/d\lambda, \tag{2.21}$$

where $d/d\lambda$ is an arbitrary tangent vector. That is, the gradient of f at any point P is that element of T^*_P whose value on an element $\bar{V}$ of T_P is the directional derivative of f along a curve whose tangent is $\bar{V}$. We must check that this is a linear function on T_P, in the sense of equation (2.15):

$$\begin{aligned}\tilde{d}f\left(a\frac{d}{d\lambda}+b\frac{d}{d\mu}\right) &= \left(a\frac{d}{d\lambda}+b\frac{d}{d\mu}\right)f\\ &= a\frac{df}{d\lambda}+b\frac{df}{d\mu}\\ &= a\,\tilde{d}f(d/d\lambda)+b\,\tilde{d}f(d/d\mu).\end{aligned}$$

So it is indeed linear. At first thought it might seem that f itself should be the one-form, since f and $d/d\lambda$ make $df/d\lambda$, a number. But this is not right; the reader is reminded that both T_P and T^*_P are defined at a point P, so all the information needed to construct $df/d\lambda$ must be present there. The value of f at P is irrelevant to $df/d\lambda$. To compute $df/d\lambda$ at P one needs to know $\partial f/\partial x^i$ at P. These are, as we shall see, the components of the gradient of f. So it is the gradient which is the one-form.

The gradient enables us to develop a picture of a one-form, complementary to the picture of a vector as an arrow. In figure 2.23 we have drawn part of a topographical map, showing contours of equal elevation. If h is the elevation, then the gradient $\tilde{d}h$ is clearly largest in an area like A, where the lines are closest together, and smallest near B, where the lines are spaced far apart. Moreover, suppose one wanted to know how much elevation a (short) walk between two points would involve. One can lay out on the map a line (vector $\Delta\bar{x}$) between

† For a nice discussion of the relation of $\tilde{d}f$ to the infinitesimal, see Spivak (1970), vol. 1.

the points. Then the number of contours the line crosses gives the change in elevation. For example, line 1 crosses $1\frac{1}{2}$ contours, while 2 crosses 2 contours. Line 3 starts near 2 but goes in a different direction, winding up only $\frac{1}{2}$ a contour higher. But these numbers are just Δh, which is a linear function of $\tilde{d}h$ and $\Delta\bar{x}$:

$$\Delta h = \sum \frac{\partial h}{\partial x^i} \Delta x^i.$$

This is the *value* of $\tilde{d}h$ on $\Delta\bar{x}$ (cf. equation (2.21) above and (2.27) below). Therefore, a one-form $\tilde{\omega}$ may be represented by a series of surfaces (figure 2.24), and its contraction with a vector $\bar{V}$ is the number of surfaces $\bar{V}$ crosses. The closer the surfaces, the larger $\tilde{\omega}$. Properly, just as a vector is straight, the one-form's surfaces are straight and parallel. This is because we deal with one-forms at a point, not over an extended region: 'tangent' one-forms, in the same sense as tangent vectors.

These pictures show why one in general *cannot* call a gradient a vector. One would like to identify the *vector* gradient as that vector pointing 'up' the slope, i.e. in such a way that it crosses the greatest number of contours per unit length. The key phrase is 'per unit length'. If there is a measure of distance on the manifold, then a vector *can* be associated with a gradient. But if one does not know how to compare the lengths of vectors that point in different directions, one cannot define a *direction* of steepest ascent, and the gradient is fundamentally different from a vector. Since we shall not assume a length (or 'metric') in

Fig. 2.23. A topographical map of a hilly region. Curves are contours of equal elevation above sea level. Arrows indicate possible paths for a walker.

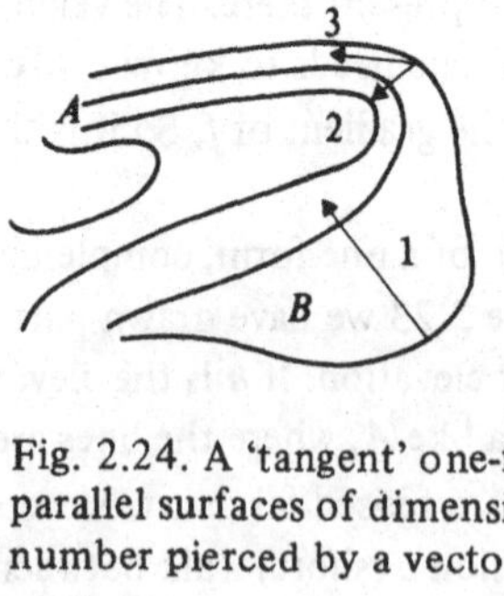

Fig. 2.24. A 'tangent' one-form $\tilde{\omega}$ represented pictorially as a series of parallel surfaces of dimension one less than that of the manifold. The number pierced by a vector $\bar{V}$ is the contraction $\langle\tilde{\omega}, \bar{V}\rangle$.

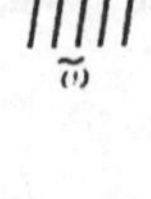

general, we must preserve the distinction between vectors and one-forms. We will return to this point in §2.29.

2.20 Basis one-forms and components of one-forms

In the vector space of one-forms at P, T^*_P, any n linearly independent one-forms constitute a basis. However, once a basis $\{\bar{e}_i, i = 1, \ldots, n\}$ has been chosen for the vectors T_P at P, this induces a preferred basis for T^*_P, called the *dual basis* $\{\tilde{\omega}^i, i = 1, \ldots, n\}$. It is defined as follows. If $\bar{V}$ is any vector in T_P then $\tilde{\omega}^i$ produces the ith component of $\bar{V}$

$$\tilde{\omega}^i(\bar{V}) = V^i. \tag{2.22}$$

It is easy to see that this is linear in the argument $\bar{V}$, since the ith component of, say, $\bar{V} + \bar{W}$ is $V^i + W^i$. So (2.22) indeed defines a linear function on T_P. In particular, since the basis vector $\bar{e}_j$ has only a jth component, all others vanishing, we have

♦ $$\tilde{\omega}^i(\bar{e}_j) = \delta^i{}_j. \tag{2.23}$$

This is the definition of $\tilde{\omega}^i$ found in most references. Note carefully that in order to define any $\tilde{\omega}^i$ *all* the vectors $\{\bar{e}_j\}$ must be known. A change in any one $\bar{e}_k$ generally changes *all* the basis one-forms $\tilde{\omega}^i$. The correspondence we have established is between one basis and its dual, not between an individual vector and an associated one-form.

We have not actually proved that the $\{\tilde{\omega}^i\}$ are linearly independent and therefore do form a basis. This follows easily from (2.23), but we will use a more indirect approach. Consider any one-form $\tilde{q}$ acting on an arbitrary vector $\bar{V}$,

$$\begin{aligned}\tilde{q}(\bar{V}) &= \tilde{q}\left(\sum_j V^j \bar{e}_j\right)\\ &= \sum_j V^j \tilde{q}(\bar{e}_j)\\ &= \sum_j \tilde{\omega}^j(\bar{V})\tilde{q}(\bar{e}_j).\end{aligned} \tag{2.24}$$

The numbers

$$q_j = \tilde{q}(\bar{e}_j) \tag{2.25}$$

are called the *components* of $\tilde{q}$ on the basis dual to $\{\bar{e}_j\}$. To see that this name is more than a mere analogy with (2.22), we rewrite (2.24) as

$$\tilde{q}(\bar{V}) = \sum_j q_j \tilde{\omega}^j(\bar{V}).$$

Since a one-form is *defined* by its values on vectors, it follows from this, equation (2.16), and the fact that $\bar{V}$ is arbitrary, that

$$\blacklozenge \quad \tilde{q} = \sum_j q_j \tilde{\omega}^j. \tag{2.26}$$

This shows that the set $\{\tilde{\omega}^j\}$ is indeed a basis, since there are only n of them and any $\tilde{q}$ is a linear combination of them. It also shows that the numbers $\{q_j\}$ are indeed the components of $\tilde{q}$ on this basis in the ordinary sense.

Most importantly, we now have a formula giving us the value of $\tilde{q}(\bar{V})$ if we know the components of $\tilde{q}$ and of $\bar{V}$:

$$\tilde{q}(\bar{V}) = \sum_j q_j V^j. \tag{2.27}$$

As remarked before, this is the contraction of $\bar{V}$ and $\tilde{q}$.

Naturally, all these considerations extend directly to one-form fields. If the set of vector fields $\{\bar{e}_i\}$ is a basis at every point in some region U of M, then the fields $\{\tilde{\omega}^j\}$ defined by (2.23) are likewise a basis at all points of U. A coordinate system on U, $\{x^i\}$, defines a natural basis for vector fields $\{\partial/\partial x^i\}$. It also defines a natural set of n one-forms, the gradients $\{\tilde{d}x^i\}$. These one-forms are in fact the basis dual to the coordinate basis vectors: by equation (2.21)

$$\tilde{d}x^i(\partial/\partial x^j) \equiv \partial x^i/\partial x^j = \delta^i{}_j, \tag{2.28}$$

the second equality following from the ordinary properties of partial derivatives.

In §2.19 we defined differentiability of one-form fields. It is now easy to prove that $\tilde{q}$ is a C^∞ one-form field if only if its components $\{q^i\}$ associated with a C^∞ basis for vector fields are C^∞ functions.

2.21 Index notation

We adopt the following conventions for the use of indices. Components of vectors, e.g. V^i, have the index written as a superscript; components of one-forms, e.g. ω_j, have subscripted indices. Members of a vector basis are labelled with subscripts ($\bar{e}_j$), those of a one-form basis with superscripts ($\tilde{\omega}^j$). (For coordinate bases, this rule means that the one-forms $\tilde{d}x^i$ have their index up, as they should, while the vectors $\partial/\partial x^j$ are considered to have their index down, since it appears in the denominator as a superscript.) These conventions are adopted for a good reason. Consider the contraction

$$\tilde{\omega}(\bar{V}) = \sum_j V^j \omega_j,$$

which is a sum of products in which one multiplier has a raised index and the other a lowered one. We shall adopt the *Einstein summation convention*: whenever an expression contains a *repeated* index, once as subscript and once as superscript, a summation over the index is understood. Thus, in the expressions

$$\tilde{\omega} = \omega_j\, \tilde{d}x^j, \quad \bar{V} = V^j \frac{\partial}{\partial x^j}, \quad \tilde{\omega}(\bar{V}) = V^j \omega_j,$$

summations are understood. In the expressions

$$V^j W^k, \qquad V^j \omega_i, \qquad V^j W^j,$$

there are no summations; in the first two there are no repeated indices, and in the last both are raised. Use of the summation convention greatly simplifies calculations in which components are used, and our rules for the placement of indices minimize the possibility that the convention will lead us into careless errors.

We are now in a position to extend our treatment of vector algebra to tensors.

2.22 Tensors and tensor fields

Tensors are a natural extension of the concepts we have already developed. Their algebra is straightforward and they have, as we shall see, many uses. The principal problem students have when they first encounter tensors is that they cannot 'visualize' them: they have no picture. We have earlier developed pictorial ways of representing vectors and one-forms; this can to some extent be extended to tensors of higher type, but the pictures rapidly become very complicated. It is perhaps better to avoid picturing most tensors directly, and to think of them in terms of the definition we shall now give, as linear operators on vectors and one-forms.

Consider a point P of M. A tensor of type $\binom{N}{N'}$ at P is defined to be a linear function which takes as arguments N one-forms and N' vectors and whose value is a real number. This is a generalization of the way we defined one-forms. By 'linear' we understand linearity on *every* argument (usually called multilinearity). For example, if $\mathbf{F}$ is a $\binom{2}{2}$ tensor then its value on the one-forms $\tilde{\omega}$ and $\tilde{\sigma}$ and the vectors $\bar{V}$ and $\bar{W}$ is

$$\mathbf{F}\,(\tilde{\omega}, \tilde{\sigma}; \bar{V}, \bar{W}).$$

As a linear function it obeys (for arbitrary numbers a, b)

$$\blacklozenge \qquad \mathbf{F}\,(a\tilde{\omega} + b\tilde{\lambda}, \tilde{\sigma}; \bar{V}, \bar{W}) = a\mathbf{F}(\tilde{\omega}, \tilde{\sigma}; \bar{V}, \bar{W}) + b\mathbf{F}(\tilde{\lambda}, \tilde{\sigma}; \bar{V}, \bar{W}), \qquad (2.29)$$

and similarly for the other arguments. If we want to speak of $\mathbf{F}$ without naming its arguments, we may sometimes write $\mathbf{F}\,(\ ,\ ;\ ,\ )$, in which empty spaces signify 'slots' into which any arguments of the appropriate type (one-forms before the semicolon, vectors after) may be placed. Naturally, the order of the arguments generally makes a difference, as is true of functions of real variables. (That is, the function $f(x, y) = 3x + 5y$ has different values for $f(1, 2)$ and $f(2, 1)$.)

As with vectors and one-forms, a $\binom{N}{N'}$ *tensor field* is a rule giving a $\binom{N}{N'}$

tensor at each point. Linearity extends to tensor fields, where the numbers a and b in equation (2.29) can have different values at each point: they are functions on M. Differentiability of the field is defined as for one-forms, §2.19.

As a special case, note that vectors are tensors of type $\binom{1}{0}$: they are linear functions of one-forms. Similarly, one-forms are tensors of type $\binom{0}{1}$. By convention, a scalar function on the manifold is taken to be a tensor of type $\binom{0}{0}$. (See §2.28 on 'Functions and scalars' below.) A $\binom{1}{1}$ tensor $\mathbf{T}$ requires two arguments. Thus, $\mathbf{T}(\tilde{\omega};\bar{V})$ is a real number; for fixed $\tilde{\omega}$, $\mathbf{T}(\tilde{\omega};\ \ )$ is a one-form, since it needs a vector argument to give a real number; $\mathbf{T}(\ \ ;\bar{V})$ is a vector. So, a $\binom{1}{1}$ tensor in particular can be thought of as a linear vector-valued function of vectors, and also as a linear form-valued function of one-forms. This game can be played with any tensor.

2.23 Examples of tensors

Although our definition of a tensor may seem rather abstract, it is in fact quite often very directly applicable to common problems. We mention three examples immediately, and later (in §2.29) devote some time to a discussion of a very important tensor, the metric tensor.

(i) We take our first example from matrix algebra. If column vectors are vectors and row vectors are one-forms, then a matrix is a $\binom{1}{1}$ tensor, since multiplying it by a vector gives a vector, and letting it operate on both in the usual way gives a number.

Exercise 2.4

A linear ('active') transformation in matrix algebra (e.g. an orthogonal rotation) transforms one matrix into another. Show that it is therefore a $\binom{2}{2}$ tensor when operating on matrices.

(ii) The second example is from the function space $C[-1, 1]$ mentioned in §2.18. A linear differential operator (e.g. $x^2\,\mathrm{d}/\mathrm{d}x$) converts functions ('vectors' in this space) into other functions (vectors). Being linear, it is therefore also a $\binom{1}{1}$ tensor in the space.

(iii) The third example is the stress tensor. Readers familiar with continuum mechanics will know the stress tensor. Given a stressed material, and given an imaginary plane passing through the material, the stress tensor gives the stress vector across that plane (the force per unit area exerted by the material on one side of the plane upon that on the other side). Now, a plane is a surface, and a surface is represented by a one-form. The stress tensor turns out to be a linear, vector-valued function of one-forms, or a $\binom{2}{0}$ tensor.

2.24 Components of tensors and the outer product

A simple $\binom{2}{0}$ tensor is the following: given two vectors $\bar{V}$ and $\bar{W}$, we form a tensor called $\bar{V} \otimes \bar{W}$ whose value on two one-forms $\tilde{p}$ and $\tilde{q}$ is the product $\bar{V}(\tilde{p})\bar{W}(\tilde{q})$:

$$\blacklozenge \qquad \bar{V} \otimes \bar{W}(\tilde{p}, \tilde{q}) \equiv \bar{V}(\tilde{p})\bar{W}(\tilde{q}). \tag{2.30}$$

The operation $\otimes$ is called the 'outer product', 'direct product', or 'tensor product'. Its generalization to arbitrary numbers and types of tensors is obvious. The outer product of a $\binom{N}{M}$ tensor with a $\binom{N'}{M'}$ tensor is a tensor of type $\binom{N+N'}{M+M'}$.

The *components* of a tensor are its values when it takes basis vectors and one-forms as arguments. If $\mathbf{S}$ is a $\binom{3}{2}$ tensor, then it has components on a basis $\{e_i\}$

$$\blacklozenge \qquad S^{ijk}{}_{lm} \equiv \mathbf{S}(\tilde{\omega}^i, \tilde{\omega}^j, \tilde{\omega}^k; \bar{e}_l, \bar{e}_m). \tag{2.31}$$

If the order of the arguments of $\mathbf{S}$ matters, then so does the order of the indices of $S^{ijk}{}_{lm}$.

The extension to components of tensor fields and their differentiability is exactly as for one-forms in §2.20.

Exercise 2.5

(a) Prove that a general $\binom{2}{0}$ tensor *cannot* be expressed as a simple outer product of two vectors. (Hint: count the number of components a $\binom{2}{0}$ tensor may have.)

(b) Prove that the $\binom{1}{1}$ tensor $\bar{V} \otimes \tilde{\omega}$ has components $V^i\omega_j$.

Exercise 2.6

Prove that the set of all $\binom{2}{0}$ tensors at P is a vector space under addition defined by analogy with equation (2.16b). Show that $\bar{e}_i \otimes \bar{e}_j$ is a basis for that space. (Thus, although a general $\binom{2}{0}$ tensor is not a simple outer product, it *can* be represented as a sum of such tensors.) This vector space is called $T_P \otimes T_P$.

2.25 Contraction

In exercise 2.5, we point out that the set $\{V^i\omega_j\}$ are components of a $\binom{1}{1}$ tensor. Now, by summing on the index, one gets $V^j\omega_j$, a number independent of the basis, the value of $\tilde{\omega}$ on $\bar{V}$, which may be thought of as a $\binom{0}{0}$ tensor. By analogy one can show that if $S^i{}_{jk}$ and P^{lm} are components, respectively, of a $\binom{1}{2}$ and $\binom{2}{0}$ tensor, then $S^i{}_{jk}\,P^{lm}$ is a component of a $\binom{3}{2}$ tensor, $S^i{}_{jk}\,P^{jm}$ of a $\binom{2}{1}$ tensor, $S^i{}_{jk}\,P^{lj}$ of a different $\binom{2}{1}$ tensor, etc. By analogy with equation (2.27), this operation is called *contraction*, and produces new tensors.

We can give a short proof of the fact that contraction is independent of the basis used. Consider the $\binom{2}{0}$ tensor **A**, the $\binom{0}{2}$ tensor **B**, and their contraction (in some basis) $A^{ij}B_{jk}$. We claim that these are the components of a $\binom{1}{1}$ tensor **C**, such that for arbitrary vector $\bar{V}$ and one-form $\tilde{\sigma}$

$$\mathbf{C}(\tilde{\sigma};\bar{V}) = \left(\sum_j A^{ij}B_{jk}\right)\sigma_i V^k = \sum_j \mathbf{A}(\tilde{\sigma},\tilde{\omega}^j)\mathbf{B}(\bar{e}_j,\bar{V}).$$

By linearity on **A**'s second argument we can write this as

$$\mathbf{C}(\tilde{\sigma};\bar{V}) = \mathbf{A}\left(\tilde{\sigma},\sum_j \mathbf{B}(\bar{e}_j,\bar{V})\tilde{\omega}^j\right),$$

since the quantities $\mathbf{B}(\bar{e}_j,\bar{V})$ are just numbers. But in §2.20 we proved in effect that, independent of the basis,

$$\sum_j \mathbf{B}(\bar{e}_j,\bar{V})\tilde{\omega}^j = \mathbf{B}(\ \ ,\bar{V}),$$

which is a one-form since (for fixed $\bar{V}$) it requires a vector as an argument (in the empty slot). This one-form occupies one of the slots in **A**, so we have proved

$$A^{ij}B_{jk} = C^i{}_k \Leftrightarrow \mathbf{C}(\tilde{\sigma};\bar{V}) = \mathbf{A}(\tilde{\sigma},\mathbf{B}(\ \ ,\bar{V})),$$

independent of any basis (cf. exercise 2.8).

Exercise 2.7

How many different $\binom{2}{1}$ tensors may be made by contraction on pairs of indices of the $\binom{3}{2}$ tensor $Q^{ijk}{}_{lm}$? How many $\binom{1}{0}$ tensors by a second contraction?

Exercise 2.8

Let **A** and **B** be two $\binom{1}{1}$ tensors, and regard them as vector-valued linear functions of vectors: if $\bar{V}$ is a vector then $\mathbf{A}(\bar{V})$ and $\mathbf{B}(\bar{V})$ are vectors. Show that if we define $\mathbf{C}(\bar{V})$ to be

$$\mathbf{C}(\bar{V}) = \mathbf{B}(\mathbf{A}(\bar{V})),$$

then **C** is a $\binom{1}{1}$ tensor as well. Show that its components are

$$C^i{}_j = B^i{}_k A^k{}_j.$$

Discuss the relation of this with the linear transformation defined in §1.6.

2.26 Basis transformations

The behavior of a tensor's components under a change of basis is at the heart of the older definition of a tensor. It has been replaced more recently by the definition we have used here in terms of linear functions, and it is a measure of how conceptually different these two approaches are that we are only now getting around to looking at basis transformations. This is not to say that these transformations are unimportant. Most practical calculations involving tensors involve working with their components, and an understanding of their transformation properties is essential.

We shall consider vectors and tensors defined at some point P of M. Suppose we begin with a vector basis $\{\bar{e}_i, i = 1, \ldots, n\}$ and wish instead to use a basis $\{\bar{e}_{j'}, j' = 1, \ldots, n\}$. (We shall use primes on the indices as our only way of distinguishing references to one basis from references to the other.) Then in T_P there is a linear transformation Λ from the old basis to the new:

$$\bar{e}_{j'} = \Lambda^i{}_{j'}\bar{e}_i. \tag{2.32}$$

The matrix $\Lambda^i{}_{j'}$ is nonsingular (otherwise $\{\bar{e}_{j'}\}$ would not be linearly independent) but otherwise arbitrary. It is not the collection of components of some tensor, since its indices refer to two different bases. It is simply called the transformation matrix.

The old one-form basis satisfies (2.23):

$$\tilde{\omega}^i(\bar{e}_k) = \delta^i{}_k.$$

Multiplying by $\Lambda^k{}_{j'}$ and using (2.32) and linearity gives

$$\tilde{\omega}^i(\bar{e}_{j'}) = \delta^i{}_k\Lambda^k{}_{j'} = \Lambda^i{}_{j'}. \tag{2.33}$$

Now the matrix $\Lambda^i{}_{j'}$ has an inverse, which we will define to be $\Lambda^{k'}{}_j$:

$$\Lambda^{k'}{}_j\Lambda^j{}_{i'} = \delta^{k'}{}_{i'}, \quad \Lambda^{k'}{}_j\Lambda^i{}_{k'} = \delta^i{}_j. \tag{2.34}$$

Multiplying (2.33) by $\Lambda^{k'}{}_i$ gives

$$\Lambda^{k'}{}_i\tilde{\omega}^i(\bar{e}_{j'}) = \delta^{k'}{}_{j'}.$$

By comparing this with (2.23)

♦ $$\tilde{\omega}^{k'} = \Lambda^{k'}{}_i\tilde{\omega}^i. \tag{2.35}$$

This is the counterpart of (2.32): basis one-forms transform *oppositely* to basis vectors (i.e. using the inverse transformation matrix) in order to satisfy (2.23) on both bases.

It is now a simple matter of transform components:

$$V^{i'} = \tilde{\omega}^{i'}(\bar{V}) = \Lambda^{i'}{}_j\tilde{\omega}^j(\bar{V}) = \Lambda^{i'}{}_jV^j, \tag{2.36}$$

$$q_{k'} = \tilde{q}(\bar{e}_{k'}) = \tilde{q}(\Lambda^j{}_{k'}\bar{e}_j) = \Lambda^j{}_{k'}\tilde{q}(\bar{e}_j) = \Lambda^j{}_{k'}q_j, \tag{2.37}$$

and similarly for tensors of higher type (cf. exercise 2.9 below). These transformation laws show that the components of vectors and the basis one-forms obey the *same* law, which is opposite (i.e. uses the matrix inverse) to the law obeyed by components of one-forms and the basis vectors. This is reasonable, in order to keep such sums as $V^i\bar{e}_i$, $V^j\sigma_j$, etc. independent of basis. This illustrates another convenience introduced by our positioning of indices and our summation convention: the position of an index automatically gives its transformation law. For example, V^i and $\tilde{\omega}^j$ obey the same law, which is

$$V^{i'} = \Lambda^{i'}{}_j V^j.$$

It could *not* use the matrix $\Lambda^i{}_{j'}$, because the summation must be on unprimed indices and must involve one index which is up and one which is down.

These opposing transformation laws gave rise to the old names, 'contravariant' and 'covariant'. What we call a vector was called contravariant because its components obey the law opposite ('contra') to the law governing the basis vectors. Similarly, one-forms were 'covariant vectors' because their components go *with* the basis vectors. The modern viewpoint emphasizes the fact that neither the vector nor the one-form is in fact changed by a basis transformation: they are coordinate-independent geometrical objects. Therefore, modern terminology has dropped the old names because they over-emphasize the coordinate-dependent descriptions of these objects.

Exercise 2.9

Show that a $\binom{2}{0}$ tensor's components transform as two vectors, i.e.

$$T^{i'j'} = \Lambda^{i'}{}_k \Lambda^{j'}{}_l T^{kl}. \tag{2.38}$$

Generalize this to type $\binom{N}{N'}$.

Exercise 2.10

Show that if a tensor's components are all zero in one basis, they are zero in all bases. (We then say the tensor is zero. It follows that if two tensors have equal components in one basis they are equal in all, and the tensors are said to be equal.)

Exercise 2.11

Associated with a particular basis $\{\bar{e}_i\}$ of a vector space of dimension n, we are given some set of numbers $\{A^i{}_j, i, j = 1, \ldots, n\}$. We define another set of numbers $A^{i'}{}_{k'} = \Lambda^{i'}{}_j \Lambda^l{}_{k'} A^j{}_l$ and call them the components of the 'tensor' **A** on the new basis $\{\bar{e}_{j'}\}$. Show that this 'tensor' is indeed a tensor as we have defined it. This shows that one can take the point of view that a tensor *is* the collection $\{A^i{}_j\}$ transforming in the given way. This is an alternative definition to the one we have used.

It is of particular interest to look at these basis transformations when they result from *coordinate transformations*, which were mentioned briefly at the end of §2.6. Suppose a region U of the manifold M has a coordinate system

$\{x^i, i = 1, \ldots, n\}$, and that we introduce new functions $\{y^{i'}, i' = 1, \ldots, n\}$ given by the equations

$$y^{i'} = f^{i'}(x^1, \ldots, x^n),\ i' = 1, \ldots, n, \tag{2.39}$$

which can be summarized as $y^{i'} = f^{i'}(x^j)$. These equations constitute a coordinate transformation if the Jacobian matrix of partial derivatives $\partial y^{i'}/\partial x^j$ has a nonvanishing determinant in U. A given point P in U can be described by two different sets of numbers, $\{x^i\}$ or $\{y^{j'}\}$. At P we likewise have two different coordinate vector bases, $\{\partial/\partial x^j\}$ and, by the chain rule of calculus,

$$\frac{\partial}{\partial y^{i'}} = \frac{\partial x^j}{\partial y^{i'}}\frac{\partial}{\partial x^j}. \tag{2.40}$$

By comparing this with (2.32) we learn

$$\Lambda^i{}_{j'} = \frac{\partial x^i}{\partial y^{j'}}. \tag{2.41}$$

Similarly, the inverse matrix is

$$\Lambda^{k'}{}_j = \frac{\partial y^{k'}}{\partial x^j}, \tag{2.42}$$

which is easily proved using the chain rule for partial derivatives:

$$\frac{\partial x^i}{\partial y^{j'}}\frac{\partial y^{j'}}{\partial x^k} = \frac{\partial x^i}{\partial x^k} = \delta^i{}_k.$$

It is important to understand that (2.42) defines only a restricted class of transformation fields $\Lambda^{k'}{}_j$ in U. At any one point P in U one can choose all n^2 elements of $\Lambda^{k'}{}_j$ arbitrarily (apart from the requirement that its determinant should not vanish), but not so in the neighborhood of P, because (2.42) implies

$$\partial\Lambda^{k'}{}_j/\partial x^l = \partial\Lambda^{k'}{}_l/\partial x^j, \tag{2.43}$$

a symmetry that an *arbitrary* field $\Lambda^{k'}{}_j$ certainly would not need to satisfy. This is another illustration that not every field of basis vectors is a coordinate basis.

2.27 Tensor operations on components

Given a tensor $\mathbf{T}$ and its components $\{T^{i\ldots}{}_{j\ldots}\}$ on some basis, suppose one multiplies each component by the number a, thereby obtaining $\{aT^{i\ldots}{}_{j\ldots}\}$. These are clearly components of the tensor $a\mathbf{T}$, and this means that the operation of multiplying all the components of $\mathbf{T}$ by a is *basis-invariant*: had we begun in coordinates $\{y^{i'}\}$ we would have obtained the components of $a\mathbf{T}$ in these new coordinates. (One could not say the same had we multiplied only *some* of $\{T^{i\ldots}{}_{j\ldots}\}$ by a.) Thus, the operation $\{T^{i\ldots}{}_{j\ldots}\} \to \{aT^{i\ldots}{}_{j\ldots}\}$ uniquely corresponds to the basis-independent statement $\mathbf{T} \to a\mathbf{T}$. Similarly, the outer product of two tensors,

$$\mathbf{A}, \mathbf{B} \to \mathbf{A} \otimes \mathbf{B},$$

has the unique component analogue (cf. exercise 2.4)

$$\{A^{i\ldots}{}_{j\ldots}\}, \{B^{k\ldots}{}_{l\ldots}\} \to \{A^{i\ldots}{}_{j\ldots}B^{k\ldots}{}_{l\ldots}\},$$

independently of what coordinate or noncoordinate bases are used. In general, an operation on components that produces components of the *same* tensor independently of the basis is called a *tensor operation*, and we will deal exclusively with them. The following list is a summary of the algebraic tensor operations (we shall consider ones involving differentiation later):

(i) Addition (and subtraction) of components of tensors of the same type.
(ii) Multiplication of all components by a number gives a tensor of the same type.
(iii) Multiplication of components of two tensors gives a tensor whose type is the sum of the two.
(iv) Contraction on pairs of indices, one of which is up and the other down.

An equation that involves components combined using only these operations is called a 'tensor equation'. It follows from exercise 2.10 that if a series of operations performed in a certain basis gives a tensor equation, then that equation is true in all bases. This often permits a convenient choice of basis for a particular calculation.

2.28 Functions and scalars

A scalar is defined as a $\binom{0}{0}$ tensor, i.e. a function on the manifold whose definition does not depend upon the choice of any particular basis. For example, the contraction $V^i\omega_i$ is a scalar, since its value is independent of the particular basis in which the components are computed. On the other hand, the component V^1 is also a function on the manifold, having a numerical value at every point; it is *not* a scalar because its value depends on the basis. Put another way, there is some (scalar) function $f(P)$ such that $V^1(P) = f(P)$ when the index '1' refers to some particular basis; when that basis is changed, the new $V^1(P)$ will not equal $f(P)$. So $f(P)$ is a scalar, whose value happens to equal that of the one-component of $\bar{V}$ in some basis. But V^1 is not a scalar since its value changes with a change in basis. You see, therefore, that whether a thing is a 'scalar' or simply a 'function' depends on its interpretation when the basis is changed, rather than on its actual value.

2.29 The metric tensor on a vector space

Most familiar vector algebras involve an inner product between vectors, as in §1.5. This is a rule which associates a number (the 'dot product') with two vectors. It is a linear function of both vectors. Therefore it is a $\binom{0}{2}$ tensor, which is called the metric tensor, g. Thus we define

◆ $$\mathsf{g}(\bar{V}, \bar{U}) = \mathsf{g}(\bar{U}, \bar{V}) \equiv \bar{U} \cdot \bar{V}. \tag{2.44}$$

The first equality above is a demand that $\bar{U} \cdot \bar{V}$ should not depend on the order of $\bar{U}$ and $\bar{V}$. We say that g is a symmetric tensor. Its components on a basis $\{\bar{e}_i\}$ are

$$g_{ij} = \mathsf{g}(\bar{e}_i, \bar{e}_j) = \bar{e}_i \cdot \bar{e}_j. \tag{2.45}$$

These components form an $n \times n$ symmetric matrix. For reasons explained later, we also demand that this matrix have an inverse. If it happens that the matrix is the unit matrix, i.e. if

$$g_{ij} = \delta_{ij},$$

we say the metric tensor is the *Euclidean metric*, and the vector space is called Euclidean space. But what can we say if g_{ij} is not this simple? Well, we are always free to try to choose a new basis $\{\bar{e}_{j'}\}$ in which the new metric components,

$$g_{i'j'} = \Lambda^k{}_{i'}\Lambda^l{}_{j'}g_{kl}, \tag{2.46}$$

are simpler. Consider this equation as a matrix equation. It is helpful to rewrite it as

$$g_{i'j'} = \Lambda^k{}_{i'}g_{kl}\Lambda^l{}_{j'}.$$

From the discussion in §1.6 it is easy to see that this is the matrix equation

$$g' = \Lambda^{\mathrm{T}} g \Lambda, \tag{2.47}$$

where Λ^{T} is the transpose of the matrix Λ whose entries are $\Lambda^k{}_{i'}$. We will now see that a clever choice of Λ will reduce the matrix g' to a very simple form. Since Λ is arbitrary, we will take it to be the product of two matrices

$$\Lambda = OD, \tag{2.48}$$

where O is an *orthogonal* matrix ($O^{\mathrm{T}} = O^{-1}$) and D is a *diagonal* matrix (so in particular $D^{\mathrm{T}} = D$). Then we have, from equation (1.41),

$$\Lambda^{\mathrm{T}} = (OD)^{\mathrm{T}} = D^{\mathrm{T}}O^{\mathrm{T}} = DO^{-1}$$

and

$$g' = DO^{-1}gOD. \tag{2.49}$$

It is well known that any symmetric matrix, such as g, can be reduced to diagonal form, g_{d}, by a similarity transformation using an orthogonal matrix, so let us choose O to do this:

$$g_{\mathrm{d}} = O^{-1}gO,$$

$$g' = Dg_{\mathrm{d}}D.$$

If g_{d} is the matrix $\mathrm{diag}(g_1, g_2, \ldots, g_n)$ and our as yet undetermined matrix D is $\mathrm{diag}(d_1, d_2, \ldots, d_n)$, then g' is

$$g' = \text{diag}(g_1 d_1^{\,2}, g_2 d_2^{\,2}, \ldots, g_n d_n^{\,2}). \tag{2.50}$$

We now choose $d_j = (|g_j|)^{-1/2}$, so that each element on the diagonal of g' is either $+1$ or -1. We cannot use d_j to change the sign of g_j, only its magnitude. Now, the diagonal elements of g_d are the *eigenvalues* of g, and are unique apart from the order in which they appear. Moreover, since g has an inverse, none of the eigenvalues is zero. If we choose O to make all the negative ones appear first, then we have proved the theorem that *any vector space with a metric tensor has a basis on which the metric tensor has the canonical form* diag$(-1, \ldots, -1, 1, \ldots, 1)$. Such a basis is said to be *orthonormal*. The sum of these diagonal elements – the trace of the canonical form – is called the *signature* of the metric.

Exercise 2.12

Find the matrices Λ which cast the following matrices into their unit diagonal form:

$$\text{(a)} \begin{pmatrix} 2 & 1 \\ 1 & 2 \end{pmatrix}, \quad \text{(b)} \begin{pmatrix} 0 & 1 \\ 1 & 0 \end{pmatrix}, \quad \text{(c)} \begin{pmatrix} 4 & 0 \\ 0 & -1 \end{pmatrix}.$$

This theorem is very important. It means that there are only a few different kinds of metric tensors on a vector space. If the metric is positive-definite, then its canonical form must have all $+1$s, and the space is Euclidean. If the metric is negative-definite it is also said to be Euclidean, since what is important for the space is whether the signs are all the same or not. If the metric is not of definite sign, it is called *indefinite*. An important case is the canonical form $(-1, 1, \ldots, 1)$, whose metric is usually called a *Minkowski metric*; special relativity has such a metric for $n = 4$, which we will discuss at length shortly.

Another consequence of this canonical form is that it picks out a preferred set of bases for the vector space, the *orthonormal bases*. In Euclidean space E^n, such a basis is called *Cartesian*. In it the metric tensor has the components $g_{ij} = \delta_{ij}$, or in matrix form $g = I$. A transformation matrix Λ_C from one such basis to another satisfies

$$I = \Lambda_C^T I \Lambda_C \Rightarrow \Lambda_C^T = \Lambda_C^{-1}. \tag{2.51}$$

So the orthogonal matrices are the transformations between Cartesian bases. These matrices form a group (the product of two orthogonal matrices is orthogonal), which is called the Euclidean *symmetry group* $O(n)$. A Minkowski metric likewise singles out its preferred *Lorentz* bases in which the metric components form the matrix

$$\eta = \text{diag}(-1, 1, \ldots, 1). \tag{2.52}$$

A transformation matrix Λ_L from one Lorentz basis to another satisfies

$$\eta = \Lambda_L^T \eta \Lambda_L. \tag{2.53}$$

Such a matrix is called a *Lorentz transformation*. It is not hard to show that these too form a group, called the *Lorentz group* $L(n)$ or $O(n-1, 1)$.

From the point of view of tensor algebra, the metric tensor's most important role is one we have not yet mentioned: it maps vectors into one-forms in a 1–1 manner. Consider a vector $\bar{V}$. Then $g(\bar{V}, \ )$ is, for fixed $\bar{V}$, a linear function of vectors into real numbers: a one-form. We denote this by

$$\tilde{V} = g(\bar{V}, \ ). \tag{2.54}$$

The fact that we demanded that the matrix g_{ij} have an inverse is what makes this map 1–1: there is only one vector $\bar{V}$ mapped to $\tilde{V}$. To see how this works, let us look at the component version of this equation. Denote the component of $\tilde{V}$ by V_i:

$$\begin{aligned} V_i &= \tilde{V}(\bar{e}_i) = g(\bar{V}, \bar{e}_i) = g(V^j\bar{e}_j, \bar{e}_i) \\ &= V^j g(\bar{e}_j, \bar{e}_i) = V^j g_{ji} = g_{ij}V^j, \end{aligned}$$

where the last equality follows from the symmetry of g. Now, the inverse matrix to g_{ij} will be called g^{ij}:

♦ $$g^{ij}g_{jk} = \delta^i{}_k. \tag{2.55}$$

Then we have

$$g^{ki}V_i = g^{ki}g_{ij}V^j = \delta^k{}_jV^j = V^k, \tag{2.56}$$

which shows that the map is invertible: the metric provides a unique pairing between one-forms and vectors. This pairing can be summarized:

♦ $$V_i = g_{ij}V^j, \tag{2.57}$$

♦ $$V^j = g^{jk}V_k. \tag{2.58}$$

Notice that we have denoted the elements of the inverse matrix by g^{ij}, and this permits (2.58) to obey the usual index conventions for a tensor equation. But for consistency one must show that the numbers g^{ij} are in fact components of a $\binom{2}{0}$ tensor. This is the object of the next exercise.

Exercise 2.13

(a) Show that $\{g^{ij}\}$ are the components of a $\binom{2}{0}$ tensor g^{-1}, either by showing that they transform properly, or that they define a bilinear function on one-forms.

(b) Show that if a vector basis $\{\bar{e}_i\}$ is orthonormal, so is its dual one-form basis $\{\tilde{\omega}^i\}$, in the sense that $g^{-1}(\tilde{\omega}^i, \tilde{\omega}^j) = \pm\,\delta^{ij}$.

In the same way, the metric can map a $\binom{2}{0}$ tensor **A** into a $\binom{1}{1}$ tensor:

$$A^i{}_j = g_{jk}A^{ik}. \tag{2.59}$$

In turn this can be mapped into a $\binom{0}{2}$ tensor

$$A_{ij} = g_{lm}A^m{}_j = g_{lm}g_{jk}A^{mk}, \tag{2.60}$$

which can be mapped back into the original tensor

$$A^{ik} = g^{il}g^{km}A_{lm}. \tag{2.61}$$

These maps are called *index raising and lowering*, and it is conventional to give all these tensors the same name (e.g. **A**), distinguishing them only by the positions of their indices. It is sometimes unimportant in vector spaces with metrics to say whether a tensor is of type $\binom{N}{N'}$ or of types $\binom{N-1}{N'+1}$, $\binom{N+1}{N'-1}$, etc. and then one speaks only of the *order* of a tensor, which is $N+N'$.

In a Euclidean vector space a Cartesian basis has $g_{ij} = \delta_{ij}$, so that $g^{ij} = \delta^{ij}$, and $U^i = U_i$: there is *no* difference between the components of a vector and of its associated one-form in this case. This is the reason that elementary discussions of Euclidean vector algebra fail to distinguish between vectors and one-forms, and also why they confine themselves to orthonormal bases. But in a nonorthonormal basis for Euclidean space and in *any* basis in an indefinite metric space, the components of a one-form can be very different from those of its vector. We will see an interesting example of this in the section on special relativity, §2.31 below.

2.30 The metric tensor field on a manifold

A metric tensor field g on a manifold is a $\binom{0}{2}$ symmetric tensor field which must have an inverse at every point. At every point P it serves as a metric on the tangent space T_P, and all the properties discussed in the previous section carry over directly. But there is much more.

The definition of a certain $\binom{0}{2}$ tensor on a manifold M as the metric of the manifold endows M with a very rich structure. It immediately becomes 'rigid': one can define such notions as distance (see below) and curvature (see chapter 6). These notions are so important in many applications, particularly in general relativity, that it is this sort of geometry that a physicist is most likely to be familiar with. But this is, from the point of view of differential geometry, a 'higher level' structure: one goes beyond the notion of a simple differentiable manifold by picking out a certain tensor field as special. In doing this one may overlook the rich geometrical structure of the ordinary manifold itself. Such important tools as Lie derivatives and differential forms have nothing to do with metrics. Accordingly we shall put the metric tensor very much in the background in this book, even in applications to manifolds on which one is defined.

In this section we take a brief look at its simplest properties. Further development of metric geometry itself is deferred to chapter 6.

The metric tensor may be as differentiable as one requires, but it must at least be continuous. This implies that its canonical form must be a constant everywhere, since it is composed only of integers, and integers cannot change continuously. So we speak of the *signature* of the field $\mathsf{g}|$. As long as one can choose the basis transformation matrix Λ freely at each point, one can transform from any given basis field to a globally orthonormal basis in which the components of $\mathsf{g}|$ are its canonical ones. But this transformation field Λ is usually not a coordinate transformation (i.e. it does not satisfy (2.43)), and in fact it is generally impossible to find a coordinate basis which is also orthonormal in any open region U of a manifold M (see exercise 2.14). The obvious exception is R^n considered as a manifold with the Euclidean metric δ_{ij} at every point. But even here only the Cartesian coordinates generate an orthonormal basis. An example of this is given in exercise 2.1 for polar coordinates in R^2. The coordinate basis is orthogonal but not normalized. The rescaled orthonormal basis is not a coordinate basis.

Exercise 2.14

Show that a C^∞ metric tensor field $\mathsf{g}|$ is *locally flat*, in the sense that any point P has a neighborhood in which there exists a coordinate system on whose basis the components g_{ij} have the following properties:

(i) $g_{ij}(P) = \pm\,\delta_{ij}$ (orthonormal form at P)

(ii) $\left.\dfrac{\partial g_{ij}}{\partial x^k}\right|_P = 0$ (orthonormal form a good approximation near P)

(iii) $\left.\dfrac{\partial^2 g_{ij}}{\partial x^k \partial x^l}\right|_P$ not necessarily all zero (no truly orthonormal coordinate system)

Exercise 2.15

In polar coordinates in the Euclidean plane, find the components of the metric on

(a) the basis $\{\partial/\partial r, \partial/\partial\theta\}$;

(b) the basis $\hat{\mathbf{r}}$, $\hat{\boldsymbol{\theta}}$ of exercise 2.1. Express $\hat{\mathbf{r}}$, $\hat{\boldsymbol{\theta}}$ in terms of $\partial/\partial r$ and $\partial/\partial\theta$.

Exercise 2.16

Find the components of $\tilde{d}f$ and the vector $\bar{d}f$ on both bases of exercise 2.15.

Here a word of caution is in order. Most treatments of vector calculus in curvilinear coordinates in Euclidean space use the components of a vector on this kind of orthonormal basis. This permits them to avoid distinguishing between vectors and one-forms. But when one compares the expressions we obtain below for, say, the divergence of a vector field in terms of its components with expressions given in other treatments, one must allow for possible differences of basis.

An important property of the metric is that it permits a definition of length on the manifold. If a curve has tangent $\bar{V} = \mathrm{d}\bar{x}/\mathrm{d}\lambda$, then a displacement $\mathrm{d}\lambda$ has squared length

$$\begin{aligned} \mathrm{d}l^2 \equiv \mathrm{d}\bar{x} \cdot \mathrm{d}\bar{x} &= (\bar{V}\mathrm{d}\lambda)\cdot(\bar{V}\mathrm{d}\lambda) = \bar{V}\cdot\bar{V}(\mathrm{d}\lambda)^2 \\ &= g(\bar{V}, \bar{V})\,\mathrm{d}\lambda^2. \end{aligned} \tag{2.62}$$

(Here the symbol 'd' *is* the infinitesimal, not the gradient.) If a metric is positive-definite, then $g(\bar{V}, \bar{V}) > 0$ for all $\bar{V} \neq 0$. In such a case $\mathrm{d}l^2$ is positive and we have

$$\mathrm{d}l = (g(\bar{V}, \bar{V}))^{1/2}\,\mathrm{d}\lambda \tag{2.63}$$

as the length of an element of the curve. In an indefinite metric, however, the squared length is not of definite sign. Curves are distinguished by having $\mathrm{d}l^2$ positive ('space-like') or negative ('time-like'). Then one defines the real number

$$\mathrm{d}l = |g(\bar{V}, \bar{V})|^{1/2}\,\mathrm{d}\lambda \tag{2.64}$$

to be 'proper distance' for space-like curves and 'proper time' for time-like curves. It is zero for 'null' curves. *When one has an indefinite metric one must be careful to distinguish a vector of zero norm from a vector which is truly zero (all its components vanishing).*

2.31 Special relativity

The vector space R^4 equipped with a metric of signature + 2 and considered as a manifold is one of the most important manifolds in physics: it is *Minkowski spacetime*, the spacetime of special relativity. Elementary treatments of special relativity often do not introduce the metric tensor explicitly, but they do provide us with all we need to see what the metric must be. In particular, we know that there exists a preferred set of coordinate systems for spacetime, called Lorentz frames, and that if two events are separated by coordinate intervals $(\Delta t, \Delta x, \Delta y, \Delta z)$ in such a frame, the number

$$\Delta s^2 = -c^2(\Delta t)^2 + (\Delta x)^2 + (\Delta y)^2 + (\Delta z)^2 \tag{2.65}$$

is independent of the Lorentz frame. (Here c is the speed of light.) Let us rescale our coordinates by defining $x^0 = ct, x^1 = x, x^2 = y, x^3 = z$. (It is a common convention to use numerical indices beginning with 0 rather than 1 in relativity.)

Let us also follow the convention of letting *Greek* letters represent spacetime indices. This will help us distinguish discussions applicable only to relativity from those of more general scope. Then equation (2.65) has the form

$$\begin{aligned}\Delta s^2 &= -(\Delta x^0)^2 + (\Delta x^1)^2 + (\Delta x^2)^2 + (\Delta x^3)^2 \\ &= \eta_{\alpha\beta}\Delta x^\alpha \Delta x^\beta, \end{aligned} \tag{2.66}$$

where $\eta_{\alpha\beta}$ is the matrix

$$\eta_{\alpha\beta} = \mathrm{diag}(-1, 1, 1, 1). \tag{2.67}$$

We now interpret (2.66) as defining the *pseudo-norm* (§1.5) of a vector $\Delta\bar{x}$ whose components are $(\Delta x^0, \Delta x^1, \Delta x^2, \Delta x^3)$. It is easy to see that this pseudo-norm satisfies axioms (Nii) and (Niv) of §1.5, which are required for an *inner product* to be defined. This is clearly

$$\bar{V} \cdot \bar{W} = \eta_{\alpha\beta} V^\alpha V^\beta, \tag{2.68}$$

and so we see that $\eta_{\alpha\beta}$ is in fact the metric tensor in canonical form and the Lorentz frame is the associated orthonormal basis.

This metric gives a good illustration of the difference between the components of a vector and its associated one-form. In a Lorentz frame

$$U_0 = \eta_{0\alpha} U^\alpha = -U^0, \tag{2.69a}$$

$$U_x = U^x, U_y = U^y, U_z = U^z. \tag{2.69b}$$

Consider the vector gradient of a function f, which is the vector mapped from the one-form $\tilde{d}f$. The gradient $\tilde{d}f$ has components $(\partial f/\partial x^0, \partial f/\partial x^1, \ldots)$ while the vector $\bar{d}f$ has $(-\partial f/\partial x^0, \partial f/\partial x^1, \ldots)$. Many treatments of special relativity introduce the gradient as a vector operator with components $(-\partial/\partial x^0, \partial/\partial x^1, \ldots)$. This odd sign is a clumsiness forced by the fact that the gradient is really a one-form.

A manifold M with a metric g is called Minkowski spacetime only if there exists a single coordinate system covering all of M in which g has components $\eta_{\alpha\beta}$. This coordinate system is a good one to work in, but it is not the only one possible on M. One can perfectly well choose others, such as those associated with accelerated observers. Provided one follows the general rules of differential geometry one will get the correct physical results.

2.32 Bibliography

For a more precise and rigorous discussion of what is meant by a manifold, and particularly of the rich and useful structure of fiber bundles, see Y. Choquet-Bruhat, C. DeWitt-Morette & M. Dillard-Bleick, *Analysis, Manifolds, and Physics* (North-Holland, Amsterdam, 1977). See M. Spivak, *A Comprehensive Introduction to Differential Geometry* (Publish or Perish, Boston, 1970) vol. 1, pp. 8–54, for a proof of the fixed-point theorem using cohomology theory (after studying §4.24

below!). The tangent bundle and related structures are also discussed in R. Hermann, *Vector Bundles in Mathematical Physics*, two volumes (Benjamin, Reading, Mass., 1970); and in R. Abraham & J. E. Marsden, *Foundations of Mechanics*, 2nd edn (Benjamin/Cummings, Reading, Mass., 1978). A discussion of fiber bundles in the context of modern research in quantum field theory and gravitation is B. Carter, Underlying mathematical structures of classical gravitation theory, in *Recent Developments in Gravitation*, ed. M. Levy & S. Deser (Plenum, New York, 1979). See also the article by A. Trautman, *Rep. Math. Phys.* **10**, 297 (1976).

Function spaces are discussed in Choquet-Bruhat *et al.* and in any textbook on functional analysis, such as F. Riesz & B. Sz.-Nagy, *Functional Analysis* (Ungar, New York, 1955). The theory of distributions (the Dirac delta function) is discussed in Choquet-Bruhat *et al.* (1977), and in G. Friedlander, *The Wave Equation on a Curved Space–Time* (Cambridge University Press, 1976).

The matrix algebra needed to reduce the metric tensor to canonical form may be found in the references quoted in chapter 1. A good reference for vector calculus in Euclidean three-space in curvilinear coordinates is *Functions of Mathematical Physics*, by W. Magnus & F. Oberhettinger, chapter 9 (Chelsea, New York, 1949). The metric structure of the manifold used in special relativity (Minkowski space) is introduced at an elementary level in *Spacetime Physics*, by E. F. Taylor & J. A. Wheeler (Freeman, San Francisco, 1963). Minkowski's own exposition of the subject is reprinted in *The Principle of Relativity*, edited and translated by W. Perrett & G. B. Jeffery (Dover, New York, 1924). As an example of a discussion of the vector gradient in special relativity without benefit of one-forms, see *The Feynman Lectures on Physics*, R. P. Feynman, R. B. Leighton & M. Sands, vol. 2, §25–3 (Addison-Wesley, Reading, Mass., 1964).

The Euler angles and other parameterizations of the rotation group are discussed at length in H. Goldstein, *Classical Mechanics* (Addison-Wesley, Reading, Mass., 1950). The student may find it helpful to read Goldstein's treatment in conjunction with our discussion of the rotation group in chapter 3 below.

3 LIE DERIVATIVES AND LIE GROUPS

3.1 Introduction: how a vector field maps a manifold into itself

In the previous sections we have developed certain aspects of index notation. This notation is often essential for dealing with actual numerical computations; but it is just as often a hindrance in developing a sound geometrical idea of what the mathematics means. We begin by defining vectors and tensors in a manner independent of any basis, and we now continue in this spirit to develop what is one of the most useful analytic tools in geometry: the Lie derivative along the congruence defined by a vector field.

We have mentioned the idea of a 'congruence' in §2.12: a set of curves that fill the manifold, or some part of it, without intersecting. Each point in the region of the manifold M is on one and only one curve. Since each curve is a one-dimensional set of points, the set of curves is $(n-1)$-dimensional. (With some suitable parameterization, the set of curves is itself a manifold.) The key point from which everything else follows is that the congruence provides a natural *mapping* of the manifold into itself. If the parameter on the curves is λ, then any sufficiently small number $\Delta\lambda$ defines a mapping in which each point is mapped into the one, a parameter distance $\Delta\lambda$ further along the same curve of the congruence (see figure 3.1). This is a 1–1 mapping, at least in any region in which the vector field is sufficiently well-behaved (a C^1 field will do). If the vector field is C^∞, the mapping is a diffeomorphism (see §2.4.). If the map exists

Fig. 3.1. The mapping of M to itself defined by mapping each point to the point on the same curve of the congruence whose parameter is some fixed number $\Delta\lambda$ larger.

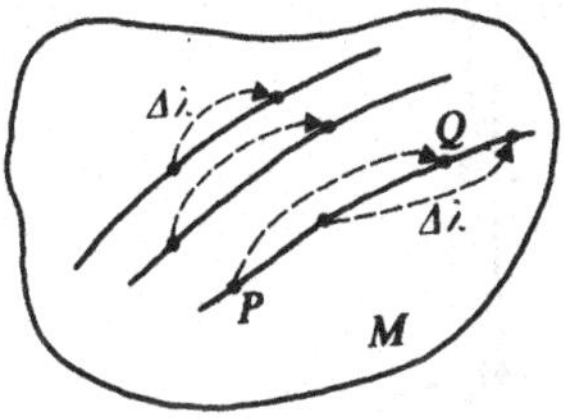

for all $\Delta\lambda$, there is a one-dimensional differentiable family of such mappings (a one-parameter Lie group, in fact, with composition law $\Delta\lambda_1 + \Delta\lambda_2$). Such a mapping is called a 'dragging' along the congruence, or a *Lie dragging*.

3.2 Lie dragging a function

If a function f is defined on the manifold, then the mapping defines a new function $f^*_{\Delta\lambda}$ by 'carrying' f along the congruence in an obvious manner: if a point P on a certain curve in figure 3.1 is mapped to the point Q, a parameter distance $\Delta\lambda$ further along the same curve, then the new field $f^*_{\Delta\lambda}$ has the same value at Q as f had at P,

♦ $$f(P) = f^*_{\Delta\lambda}(Q).$$

(Here the asterisk on $f^*_{\Delta\lambda}$ simply means 'new'.) If it happens that the value $f^*_{\Delta\lambda}(Q)$ in fact equals the old value at the point Q, $f(Q)$, for all Q,

$$f = f^*_{\Delta\lambda},$$

then the function is invariant under the mapping. If the function is invariant for *all* $\Delta\lambda$ then it is said to be *Lie dragged*. Clearly, a function that is Lie dragged must be constant along any curve of the congruence: $df/d\lambda = 0$.

3.3 Lie dragging a vector field

To see the effect this map has on vector fields, recall that any vector field is defined by the congruence of curves for which it is the tangent field. In

Fig. 3.2. How a new vector field $d/d\mu^*_{\Delta\lambda}$ is defined by Lie dragging its path and its parameter μ. Curves (1)–(4) are members of the λ-congruence. Curve (A) is a μ-curve passing through P and is mapped to curve (A') by being Lie dragged a parameter distance $\Delta\lambda$. Curve (B) is a μ-curve of the old congruence also passing through Q. The image of (B) under the dragging is not shown. In general (B) and (A') will be different curves. If they are the same the μ-congruence is said to be Lie dragged.

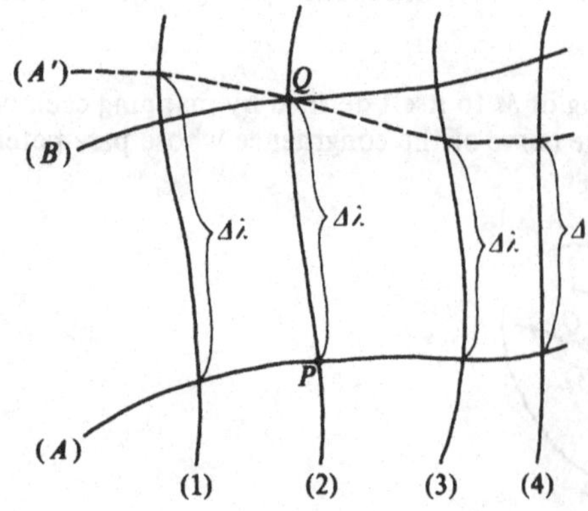

figure 3.2, we show two congruences: one, for $d/d\lambda$, generates a map of the manifold; the other, which defines the arbitrary field $d/d\mu$, will be acted on by the map. This action is very simple: any curve of the μ-congruence is mapped into a new curve which runs through the images under the Lie dragging of the points it used to run through, and the parameter values μ are carried along to the new points as well. This defines a new congruence with parameter $\mu^*_{\Delta\lambda}$. This new congruence has a tangent vector field $d/d\mu^*_{\Delta\lambda}$, which is called the image of $d/d\mu$ under the Lie dragging.

In general the $\mu^*_{\Delta\lambda}$-congruence is different from the μ-congruence. If it is the same, then $d/d\mu^*_{\Delta\lambda} = d/d\mu$ everywhere and we say the vector field and congruence are invariant under the map. If they are invariant for *all* $\Delta\lambda$ then we say they are *Lie dragged* by the vector field $d/d\lambda$.

A Lie dragged vector field has a simple geometric interpretation, illustrated by figure 3.3. It is clear that (in the limit of infinitesimal $\Delta\lambda$ and infinitesimal separation between curves (2) and (3)) if $d/d\mu$ at P 'stretches' exactly from P to R on curve (A), then $d/d\mu^*_{\Delta\lambda}$ stretches exactly from Q to S on (A'). If $d/d\mu$ is Lie dragged, then curve (B) of figure 3.2 coincides with (A') and $(d/d\mu^*_{\Delta\lambda})_Q = (d/d\mu)_Q$, so $d/d\mu$ also stretches from Q to S. Referring to our discussion of Lie brackets in §2.14, we find that this implies $[d/d\lambda, d/d\mu] = 0$: a vector field is Lie dragged if its Lie bracket with the dragging field vanishes:

$$[d/d\lambda, d/d\mu] = 0. \tag{3.1}$$

There is another way to see the same thing. Suppose we look at figure 3.3 differently, as if we were given only a single curve (A) with parameter μ, not a whole congruence. Then we can generate from this curve a whole congruence by Lie dragging it for all possible values of $\Delta\lambda$. One such curve is (A'). Let us call this field $d/d\mu_L$ with parameter μ_L. By this construction, the derivative $d/d\lambda$ is

Fig. 3.3. The central section of figure 3.2, with curve (B) omitted and the tangent vectors to (A) and (A') at P and Q respectively drawn in.

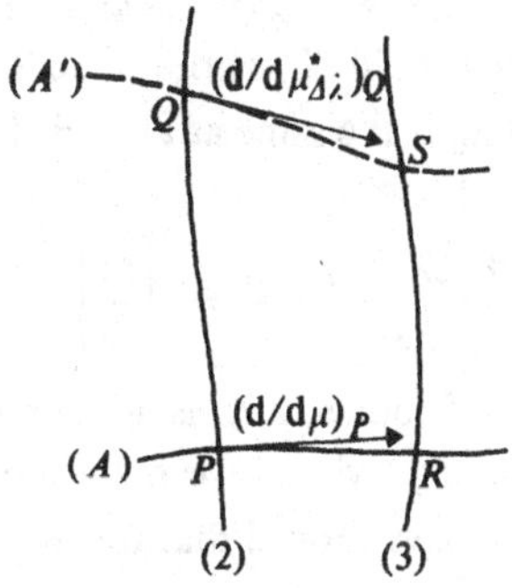

always on a curve of fixed μ and the derivative $d/d\mu_L$ is always on a curve of fixed λ. Therefore they must commute.

3.4 Lie derivatives

The concept of dragging permits the definition of a derivative along the congruence. There is a difficulty inherent in any attempt to define derivatives of vector and tensor fields. Consider trying to define a vector field's derivative as the limit of the difference between the vectors at different points divided by the distance between the points. One problem is defining 'distance' between points; if one has a curve between the points, one can take this to be the difference between the parameter values at the points. (This gives a derivative with respect to the parameter, and on manifolds without metrics this is all one can hope for.) A more serious problem is the comparison of vectors at different points: are two vectors at different points 'parallel' or not? In the Euclidean plane this is a simple question to answer. On a curved surface it may not have a unique answer. On a simple differentiable manifold the question of parallelism at different points does not even make sense, since there are no 'markers' or rules for moving vectors around in a parallel manner. One must add more structure – called an 'affine connection' – to the manifold in order to define an absolute parallelism. This is treated in chapter 6 on Riemannian geometry. What we shall consider here is an alternative that one should expect to find useful in any problem in which a congruence plays a central role. The congruence itself can provide a substitute for the concept of parallelism at different points. That is, when comparing vectors at points λ and $\lambda + \Delta\lambda$ on a certain curve, one can Lie drag the vector at $\lambda + \Delta\lambda$ back to the point λ. This defines a new vector at λ, which can be subtracted from the old one to define the difference between them. Notice that this is a *unique* difference, and hence a unique derivative, given the congruence. But it does depend on the congruence.

Let us derive analytic expressions for this. First consider a scalar function. Evaluate the scalar at the point $\lambda_0 + \Delta\lambda$, drag it back to λ_0, subtract the value of the scalar at λ_0, divide by $\Delta\lambda$ and take the limit $\Delta\lambda \to 0$. Its value at $\lambda_0 + \Delta\lambda$ is $f(\lambda_0 + \Delta\lambda)$. By dragging one defines a new scalar field f^*, whose value is defined by the rule $df^*/d\lambda = 0$. Therefore its value at λ_0 is the same as at $\lambda_0 + \Delta\lambda$: $f^*(\lambda_0) = f(\lambda_0 + \Delta\lambda)$. The derivative so defined is

$$\lim_{\Delta\lambda\to 0} \frac{f^*(\lambda_0) - f(\lambda_0)}{\Delta\lambda} = \lim_{\Delta\lambda\to 0} \frac{f(\lambda_0 + \Delta\lambda) - f(\lambda_0)}{\Delta\lambda} = \left[\frac{df}{d\lambda}\right]_{\lambda_0}. \quad (3.2)$$

The result for the Lie derivative of f is not, of course, surprising. There is a special notation for the Lie-derivative operator: $£_{\bar{V}}$, where $\bar{V}$ is the vector field generating the mappings ($d/d\lambda$ in our case). We have proved that for functions

♦ $$£_{\bar{V}}f = \bar{V}(f) = \mathrm{d}f/\mathrm{d}\lambda. \tag{3.3}$$

Now we do the same for a vector field $\bar{U} = \mathrm{d}/\mathrm{d}\mu$. Since a vector is defined by its effect on functions, we use an *arbitrary* function f in what follows. At λ_0 the field $\bar{U}$ gives the derivative $(\mathrm{d}f/\mathrm{d}\mu)_{\lambda_0}$, while at $\lambda_0 + \Delta\lambda$ it gives $(\mathrm{d}f/\mathrm{d}\mu)_{\lambda_0+\Delta\lambda}$. By dragging $\bar{U}(\lambda_0 + \Delta\lambda)$ in the sense of §3.3, one gets a new field $\bar{U}^* = \mathrm{d}/\mathrm{d}\mu^*$, defined by $[\bar{U}^*, \bar{V}] = 0$ and by $\bar{U}^*(\lambda_0 + \Delta\lambda) = \bar{U}(\lambda_0 + \Delta\lambda)$. The vanishing of the commutator implies

$$\frac{\mathrm{d}}{\mathrm{d}\lambda}\frac{\mathrm{d}}{\mathrm{d}\mu^*}f = \frac{\mathrm{d}}{\mathrm{d}\mu^*}\frac{\mathrm{d}}{\mathrm{d}\lambda}f \tag{3.4}$$

everywhere. Therefore we have (for analytic vector fields)

$$\begin{aligned}
\left[\frac{\mathrm{d}}{\mathrm{d}\mu^*}f\right]_{\lambda_0} &= \left[\frac{\mathrm{d}}{\mathrm{d}\mu^*}f\right]_{\lambda_0+\Delta\lambda} - \Delta\lambda\left[\frac{\mathrm{d}}{\mathrm{d}\lambda}\left(\frac{\mathrm{d}}{\mathrm{d}\mu^*}f\right)\right]_{\lambda_0} + O(\Delta\lambda^2) \\
&= \left[\frac{\mathrm{d}}{\mathrm{d}\mu}f\right]_{\lambda_0+\Delta\lambda} - \Delta\lambda\left[\frac{\mathrm{d}}{\mathrm{d}\mu^*}\left(\frac{\mathrm{d}}{\mathrm{d}\lambda}f\right)\right]_{\lambda_0} + O(\Delta\lambda^2) \\
&= \left[\frac{\mathrm{d}}{\mathrm{d}\mu}f\right]_{\lambda_0} + \Delta\lambda\left[\frac{\mathrm{d}}{\mathrm{d}\lambda}\left(\frac{\mathrm{d}}{\mathrm{d}\mu}f\right)\right]_{\lambda_0} \\
&\quad - \Delta\lambda\left[\frac{\mathrm{d}}{\mathrm{d}\mu^*}\left(\frac{\mathrm{d}}{\mathrm{d}\lambda}f\right)\right]_{\lambda_0} + O(\Delta\lambda^2).
\end{aligned}$$

We define the Lie derivative $£_{\bar{V}}\bar{U}$ as the vector field which operates on f to give

$$\begin{aligned}
[£_{\bar{V}}\bar{U}](f) &= \lim_{\Delta\lambda\to 0}\left[\frac{\bar{U}^*(\lambda_0) - \bar{U}(\lambda_0)}{\Delta\lambda}\right](f) \qquad (3.5)\\
&= \lim_{\Delta\lambda\to 0}\left[\left(\frac{\mathrm{d}}{\mathrm{d}\mu^*}f\right)_{\lambda_0} - \left(\frac{\mathrm{d}}{\mathrm{d}\mu}f\right)_{\lambda_0}\right]\Big/\Delta\lambda \\
&= \lim_{\Delta\lambda\to 0}\left(\frac{\mathrm{d}}{\mathrm{d}\lambda}\frac{\mathrm{d}}{\mathrm{d}\mu}f - \frac{\mathrm{d}}{\mathrm{d}\mu^*}\frac{\mathrm{d}}{\mathrm{d}\lambda}f\right).
\end{aligned}$$

Now, the difference between μ^* and μ is clearly a term of first order in $\Delta\lambda$, which means we can replace μ^* by μ in the last equation above. Since this equation is true for all f, we have

♦ $$£_{\bar{V}}\bar{U} = \frac{\mathrm{d}}{\mathrm{d}\lambda}\bar{U} - \frac{\mathrm{d}}{\mathrm{d}\mu}\bar{V} = [\bar{V}, \bar{U}]. \tag{3.6}$$

This is again a sensible result. By definition of the Lie derivative along $\bar{V}$, a vector field has a zero Lie derivative if it is Lie dragged, i.e. if it has zero Lie bracket with $\bar{V}$. Therefore it makes sense that its derivative is in fact its Lie bracket. By the antisymmetry of the Lie bracket we find

$$£_{\bar{V}}\bar{U} = -£_{\bar{U}}\bar{V}. \tag{3.7}$$

Exercise 3.1

(a) Show that, on functions and fields,

$$[\pounds_{\bar{V}}, \pounds_{\bar{W}}] = \pounds_{[\bar{V},\bar{W}]} \tag{3.8}$$

for any two twice-differentiable vector fields $\bar{V}$ and $\bar{W}$.

(b) Prove the Jacobi identity for Lie derivatives on functions and vector fields:

$$[[\pounds_{\bar{X}}, \pounds_{\bar{Y}}], \pounds_{\bar{Z}}] + [[\pounds_{\bar{Y}}, \pounds_{\bar{Z}}], \pounds_{\bar{X}}] + [[\pounds_{\bar{Z}}, \pounds_{\bar{X}}], \pounds_{\bar{Y}}] = 0, \tag{3.9}$$

where $\bar{X}, \bar{Y}, \bar{Z}$ are any three-times-differentiable vector fields. (Hint: for (a) on vectors, show that (3.8) is equivalent to (2.14). For (b) on vectors, use (3.8) and the fact that, as is obvious from its definition, $\pounds_{\bar{A}} + \pounds_{\bar{B}} = \pounds_{\bar{A}+\bar{B}}$.)

Exercise 3.2

(a) Deduce the Leibniz rule

$$\pounds_{\bar{V}}(f\bar{U}) = (\pounds_{\bar{V}}f)\bar{U} + f\pounds_{\bar{V}}\bar{U} \tag{3.10}$$

from the definitions of $\pounds_{\bar{V}}$ on functions and vector fields.

(b) From (2.7) we know that the components of $\pounds_{\bar{V}}\bar{U}$ on a *coordinate basis* are

$$(\pounds_{\bar{V}}\bar{U})^i = V^j \frac{\partial}{\partial x^j} U^i - U^j \frac{\partial}{\partial x^j} V^i. \tag{2.7}$$

Given an arbitrary basis $\{\bar{e}_i\}$ for vector fields, show from (a) that

$$(\pounds_{\bar{V}}\bar{U})^i = V^j\bar{e}_j(U^i) - U^j\bar{e}_j(V^i) + V^jU^k(\pounds_{\bar{e}_j}\bar{e}_k)^i, \tag{3.11}$$

where $\bar{e}_j(U^i)$ means the derivative of the function U^i with respect to the vector field $\bar{e}_j$.

Exercise 3.3

Show that if one chooses a coordinate system in which $\bar{V}$ is a coordinate basis vector, say $\partial/\partial x^1$, then for any vector field $\bar{W}$

♦ $$(\pounds_{\bar{V}}\bar{W})^i = \partial W^i/\partial x^1. \tag{3.12}$$

That is, the Lie derivative is the coordinate-independent form of the partial derivative.

3.5 Lie derivative of a one-form

Since fields of one-forms and tensors of higher rank are defined in terms of vector fields and scalar functions, one can deduce the Lie derivatives of one-forms from the Lie derivatives of vectors and scalars. Conceptually, the definition is the same: a one-form field is said to be Lie dragged if its value on any Lie dragged vector field is constant. The Lie derivative is found by dragging the one-form at $\lambda_0 + \Delta\lambda$ back to λ_0 and taking the difference. The result is that if $\tilde{\omega}$ is a one-form, then $\pounds_{\bar{V}}\tilde{\omega}$ is the one-form field which is the Lie derivative of $\tilde{\omega}$ along $\bar{V}$ defined by the product rule (just the Leibniz rule for first-order derivatives):

$$\pounds_{\bar{V}}[\tilde{\omega}(\bar{W})] = (\pounds_{\bar{V}}\tilde{\omega})(\bar{W}) + \tilde{\omega}(\pounds_{\bar{V}}\bar{W}) \tag{3.13}$$

for all vector fields $\bar{W}$. Since $\tilde{\omega}(\bar{W})$ is simply a function, this defines $\pounds_{\bar{V}}\tilde{\omega}$ in terms of known operations, the Lie derivative of functions and vector fields.

Exercise 3.4

From (3.13) and the expression (2.7) for the components of $\pounds_{\bar{V}}\bar{W} = [\bar{V}, \bar{W}]$, deduce that $\pounds_{\bar{V}}\tilde{\omega}$ has components, on a coordinate basis,

♦ $$(\pounds_{\bar{V}}\tilde{\omega})_i = V^j \frac{\partial}{\partial x^j}\omega_i + \omega_j \frac{\partial}{\partial x^i} V^j. \tag{3.14}$$

The natural extension of (3.13) to tensors of higher type gives the Lie derivative the properties

$$\pounds_{\bar{V}}(\mathbf{A} \otimes \mathbf{B}) = (\pounds_{\bar{V}}\mathbf{A}) \otimes \mathbf{B} + \mathbf{A} \otimes (\pounds_{\bar{V}}\mathbf{B}) \tag{3.15}$$

and

$$\begin{aligned} \pounds_{\bar{V}}(\mathbf{T}(\tilde{\omega}, \ldots; \bar{U}, \ldots)) = {} & (\pounds_{\bar{V}}\mathbf{T})(\tilde{\omega}, \ldots; \bar{U}, \ldots) \\ & + \mathbf{T}(\pounds_{\bar{V}}\tilde{\omega}, \ldots; \bar{U}, \ldots) + \ldots \\ & + \mathbf{T}(\tilde{\omega}, \ldots; \pounds_{\bar{V}}\bar{U}, \ldots) + \ldots, \end{aligned} \tag{3.16}$$

where $\mathbf{A}, \mathbf{B}, \mathbf{T}$ are arbitrary tensors and $\tilde{\omega}$ and $\bar{U}$ arbitrary one-form and vector, respectively.

3.6 Submanifolds

A submanifold of a manifold M is a manifold which is a smooth subset of M. If M is ordinary three-dimensional Euclidean space, then ordinary smooth surfaces and curves are submanifolds. In four-dimensional Minkowski spacetime (§2.31), the three-dimensional space of events simultaneous to a given event in the view of a particular observer (same time coordinate t) is a submanifold, and so is the hyperboloid of all events at constant interval Δs^2 from a given event. The word 'hypersurface' is sometimes used instead of 'submanifold', but some textbooks use 'hypersurface' only to describe a submanifold whose dimension is one less than that of M.

Although the idea of a submanifold is easy enough to visualize in simple cases, the word 'smooth' in the definition given above needs to be made more precise, and different textbooks give different (and inequivalent) definitions. We shall use the one which guarantees the greatest smoothness and is closest to our definition of a manifold. An m-dimensional *submanifold* S of an n-dimensional manifold M is a set of points of M which have the following property: in some open neighborhood in M of any point P of S there exists a coordinate system for M in which the points of S in that neighborhood are the points characterized by $x^1 = x^2 = \ldots = x^{n-m} = 0$ (see figure 3.4). A one-dimensional submanifold is a kind of *curve*, and its smoothness requirement is illustrated in figure 3.5. It is clear that the definition of S guarantees it is itself a manifold, since it has the requisite coordinate patches (charts). A special case is $m = n$: any open set of M is a submanifold of M.

Our interest in submanifolds stems mostly from the fact that solutions of differential equations are usually relations, say $\{y_i = f_i(x^1, \ldots, x^m), i = 1, \ldots, p\}$, which can be thought of as submanifolds with coordinates $\{x^1, \ldots, x^m\}$ of a larger manifold whose coordinates are $\{y_1, \ldots, y_p, x^1, \ldots, x^m\}$. We shall begin our investigation of submanifolds from a different perspective, however, and the tie-in with differential equations will not come until chapter 4.

Suppose P is a point of a submanifold S (dimension m) of M (dimension n). A curve in S through P is also a curve in M through P, so naturally a tangent vector to such a curve at P is an element of both T_P, the tangent space to M at P, and V_P, the tangent space to S at P. In fact, V_P is a vector subspace of T_P of dimension m. On the other hand, an arbitrary vector of T_P not in V_P has no unique 'projection' onto V_P (recall there is no notion of orthogonality in general).

Fig. 3.4. A two-dimensional submanifold S of a three-dimensional manifold M is shown, along with coordinates near a point P which satisfy the definition given in the text. The coordinate line of x^1 intersects S only at P.

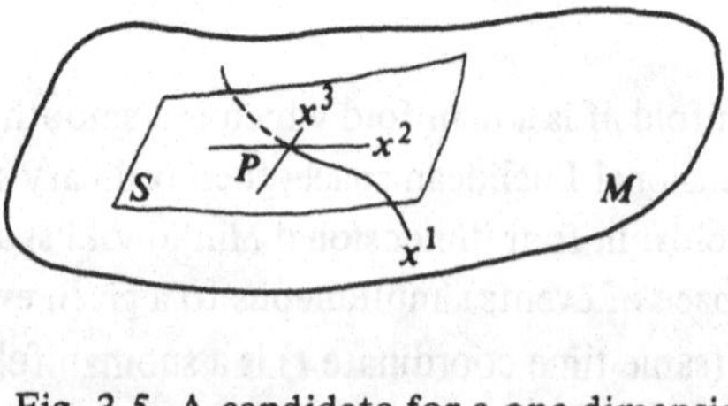

Fig. 3.5. A candidate for a one-dimensional submanifold of a two-dimensional manifold, which fails because it crosses itself at P. At P one cannot construct the necessary coordinates. Only some curves, therefore, are submanifolds.

The situation for one-forms at P is just the reverse. Let T^*_P be the dual of T_P, the set of one-forms at P which are functions defined on all of T_P. Similarly, let V^*_P be the dual of V_P, the one-forms S itself has at P. Any one-form in T^*_P defines one in V^*_P: this only involves restricting its domain from all of T_P down to its subspace V_P. But there is no unique element of T^*_P corresponding to a given element of V^*_P, since simply knowing the values of a one-form on V_P does not tell us what its value will be on a vector not in V_P.

In summary, then, a vector defined on a submanifold S is also a vector on M, and a one-form on M is also a one-form on S. But neither statement is reversible. We will discuss one-forms and submanifolds again in chapter 4. Here we shall concentrate of vector fields.

3.7 Frobenius' theorem (vector field version)

In any coordinate patch of S there are coordinates $\{y^a, a = 1, \ldots, m\}$ and basis vectors $\{\partial/\partial y^a\}$ for vector fields on S. All these basis fields naturally commute:

$$[\partial/\partial y^a, \partial/\partial y^b] = 0. \tag{3.17}$$

Exercise 3.5

(a) Show that if $\bar{V}$ and $\bar{W}$ are linear combinations (not necessarily with constant coefficients) of m vector fields that all commute with one another, then the Lie bracket of $\bar{V}$ and $\bar{W}$ is a linear combination of the same m fields.

(b) Prove the same result when the m vector fields have Lie brackets which are nonvanishing linear combinations of the m fields.

From exercise 3.5(a) it follows that *any* two vector fields on S have a Lie bracket which is also tangent to S, since these fields are certainly linear combinations of the commuting fields $\{\partial/\partial y^a\}$. The important statement is the converse: *if a set of m C^∞ vector fields defined in a region U of M have Lie brackets with one another, all of which are linear combinations of the m vector fields, then the integral curves of the fields mesh to form a family of submanifolds.* Each submanifold has dimension equal to the dimension of the vector space these fields define at any point, which is at most m, but which may be smaller (as in §3.9 below). Each point of U is on one and only one such submanifold, provided that the dimension of the vector space defined by the fields is the same everywhere in U. This family of submanifolds fills U in much the same way as a congruence of curves does (§2.12), and it is called a *foliation* of U. Each submanifold is a *leaf* of the foliation. Two foliations are illustrated in figure 3.6.

This result is called *Frobenius' theorem*. The proof is sketched in the next section, but it is easy to see the central idea. If the integral curves of the various fields are to define a submanifold, they must remain tangent to it: no curve can start 'sticking out' off it. This tangency is guaranteed if all the Lie brackets are themselves tangent, since the Lie brackets are simply the derivatives of the various vector fields along one another. If no vector field has a derivative with a component off the hypersurface, then no integral curve can leave the hypersurface. See figure 3.7 for some examples. When we come to the study of differential forms, we shall encounter another version of Frobenius' theorem, which will show us that it is the fundamental theorem giving conditions for the existence of solutions to partial differential equations ('integrability conditions').

Fig. 3.6. (*a*) A foliation of R^3 by parallel planes. Each point of R^3 is on one plane of the foliation. Only a few such planes are shown. (*b*) A foliation of R^3 by concentric spheres S^2. The centre is a degenerate point of the foliation.

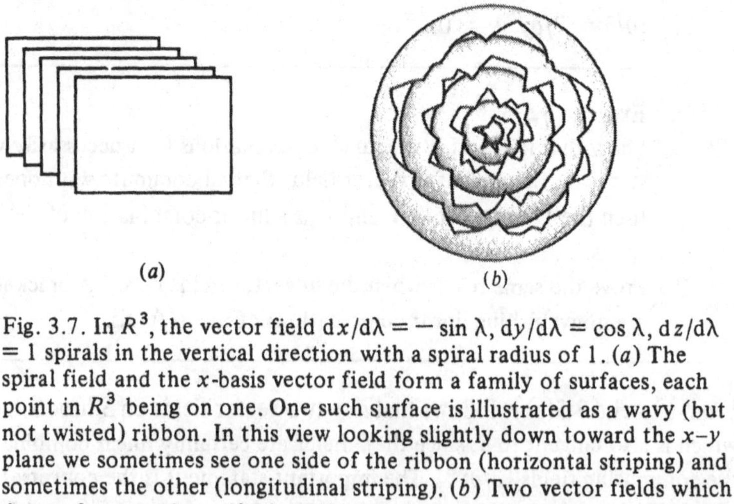

Fig. 3.7. In R^3, the vector field $dx/d\lambda = -\sin\lambda$, $dy/d\lambda = \cos\lambda$, $dz/d\lambda = 1$ spirals in the vertical direction with a spiral radius of 1. (*a*) The spiral field and the *x*-basis vector field form a family of surfaces, each point in R^3 being on one. One such surface is illustrated as a wavy (but not twisted) ribbon. In this view looking slightly down toward the x–y plane we sometimes see one side of the ribbon (horizontal striping) and sometimes the other (longitudinal striping). (*b*) Two vector fields which do not form a submanifold are the spiral one and the *z*-basis vector field. The plane defined by the two at any point is not tangent to the 'next' spiral curve above or below it.

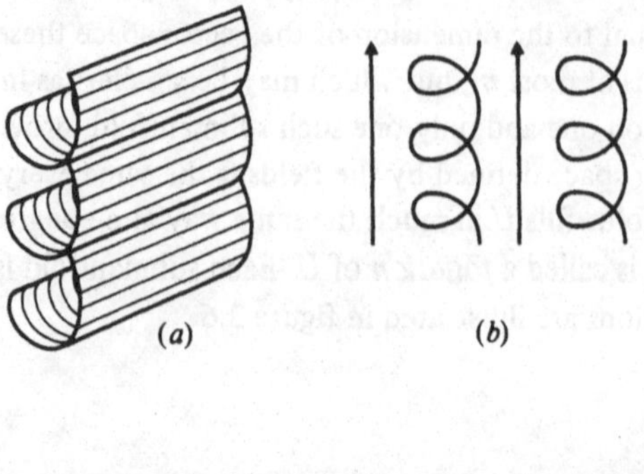

3.8 Proof of Frobenius' theorem

Suppose in some open region U' of M we are given m' vector fields which at every point P of U' span a subspace of T_P of dimension $m \leqslant m'$. (The set of all these subspaces is called an *m-dimensional distribution* on M. This has no relation to the delta-function distributions of §2.18.) At least in some neighborhood U of any point P in U' we can choose m of the fields as a linearly independent basis for the set, and these fields $\{\bar{V}_{(a)}, a = 1, \ldots, m\}$ will (by exercise 3.5(b)) have the property

$$[\bar{V}_{(a)}, \bar{V}_{(b)}] = \sum_c \alpha_{abc} \bar{V}_{(c)} \tag{3.18}$$

in U. So we never really need to consider the case where the fields are not linearly independent: such a set reduces locally to a linearly independent set of smaller dimension. Let the manifold M have dimension n.

The theorem is trivial when there is only one vector field $\bar{V}$ (i.e. $m = 1$). The integral curves clearly exist in U if $\bar{V}(P) \neq 0$. Each curve is a one-dimensional manifold, a submanifold of M.

The theorem for $m \geqslant 2$ will be proved by induction. First we will establish a formula we will find useful in the proof. From equation (3.14) it is easy to prove that for any function f and vector field $\bar{V}$,

$$£_{\bar{V}}(\tilde{d}f) = \tilde{d}(£_{\bar{V}}f). \tag{3.19}$$

Moreover, equation (3.15) implies, for any vector field $\bar{W}$,

$$£_{\bar{V}}\langle \tilde{d}f, \bar{W}\rangle = \langle £_{\bar{V}}\tilde{d}f, \bar{W}\rangle + \langle \tilde{d}f, £_{\bar{V}}\bar{W}\rangle. \tag{3.20}$$

Combining these two equations and remembering that $£_{\bar{V}}\bar{W} = [\bar{V}, \bar{W}]$, we find the result we shall need:

$$\langle \tilde{d}f, [\bar{V}, \bar{W}]\rangle = £_{\bar{V}}\langle \tilde{d}f, \bar{W}\rangle - \langle \tilde{d}(£_{\bar{V}}f), \bar{W}\rangle. \tag{3.21}$$

Returning now to the main proof, we first note that if the m vector fields all actually *commute* (have zero Lie brackets with one another), then the construction of §2.15 shows that they define a coordinate system for the points on their integral curves, hence the required family of submanifolds of M. We shall prove that the submanifolds exist in the general case (Lie brackets linearly dependent on the fields) by constructing m linearly independent linear combinations of the original fields which do commute. Thus, suppose we have m linearly independent vector fields $\bar{V}_{(a)}$ whose Lie brackets are linearly dependent on the fields. We select any one, say $\bar{V}_{(m)} = d/d\lambda_{(m)}$. Now the parameter $\lambda_{(m)}$ along the $\bar{V}_{(m)}$ congruence is a number defined at every point, so it is a function on the region U of M we are looking at. Accordingly, its gradient $\tilde{d}\lambda_{(m)}$ exists, and we use it as follows. We define $(m - 1)$ vector fields $\bar{X}_{(a)}$ which are linear combinations of all the original $\bar{V}_{(a)}$s and which satisfy

$$\langle \tilde{d}\lambda_{(m)}, \bar{X}_{(a)} \rangle = 0, \quad a = 1, \ldots, m-1. \tag{3.22}$$

This determines the set $\{\bar{X}_{(k)}\}$ up to linear combinations of themselves. Now we write (again by exercise 3.5(b))

$$[\bar{X}_{(a)}, \bar{X}_{(b)}] = \sum_{c=1}^{m-1} \beta_{abc} \bar{X}_{(c)} + \gamma_{ab} \bar{V}_{(m)}, \tag{3.23}$$

$$[\bar{V}_{(m)}, \bar{X}_{(a)}] = \sum_{b=1}^{m-1} \mu_{ab} \bar{X}_{(b)} + \nu_a \bar{V}_{(m)}, \tag{3.24}$$

where β_{abc}, γ_{ab}, μ_{ab}, and ν_a are all functions on U. We contract these equations with $\tilde{d}\lambda_m$ and use (3.21), (3.22), and the following simple identity

$$\langle \tilde{d}\lambda_m, \bar{V}_{(m)} \rangle = £_{\bar{V}_{(m)}} \lambda_{(m)} = d\lambda_{(m)}/d\lambda_{(m)} = 1 \Longrightarrow \tilde{d}(£_{\bar{V}_{(m)}} \lambda_{(m)}) = 0. \tag{3.25}$$

The left-hand sides of both (3.23) and (3.24) contract to zero and the resulting equations imply $\gamma_{ab} = \nu_a = 0$. So, in particular, the Lie brackets of the $\bar{X}_{(a)}$s do not involve $\bar{V}_{(m)}$ at all. This was the purpose of imposing (3.22) on their construction.

We now invoke the inductive hypothesis, that any set of $(m-1)$ vector fields having Lie brackets linearly dependent on them form an $(m-1)$-dimensional submanifold. This applies to the set $\{\bar{X}_{(a)}, a = 1, \ldots, m-1\}$, which is therefore assumed to form a family of $(m-1)$-dimensional submanifolds filling U. Define a set of vector fields $\{\bar{Y}_{(a)}, a = 1, \ldots, m-1\}$ which form a coordinate basis for one of the submanifolds, say S', so that these fields commute on S'. We shall define fields $\{\bar{Z}_{(a)}, a = 1, \ldots, m-1\}$ off S' by Lie dragging along $\bar{V}_{(m)}$:

$$\left.\begin{array}{lll} \bar{Z}_{(a)} = \bar{Y}_{(a)} & \text{on } S' \\ [\bar{V}_{(m)}, \bar{Z}_{(a)}] = 0 & \text{in } U \text{ along any curve} \\ & \bar{V}_m \text{ passing through } S' \end{array}\right\} \text{for } a = 1, \ldots, m-1. \tag{3.26}$$

What we aim to prove is that the $\bar{Z}_{(a)}$s commute among themselves everywhere, as they do on S'. Then we will have constructed the fully commuting set $\{\bar{V}_{(m)}, \bar{Z}_{(a)}, a = 1, \ldots, m-1\}$ and proved the theorem. But first we must establish that each $\bar{Z}_{(a)}$ is still a linear combination of the $\bar{V}_{(a)}$s. In fact we will prove that it is a linear combination of the $\bar{X}_{(a)}$s alone, without $\bar{V}_{(m)}$. Each field $\bar{Z}_{(a)}$ is certainly unique, so let us see whether we can satisfy (3.26) with a linear combination

$$\bar{Z}_{(a)} = \sum_b \alpha_{ab} \bar{X}_{(b)}.$$

Then we must have (all sums running from 1 to $m-1$)

$$0 = [\bar{V}_{(m)}, \bar{Z}_{(a)}] = £_{\bar{V}_{(m)}} \bar{Z}_{(a)}$$

$$= \sum_b (£_{\bar{V}_{(m)}} \alpha_{ab}) \bar{X}_{(b)} + \sum_b \alpha_{ab} [\bar{V}_{(m)}, \bar{X}_{(b)}]$$

$$= \sum_b \frac{d\alpha_{ab}}{d\lambda_m}\bar{X}_{(b)} + \sum_{bc} \alpha_{ab}\mu_{bc}\bar{X}_{(c)}, \tag{3.27}$$

to achieve which, we have used (3.24) with $\nu_a = 0$. Redefining the summation indices in the last sum ($b \to c$ and $c \to b$) gives

$$0 = \sum_b \left(d\alpha_{ab}/d\lambda_m + \sum_c \alpha_{ac}\mu_{cb} \right) \bar{X}_b.$$

Since the $\bar{X}_{(a)}$s are linearly independent, this requires

$$\frac{d\alpha_{ab}}{d\lambda_m} + \sum_c \alpha_{ac}\mu_{cb} = 0, \tag{3.28}$$

which is a set of ordinary differential equations. The initial conditions (at S'), that α_{ab} give the appropriate combination of $\bar{X}_{(b)}$s to form $\bar{Y}_{(a)}$, determine a unique solution, which always exists. Therefore, at every point the $\bar{Z}_{(a)}$s are linear combinations of the $\bar{X}_{(a)}$s.

The final step is to observe that the Lie dragging preserves the fact that they commute:

$$[\bar{Z}_{(a)}, \bar{Z}_{(b)}] = 0, \quad a, b = 1, \ldots, m-1. \tag{3.29}$$

This can be proved by using the Jacobi identity, exercise 2.3, among the three fields $\bar{V}_{(m)}$, $\bar{Z}_{(a)}$, and $\bar{Z}_{(b)}$. By construction we now have m fields $\{\bar{V}_{(m)}, \bar{Z}_{(a)}, a = 1, \ldots, m-1\}$ which all commute and which therefore form a coordinate basis for a submanifold of dimension m. Since the original fields $\{\bar{V}_{(a)}\}$ are linear combinations of these, we have proved the theorem.

3.9 An example: the generators of S^2

Readers familiar with angular momentum in quantum mechanics may have found many of the ideas presented so far familiar. Consider the (unnormalized) ϕ-basis vector of spherical coordinates, sometimes called $\bar{e}_\phi$:

$$\bar{e}_\phi = -y\bar{e}_x + x\bar{e}_y,$$

where $\bar{e}_x$ and $\bar{e}_y$ are the usual Cartesian basis vectors. In our notation this becomes

$$\frac{\partial}{\partial\phi} = -y\frac{\partial}{\partial x} + x\frac{\partial}{\partial y},$$

which we shall call $\bar{l}_z$, the 'angular momentum operator' for the z-direction:

$$\bar{l}_z = \frac{\partial}{\partial\phi}.$$

(This differs from the usual definition in quantum mechanics by a factor $\hbar/\mathrm{i}$.) One can define $\bar{l}_x$ and $\bar{l}_y$ in analogous ways, and one finds the commutation relations (Lie brackets)

$$[\bar{l}_x, \bar{l}_y] = -\bar{l}_z,$$
$$[\bar{l}_y, \bar{l}_z] = -\bar{l}_x, \qquad (3.30)$$
$$[\bar{l}_z, \bar{l}_x] = -\bar{l}_y.$$

Therefore the three vectors determine a submanifold. However, it would appear that this submanifold need only have dimension three – i.e. be all of the space – because there are three vectors. That it is really two-dimensional, we can see by realizing that if we define $r \equiv (x^2 + y^2 + z^2)^{1/2}$, then $\bar{l}_x(r) = \bar{l}_y(r) = \bar{l}_z(r) = 0$. Put another way,

$$\tilde{d}r(\bar{l}_x) = \tilde{d}r(\bar{l}_y) = \tilde{d}r(\bar{l}_z) = 0. \qquad (3.31)$$

From our picture of $\tilde{d}r$ as a set of surfaces of constant r, and our interpretation of its contraction with, say, $\bar{l}_x$ as the number of such surfaces $\bar{l}_x$ pierces, we see that (3.31) means that $\bar{l}_x$, $\bar{l}_y$, and $\bar{l}_z$ are all *tangent* to the sphere $r = \text{const}$. Therefore, at any point they are linearly dependent, and they generate a submanifold of dimension two: the sphere, of course.

Exercise 3.6

Show that exercise 3.3 is valid when $\bar{W}$ is replaced by *any* tensor field.

Exercise 3.7

Define the operator

$$L^2 = £_{\bar{l}_x} £_{\bar{l}_x} + £_{\bar{l}_y} £_{\bar{l}_y} + £_{\bar{l}_z} £_{\bar{l}_z}. \qquad (3.32)$$

Show that $£_{\bar{l}_z}$ and L^2 commute. By symmetry this also implies that L^2 commutes with $£_{\bar{l}_x}$ and $£_{\bar{l}_y}$. Show that if f is a scalar function, then

$$L^2 f = \frac{1}{\sin\theta} \frac{\partial}{\partial\theta}\left(\sin\theta \frac{\partial f}{\partial\theta}\right) + \frac{1}{\sin^2\theta} \frac{\partial^2 f}{\partial\phi^2}, \qquad (3.33)$$

where θ and ϕ are the usual spherical coordinates. That is, $L^2 f$ is the angular part of $\nabla^2 f$ on the unit sphere.

3.10 Invariance

One of the principal uses of Lie derivatives in physics is to express the notion that a tensor field is invariant under some transformation. We say that a tensor field T is *invariant under a vector field* $\bar{V}$ if

$$£_{\bar{V}} \mathsf{T} = 0. \qquad (3.34)$$

If T has physical importance – e.g. it might be the metric tensor, or a scalar field describing the potential energy of a particle, or a vector field of force – then those special vector fields (if any) under which T is invariant will also be important. For example, in the preceding section we discussed the vector fields associated with rotations of the sphere. One knows that angular momentum will be important in a physical problem only if the problem is invariant under the rotations associated with at least one of the vector fields. For instance, if the system is invariant under rotations in some plane, it is said to be *axially symmetric* (or axisymmetric) and the angular momentum associated with the vector generating those rotations is conserved. How this comes about will be discussed in exercise 5.8. Here we shall look at invariance generally.

The following theorem is of central importance to the whole theory of invariance. Suppose we have a set $F = \{\mathbf{T}_1, \mathbf{T}_2, \ldots\}$ of tensor fields whose invariance properties are being studied. Then *the set of all vector fields $\bar{V}$ under which all fields in F are invariant is a Lie algebra*, as defined in §2.14. The proof of this theorem has two steps. The first step is supplied by exercise 3.8, which shows that the set of fields is a vector space over the real numbers.

Exercise 3.8

Show that if a tensor $\mathbf{T}$ is invariant under both $\bar{V}$ and $\bar{W}$ then it is invariant under $a\bar{V} + b\bar{W}$, where a and b are *constants*.

The second step relies on the result of exercise 3.1(a), which applies as well to all tensor fields by (3.13) and (3.15). If $\bar{V}$ and $\bar{W}$ are vector fields in the set then for any tensor field $\mathbf{T}_i$ in F

$$£_{\bar{V}}\mathbf{T}_i = £_{\bar{W}}\mathbf{T}_i = 0 \Rightarrow [£_{\bar{V}}, £_{\bar{W}}]\mathbf{T}_i = 0 \Rightarrow £_{[\bar{V},\bar{W}]}\mathbf{T}_i = 0. \quad (3.35)$$

Therefore $[\bar{V}, \bar{W}]$ is in the set if $\bar{V}$ and $\bar{W}$ are. This proves the theorem. We will shortly see that Lie algebras are very closely related to Lie groups, and this theorem then explains some of the usefulness of Lie groups in physics. In the next sections we will study some examples of invariance.

It is important to understand what sort of vector space this Lie algebra is. One usually thinks of a linear combination of vector fields $\bar{V}$ and $\bar{W}$ as another vector field $a\bar{V} + b\bar{W}$, where a and b are functions on the manifold. The linear combinations permitted by exercise 3.8, however, use only *constants* for a and b. The vector space we have constructed has the *fields* $\bar{V}$ and $\bar{W}$ as *single elements*; it is not a fiber bundle of which $\bar{V}$ and $\bar{W}$ are cross-sections. It is more like a finite-dimensional function space (see §2.3). This point may seem subtle, but it is important for understanding the dimension of the vector space. For example, the three vector fields $\bar{l}_x$, $\bar{l}_y$, and $\bar{l}_z$ of the previous section are linearly dependent as vector fields on R^3, since they are all tangent to S^2. But to express one in terms of the other two one must use a linear combination with *variable* coefficients. Therefore the three fields are *linearly independent* elements of the Lie algebra: no linear combination of them with constant coefficients (not all zero) equals the zero element of the algebra, which is the zero vector field. We therefore say that these vector fields are a basis for a three-dimensional Lie algebra. The only other Lie algebra we have encountered so far is the algebra of *all* tangent vector fields to a manifold or submanifold. No finite number of such fields can be a basis for linear combinations using constant coefficients, so we say that this Lie algebra is infinite-dimensional.

3.11 Killing vector fields

Many manifolds of interest in physics have metrics, and it is therefore of considerable interest whenever the metric is invariant with respect to some vector field. A *Killing vector field* is defined to be a vector field $\bar{V}$ such that

♦ $$\pounds_{\bar{V}}\mathsf{g} = 0. \tag{3.36}$$

It can be deduced that the component form of this equation in a coordinate system is (cf. equation (3.14))

$$(\pounds_{\bar{V}}\mathsf{g})_{ij} = V^k \frac{\partial}{\partial x^k} g_{ij} + g_{ik} \frac{\partial}{\partial x^j} V^k + g_{kj} \frac{\partial}{\partial x^i} V^k = 0. \tag{3.37}$$

It is often convenient to use a coordinate system in which the integral curves of $\bar{V}$ are one family of coordinate lines, say for the x^1 coordinate. Then, from exercise 3.6 we find

$$(\pounds_{\bar{V}}\mathsf{g})_{ij} = \frac{\partial}{\partial x^1} g_{ij} = 0, \tag{3.38}$$

and so the metric components are independent of the coordinate x^1. Conversely, if there exists a coordinate system in which the components of the metric are independent of a certain coordinate, then the basis vector for that coordinate is a Killing vector. This is often a convenient way of identifying Killing vectors.

As an example, let us find the Killing vector fields of three-dimensional Euclidean space. The metric in Cartesian coordinates has components

$$g_{ij} = \delta_{ij}, \tag{3.39}$$

which is independent of x, y, and z. Therefore $\partial/\partial x$, $\partial/\partial y$, and $\partial/\partial z$ are Killing vectors. The same metric in spherical polar coordinates has components

$$\begin{aligned} g_{rr} &= \frac{\partial}{\partial r} \cdot \frac{\partial}{\partial r} = 1, \\ g_{\theta\theta} &= \frac{\partial}{\partial \theta} \cdot \frac{\partial}{\partial \theta} = r^2, \\ g_{\phi\phi} &= \frac{\partial}{\partial \phi} \cdot \frac{\partial}{\partial \phi} = r^2 \sin^2\theta. \end{aligned} \tag{3.40}$$

Therefore $\partial/\partial\phi$ is a Killing vector: $\bar{l}_z$ in fact. Clearly, $\bar{l}_x$ and $\bar{l}_y$ will also be Killing vectors. These six Killing vectors turn out to be a basis for the Lie algebra of Killing vector fields. We shall prove this in chapter 5, part E, where we undertake a more thorough study of spaces with high symmetry.

3.12 Killing vectors and conserved quantities in particle dynamics

It is well-known in classical mechanics that if a force is the gradient of a potential which is axially symmetric, then the angular momentum of a particle about the axis of symmetry is constant on the particle's trajectory. Similarly, if the potential is independent of one of the Cartesian coordinates, say x, then the x-component of momentum is conserved. However, it is not often remarked that if the potential has some other sort of symmetry (constant, say, on a family of similar ellipsoids) then there is *not* a conserved momentum associated with that symmetry. That is, conserved quantities in particle dynamics do not follow simply from invariance of the potential under some motion (circular, linear, or elliptical in our three examples), but also require that the motion be along a *Killing vector field* of the Euclidean space in which the dynamics takes place. Although we do not yet have quite enough mathematical machinery to prove this assertion (deferred to exercise 5.8), its reasonableness can be seen from the equation of motion. Written in ordinary vector calculus notation, this is

$$m\dot{\mathbf{V}} = -\nabla\Phi, \text{ or } m\dot{V}^i = -\nabla^i\Phi. \tag{3.41}$$

But with our understanding that the vector gradient involves the metric, we know that this is really

$$m\dot{V}^i = -g^{ij}\frac{\partial}{\partial x^j}\Phi. \tag{3.42}$$

Any invariants derived from this equation clearly must involve not only the invariance of Φ but also of g.

3.13 Axial symmetry

To illustrate the natural way in which Lie derivatives enter problems with symmetry, we consider the case of axial symmetry. Axial symmetry is invariance under rotations about some fixed axis. (It should not be confused with cylindrical symmetry, which has the added assumption of invariance under translation along the axis of symmetry.) Let the angle about the axis be ϕ. Situations often arise in which a problem has a certain 'background' axial symmetry. One may be dealing with a particle orbiting in an axially symmetric potential, or with small perturbations of an axially symmetric system. In such a case, one gets a linear equation for the unknown ψ,

$$L(\psi) = 0, \tag{3.43}$$

where L is some operator which is independent of the coordinate transformation $\phi \rightarrow \phi + \text{const}$. Solutions to (3.43) are not necessarily axially symmetric: the particle is at one angle at one time and another angle a moment later, or the perturbation has nonaxisymmetric initial values. But scalar solutions do have the nice property that, when Fourier-analyzed in ϕ as

$$\psi(\phi, x^j) = \sum_{m=-\infty}^{\infty} \psi_m(x^j)\,\mathrm{e}^{im\phi}, \tag{3.44}$$

the functions $\psi_m(x^j)$ (the index j runs over all coordinates but ϕ) satisfy the related differential equation

$$0 = L_m(\psi_m) = \mathrm{e}^{-im\phi} L(\psi_m\,\mathrm{e}^{im\phi}). \tag{3.45}$$

The operators L and L_m are not usually identical because L can contain derivatives with respect to ϕ, but L_m must not. For example, consider the operator

$$\nabla^2 = \frac{1}{r^2}\frac{\partial}{\partial r} r^2 \frac{\partial}{\partial r} + \frac{1}{r^2 \sin\theta}\frac{\partial}{\partial\theta}\sin\theta\frac{\partial}{\partial\theta} + \frac{1}{r^2\sin^2\theta}\frac{\partial^2}{\partial\phi^2},$$

which is clearly unchanged by the transformation $\phi \rightarrow \phi + \text{const}$. Then when it operates on a function $f(r, \theta)\,\mathrm{e}^{im\phi}$ it gives

$$\nabla^2(f(r,\theta)\,\mathrm{e}^{im\phi}) = \mathrm{e}^{im\phi}\left\{\frac{1}{r^2}\frac{\partial}{\partial r}r^2\frac{\partial}{\partial r} + \frac{1}{r^2\sin\theta}\frac{\partial}{\partial\theta}\sin\theta\frac{\partial}{\partial\theta} - \frac{m^2}{r^2\sin^2\theta}\right\} f(r,\theta). \tag{3.46}$$

The operator in curly brackets is ∇_m^2, as defined in (3.45). This Fourier decomposition of the function f is not usually useful in the case of particle motion, where the particle's position is a delta function in ϕ, but it is very helpful for continuous systems, like waves on an axially symmetric background. The key functions, $\mathrm{e}^{im\phi}$, may be called scalar axial harmonics.

We say that a solution ψ to (3.43) has *axial eigenvalue m* if

$$\blacklozenge \qquad £_{\bar{e}_\phi}\psi = im\psi, \tag{3.47}$$

where $\bar{e}_\phi \equiv \partial/\partial\phi$ is the tangent to the circles of symmetry. None of this is difficult if ψ is a scalar function, but suppose one is dealing with a vector equation, such as the one for the vector potential of electromagnetism. It will again be useful to have axial harmonics here, but they must be *vector* axial harmonics. We proceed to construct these.

Consider the submanifold $\phi = 0$ (really a submanifold with boundary, this being on the axis of symmetry). At each point choose a basis $\{\bar{e}_j\}$ for vectors tangent to the submanifold. Supplement this basis by $\bar{e}_\phi$ so that $\{\bar{e}_\phi, \bar{e}_j\}$ is a basis for all vectors tangent to the manifold at the points of the submanifold. Now generate a basis for the entire manifold by Lie dragging this basis along $\bar{e}_\phi$ all the way around the axis of symmetry, as shown in figure 3.8. The resulting fields all satisfy

$$\pounds_{\bar{e}_\phi} \bar{e}_j = 0, \tag{3.48}$$

i.e. they are all axially symmetric. Notice that in conventional Cartesian coordinates the components of $\bar{e}_j$ change on going around the axis. Axial symmetry for a vector field does *not* mean that its Cartesian components are independent of ϕ, but rather that its components in a coordinate system that includes ϕ are independent of ϕ.

We now have a basis which has axial eigenvalue 0, by equation (3.48). Clearly, a basis which has axial eigenvalue m is

$$\bar{e}_{(m)j} = \bar{e}_j \, \mathrm{e}^{\mathrm{i}m\phi}, \bar{e}_{(m)\phi} = \bar{e}_\phi \, \mathrm{e}^{\mathrm{i}m\phi}. \tag{3.49}$$

Any vector field satisfying

$$\pounds_{\bar{e}_\phi} \bar{V} = \mathrm{i}m\bar{V}$$

can be expressed as a linear combination of the vector axial harmonics of eigenvalue m given in (3.49), which coefficients which are independent of ϕ.

Exercise 3.9

In Euclidean three-space, construct axial vector harmonics for rotations about the z-axis by choosing the basis in the plane $\phi = 0$ to be $\{\bar{e}_x, \bar{e}_z\}$. Find the Cartesian components of the three vector harmonics for $m = 2$. In a similar way, find the basis one-form axial harmonics for $m = 2$, beginning with $\{\tilde{\mathrm{d}}x, \tilde{\mathrm{d}}z\}$ in the plane $\phi = 0$. If f is a scalar function of axial eigenvalue 2, show that the gradient, $\tilde{\mathrm{d}}f$, is a one-form of axial eigenvalue 2. Show that $\bar{e}_x \pm \mathrm{i}\bar{e}_y$ has axial eigenvalue $\pm$ 1.

Although we have not yet exploited it, there is clearly a close relationship here with group theory. The existence of axial symmetry means that the

Fig. 3.8. View down the axis of symmetry of a basis $\{\bar{e}_\phi, \bar{e}_j\}$ formed by Lie dragging along $\bar{e}_\phi$.

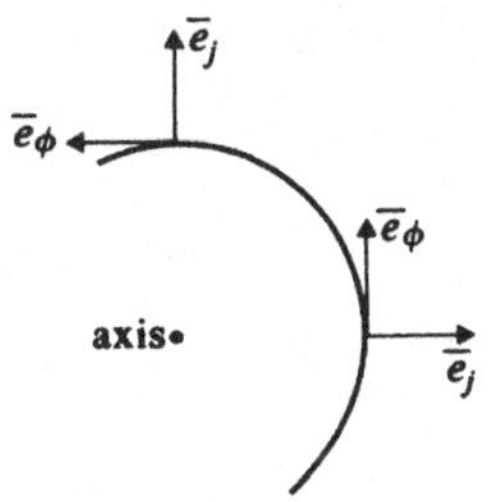

'background' physical situation is invariant under Lie draggings along $\partial/\partial\phi$. These draggings form a Lie group, as described in §3.1. The group involved, $SO(2)$, is particularly simple. The more important example of the rotation group (whose associated symmetry is called spherical symmetry) is more complicated because the various Lie derivatives along $\bar{l}_x$, $\bar{l}_y$, and $\bar{l}_z$ do not commute. We can deal with this problem only by studying Lie groups themselves systematically, and that will occupy the rest of this chapter.

3.14 Abstract Lie groups

We have touched on Lie groups or Lie algebras several times. Now we shall study them more systematically. The main reason they are interesting in physics is, as we have seen, that they express the invariance properties of important tensors. We will explore that aspect in later sections. Here our intention is to study the group manifold itself. This is an important distinction which must not be blurred: the group manifold is quite separate from whatever manifold contains the tensor whose invariance properties the group expresses. The manifold of all rotations ($SO(3)$) is different from the manifold whose coordinate systems are rotated (E^3).

Let us assume we have a finite-dimensional Lie group, i.e. a C^∞ manifold G of dimension n, which has the following C^∞ maps (diffeomorphisms): any element g of G maps $h \mapsto gh$ (*left translation by g*) or $h \mapsto hg$ (*right translation by g*). We do not assume the group is Abelian (i.e. $hg \neq gh$ in general), and we shall denote the identity element by e. Any neighborhood of e is mapped by left translation along a particular g onto a neighborhood of g, as shown in figure 3.9. Because the map carries curves into curves it maps tangent vectors at e (elements of T_e) to those at g. This is a map called $L_g: T_e \to T_g$, which is also illustrated in figure 3.9. (The concept is the same as for the Lie dragging map, §3.3.) A vector

Fig. 3.9. The left translation along g maps a neighborhood of e onto one of g. There is a natural map of a vector at e to one at g.

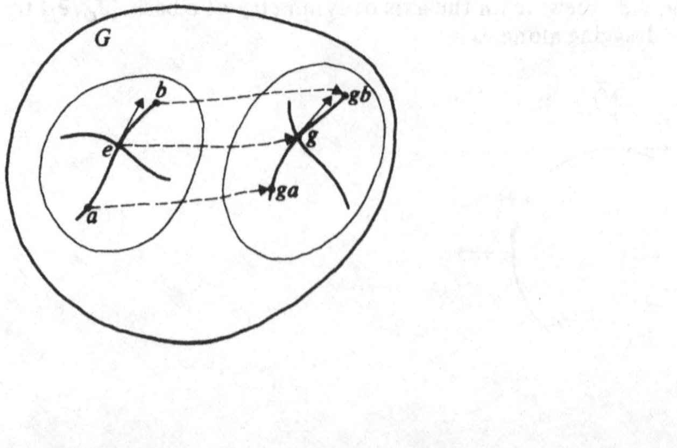

field $\bar{V}$ on G is said to be *left-invariant* if L_g maps $\bar{V}$ at e to $\bar{V}$ at g (L_g: $\bar{V}(e) \mapsto \bar{V}(g)$) for *all* g. By the group composition law it follows that L_g maps $\bar{V}(h) \mapsto \bar{V}(gh)$ for any h in G, so that what we have is a natural definition of a 'constant' vector field on G. It is also clear that each vector in T_e defines a unique left-invariant vector field, so that the left-invariant vector fields form an n-dimensional vector space. (As in §3.10, linear combinations of these fields use *constants*, not functions on G.) In fact it is easy to see (figure 3.10) that if $\bar{V}$ and $\bar{W}$ are any two left-invariant vector fields, then L_g maps $[\bar{V}, \bar{W}]$ at e to $[\bar{V}, \bar{W}]$ at g: the field $[\bar{V}, \bar{W}]$ is also left-invariant. (The reader who is not convinced by the diagram is invited to use coordinates on G to prove the result.) This is important, because it means that *the left-invariant vector fields form a Lie algebra*. This is called the *Lie algebra* of G, denoted by $\mathfrak{L}(G)$. (Some authors use $\mathfrak{g}$.) This Lie algebra is completely characterized by its *structure constants* c^j_{kl}, defined as follows. Let $\{\bar{V}_{(i)}, i = 1, \ldots, n\}$ be a basis for the Lie algebra, a linearly independent set of left-invariant vector fields. (If they are linearly independent at one point, say e, the map shows they are independent everywhere.) Then we can always write

$$[\bar{V}_{(k)}, \bar{V}_{(l)}] = c^j_{kl}\bar{V}_{(j)} \tag{3.50}$$

(summation convention assumed). If all the structure constants vanish, the Lie algebra is said to be *Abelian*. We shall see that it implies G is also Abelian. Naturally the basis $\{\bar{V}_{(k)}\}$ is not unique, and under a change of basis the numbers c^j_{kl} transform as components of a $\binom{1}{2}$ tensor. Every Lie group and algebra has a unique 'structure tensor' **C**. There is a limited converse to this, that a given set of structure constants 'almost' determines the Lie group whose Lie algebra they embody. This will be discussed in §3.16 below.

Fig. 3.10. The mapping of figure 2.21 for left-invariant vector fields. Because they are left-invariant, translations by parameter distance ϵ near e map into the same ones near g, and so the 'gap' near e that represents their Lie bracket is mapped to that near g, which is the bracket of the translated fields.

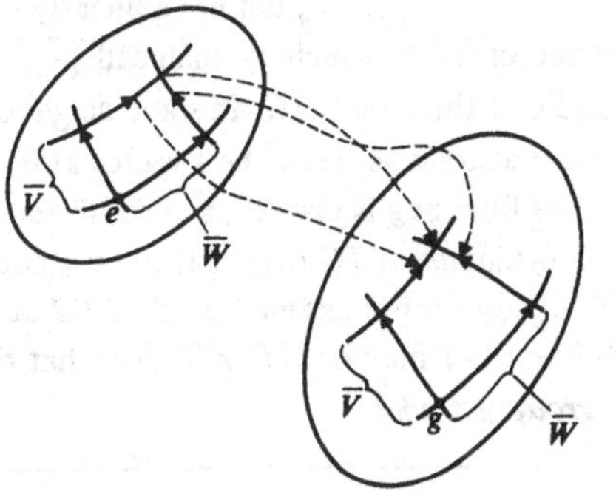

Consider the integral curve of a left-invariant vector field $\bar{V}$ which passes through e. It has a unique tangent vector $\bar{V}_e$ at e and a unique parameter t for which e corresponds to $t = 0$. As in §2.13, the points on the curve may be located by exponentiation of $\bar{V}$, $\exp(t\bar{V})$. This just involves the diffeomorphism of G onto itself generated by $\bar{V}$, as discussed in §3.1. Unlike an arbitrary vector field, $\bar{V}$ is determined completely by $\bar{V}_e$, so we can denote the points of G on this curve by

$$g_{\bar{V}_e}(t) = \exp(t\bar{V})|_e. \tag{3.51}$$

Because exponentiation has, by definition, the property

$$\exp(t_2\bar{V})\exp(t_1\bar{V})|_e = \exp[(t_1 + t_2)\bar{V}]|_e,$$

the points on these integral curves form a group:

$$\begin{aligned} g_{\bar{V}_e}(t_1 + t_2) &= \exp[(t_1 + t_2)\bar{V}]|_e = \exp(t_2\bar{V})\exp(t_1\bar{V})|_e \\ &= g_{\bar{V}_e}(t_2)g_{\bar{V}_e}(t_1). \end{aligned} \tag{3.52}$$

This is called a *one-parameter subgroup* of G. It is obviously always Abelian, $g_{\bar{V}_e}(t_1 + t_2) = g_{\bar{V}_e}(t_2 + t_1)$, simply because the group operation corresponds to addition of parameter values. To each vector in T_e there corresponds a unique subgroup. Moreover, since every one-parameter subgroup must be a C^∞-curve in G which passes through e (a subgroup must always contain the identity element) there is a one-to-one correspondence between the one-parameter subgroups of G and the elements of the Lie algebra of G.

Exercise 3.10

Define right-invariant vector fields. Show that they form a Lie algebra. Show that their integral curves through e coincide with those of the left-invariant fields. Show that their integral curves through other elements do not coincide with those of the left-invariant fields in general unless the group is Abelian.

Exercise 3.11

(a) Show any basis $\{\bar{V}_i(e), i = 1, \ldots, n\}$ for T_e defines a linearly independent set of left-invariant vector fields, which we shall call $\{\bar{V}_i\}$.

(b) Consider the tangent bundle of the group, TG. In some neighborhood U of e adopt coordinates for it as follows. Let $\bar{X}$ be a vector at point g of U, with $\bar{X} = \Sigma_i \alpha_i \bar{V}_i(g)$. The fiber at g is just R^n, so take the coordinates of $\bar{X}$ to be $\{\alpha_i\}$. Let the coordinates of TG over U then be {{coordinates of g}, $\{\alpha_i\}$}. Show that this prescription extends to all of TG in such a way as to prove that TG has a 1–1 map onto $G \times R^n$, *i.e.* that *the tangent bundle of a Lie group is trivial.*

3.15 Examples of Lie groups

(i) The simplest example is R^n, which is a manifold and a group under vector addition. This is an Abelian group. The one-paramenter subgroups are the 'rays' (straight lines through the origin). The left-invariant vector fields are parallel to the rays, so they all commute. The Lie algebra is thus the vector space T_e equipped with the trivial Abelian bracket: $[\bar{V}, \bar{W}] = 0$ for all $\bar{V}$ and $\bar{W}$ in T_e.

(ii) For physics, one of the most important Lie groups is the group of all $n \times n$ real matrices with nonvanishing determinant, called $GL(n, \mathrm{R})$ or the General Linear group in n Real dimensions, which is a Lie group for the following reasons. First, it is a group with the operation of matrix multiplication, the unit matrix being the identity element. (The restriction to nonvanishing determinant is necessary to ensure the existence of an inverse element for any matrix.) Second, it is a Lie group because it is a manifold. Any matrix A in $GL(n, \mathrm{R})$ with entries $\{a^i{}_j, i, j = 1, \ldots, n\}$ has a neighborhood of radius ϵ defined as those matrices B for which $|b^i{}_j - a^i{}_j| < \epsilon$ for all i and j, and ϵ can be chosen small enough so that every B has nonvanishing determinant. The numbers $x^i{}_j = b^i{}_j - a^i{}_j$ are coordinates for this neighborhood, and as there are n^2 of them, all independent, the dimension of $GL(n, \mathrm{R})$ is n^2. In fact it is a submanifold of R^{n^2}. Since R^{n^2}, like any R^m, is identical with the tangent space of any of its points, the tangent space of the identity e of $GL(n, \mathrm{R})$ is R^{n^2}, and any tangent vector can be represented as a matrix. For instance, the curve in $GL(n, \mathrm{R})$ comprising the matrices diag $(1 + \exp(\lambda), 1, 1, \ldots, 1)$, which has parameter λ, has tangent diag $(1, 0, 0, \ldots, 0)$ at $\lambda = 0$. This matrix has zero determinant, illustrating the fact that *any* matrix is in T_e and therefore any matrix generates a one-parameter subgroup, a left-invariant vector field,† and an element of the Lie algebra.

The one-parameter subgroup generated by any matrix A is the integral curve through e of the left-invariant vector field whose tangent at e is A. If we denote these matrices by $g_A(t)$ with $\mathrm{d}g_A(t)/\mathrm{d}t|_0 = A$ (which simply means $\mathrm{d}(g_A)^i{}_j/\mathrm{d}t|_0 = a^i{}_j$ for all i, j), then by (3.52) we have

$$g_A(t + \Delta t) = g_A(t)g_A(\Delta t)$$

$$\Rightarrow \mathrm{d}g_A(t)/\mathrm{d}t = g_A(t)A \tag{3.53}$$

$$\Rightarrow g_A(t) = \exp(tA) \tag{3.54}$$

$$= 1 + tA + \frac{1}{2!}t^2A^2 + \frac{1}{3!}t^3A^3 + \ldots. \tag{3.55}$$

Equation (3.55) is the definition of the exponential of a matrix, and with (3.54) gives concreteness of the formal expression (3.51). So the one-parameter

† The reader should bear in mind that a vector tangent to G is in fact a matrix, not to be confused with a 'column vector', which plays no role here.

subgroups of $GL(n, R)$ are the exponentials of arbitrary $n \times n$ matrices. The matrix A is often called by physicists the *infinitesimal generator* of the subgroup $g_A(t)$. Exercise 3.12 explores properties of $\exp(tA)$.

Exercise 3.12

(a) Show that (3.55) satisfies (3.53).

(b) Show that (3.55) implies

$$\exp(B^{-1}AB) = B^{-1}\exp(A)B. \tag{3.56}$$

(c) It can be shown (see Hirsch & Smale, 1974) that for any real matrix A, a real matrix B can be chosen so that $B^{-1}AB$ has the following *canonical form* (called block-diagonal form, since the nonzero elements fall in square blocks along the main diagonal)

$$B^{-1}AB = \begin{pmatrix} P_1 & 0 & 0 & \cdots \\ 0 & P_2 & 0 & \cdots \\ 0 & 0 & P_3 & \\ \vdots & \vdots & \vdots & \ddots \end{pmatrix} \tag{3.57}$$

where each P_j is a square matrix having one of the forms

(i) P_j is a 1×1 matrix

$$(\lambda_j), \tag{3.58a}$$

or (ii) P_j is a 2×2 nondiagonal matrix given by

$$\begin{pmatrix} r_j & s_j \\ -s_j & r_j \end{pmatrix}, \tag{3.58b}$$

or (iii) P_j is an $n_j \times n_j$ nondiagonal matrix ($n_j \geqslant 2$) given by

$$\begin{pmatrix} \mu_j & 1 & 0 & \cdots & 0 & 0 \\ 0 & \mu_j & 1 & \ddots & 0 & 0 \\ 0 & 0 & \mu_j & \ddots & 1 & 0 \\ 0 & 0 & 0 & \ddots & \mu_j & 1 \\ 0 & 0 & 0 & \cdots & 0 & \mu_j \end{pmatrix}. \tag{3.58c}$$

Moreover, the numbers λ_j, μ_j, and $r_j \pm \mathrm{i}s_j$ are the *eigenvalues* of A. Show from this and (3.55) that $\exp(tB^{-1}AB)$ similarly has block-diagonal form with the corresponding blocks:

(i) $$(\mathrm{e}^{t\lambda_j}), \tag{3.59a}$$

(ii) $$\mathrm{e}^{tr_j}\begin{pmatrix} \cos ts_j & \sin ts_j \\ -\sin ts_j & \cos ts_j \end{pmatrix}, \tag{3.59b}$$

(iii)

$$e^{t\mu_j} \begin{pmatrix} 1 & t & \frac{1}{2!}t^2 & \frac{1}{3!}t^3 & \cdots \\ 0 & 1 & t & \frac{1}{2!}t^2 & \cdots \\ 0 & 0 & 1 & t & \cdots \\ 0 & 0 & 0 & 1 & \cdots \\ \vdots & \vdots & \vdots & \vdots & \ddots \end{pmatrix} \quad (3.59c)$$

It follows from (a) that a matrix B which puts A into canonical form also puts $\exp(tA)$ into canonical form in cases (i) and (ii), but that the transformation to canonical form in (iii) is a function of t.

Note that not every element of $GL(n, \mathrm{R})$ is a member of a one-parameter subgroup. One reason is that such a subgroup is a continuous curve in $GL(n, \mathrm{R})$, on which the determinant must change continuously. Since the determinant is 1 at e and cannot be zero, there is *no* continuous curve linking e to a matrix with negative determinant. (The reader can easily see that (3.59) represents only matrices with positive determinants.) This is an example of a *disconnected group* and illustrates the interesting global properties Lie groups can have: one does not usually learn everything about a Lie group just by studying its one-parameter subgroups or even its Lie algebra. Those elements which can be joined to e by a *continuous path* (not necessarily a one-parameter subgroup) are called the *component of the identity* of the group.

Exercise 3.13

Show that the matrix

$$\begin{pmatrix} -1 & 1 \\ 0 & -1 \end{pmatrix}$$

is in the component of the identity of $GL(2, \mathrm{R})$, but is not in any one-parameter subgroup. (*Hint*: construct a continuous path joining it to $e = \left(\begin{smallmatrix} 1 & 0 \\ 0 & 1 \end{smallmatrix}\right)$.)

What is the Lie algebra of $GL(n, \mathrm{R})$? Given a tangent vector $\bar{A}_e$ at e and its one-parameter subgroup $g_{\bar{A}_e}(t)$, the left-translation $fg_{\bar{A}_e}(t)$ of this curve by any matrix f of $GL(n, \mathrm{R})$ produces a curve of the congruence of the left-invariant vector field corresponding to $\bar{A}_e$, as in figure 3.11. This is how $\bar{A}_e$ generates its left-invariant field, which we call simply $\bar{A}$. If in fact f is on the curve $g_{\bar{B}_e}(t)$

through e generated by any matrix $\bar{B}_e$ in T_e then the Lie bracket of the two vector fields at e, $[\bar{A}, \bar{B}]|_e$, is by (2.12) just

$$\lim_{t \to 0} \frac{1}{t^2} [g_{\bar{A}_e}(t) g_{\bar{B}_e}(t) - g_{\bar{B}_e}(t) g_{\bar{A}_e}(t)],$$

which is easily evaluated using (3.55):

$$[\bar{A}, \bar{B}]|_e = \bar{A}_e \bar{B}_e - \bar{B}_e \bar{A}_e. \tag{3.60}$$

That is, the Lie bracket of any two left-invariant vector fields at e in $GL(n, \mathrm{R})$ is just the ordinary *matrix commutator* of the two matrices which generate the fields. The left-invariant vector field generated by this commutator is the element of the Lie algebra $\mathfrak{L}(GL(n, \mathrm{R}))$ which is the bracket of the original fields.

(iii) We have seen that the rotation group is a Lie group (§2.3(vi)). We will study it closely below, but here we examine it as a subgroup of $GL(n, \mathrm{R})$. In §2.29 we saw that the matrices A for which $A^{-1} = A^{\mathrm{T}}$ are elements of the Euclidean symmetry group $O(n)$. (The symbol $O(n)$ means the Orthogonal group in n dimensions.) Since the determinant of any matrix obeys the rules (§1.6)

$$\det(A) = 1/\det(A^{-1}), \quad \det(A) = \det(A^{\mathrm{T}}), \tag{3.61}$$

matrices in $O(n)$ have determinant ± 1. Those with determinant $+1$ form a subgroup called $SO(n)$ – the Special Orthogonal group – and we shall now demonstrate that this is the group of rotations. (The matrices in $O(n)$ which have determinant -1 are not a subgroup since they do not include the identity matrix. Like $GL(n, \mathrm{R})$, $O(n)$ is disconnected.)

Exercise 3.14

(a) Show that if A is in $O(n)$ its eigenvalues equal those of A^{-1}. (Use the fact that $\det B = \det B^{\mathrm{T}}$ for any B.) The reader not experienced with eigenvalues should look up their definition in §1.6.

Fig. 3.11. Left-translation of $g_{A_e}(t)$ by f.

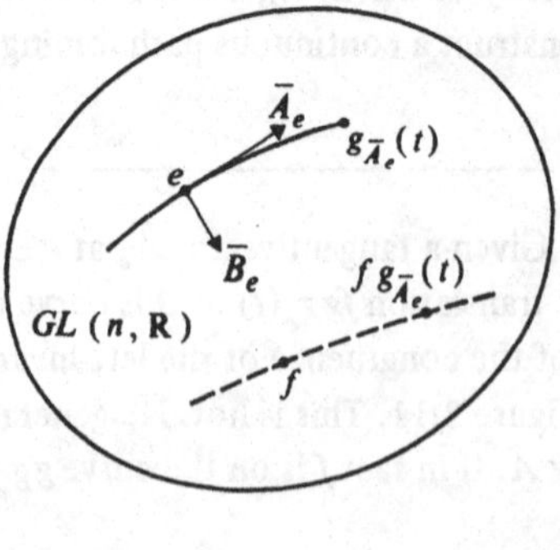

(b) Show that for any nonsingular matrix A the eigenvalues of A are the reciprocals of those of A^{-1}. (Use the fact that $\det(AB) = \det A \det B$.) Conclude that there are two types of eigenvalues $(\lambda_1, \ldots, \lambda_n)$ of A in $O(n)$: either (i) $\lambda_j = \pm 1$ or (ii) $\lambda_j\lambda_k = 1$ for $j \neq k$. Show that case (ii) implies the eigenvalues come in pairs $(e^{i\theta}, e^{-i\theta})$ for real θ.

(c) It can also be shown that the canonical form of a matrix A in $O(n)$ can be achieved by a transformation $B^{-1}AB$, where B is a matrix in $SO(n)$. Use this to conclude that a matrix in $O(n)$ has canonical form consisting of blocks

$$\text{(i)}\ (1), \tag{3.62a}$$

$$\text{or (ii)}\ (-1), \tag{3.62b}$$

$$\text{or (iii)}\ \begin{pmatrix} \cos\theta & \sin\theta \\ -\sin\theta & \cos\theta \end{pmatrix}. \tag{3.62c}$$

(d) Show that the Lie algebra of $O(n)$ consists of all antisymmetric matrices. Show from this that $O(n)$ has dimension $\frac{1}{2}n(n-1)$.

Now, a matrix A in $GL(n, R)$ may be regarded as an invertible $\binom{1}{1}$ tensor on R^n, mapping a column vector $\bar{V}$ of R^n to $A\bar{V}$, obtained by matrix multiplication. The transformation $B^{-1}AB$ is nothing more than the transformation of the *components* of this tensor (§2.26) when the basis $\{\bar{e}_1, \ldots, \bar{e}_n\}$ is transformed to $\{B^{-1}\bar{e}_1, \ldots, B^{-1}\bar{e}_n\}$. We can therefore take the view that any matrix of $SO(n)$ is equivalent to successive rotations in independent two-dimensional planes, since the canonical form (3.62c) obviously does that, while the form (3.62b) must occur an even number of times (for the determinant to be positive), which directions can be paired to give (3.62c) with $\theta = \pi$. Thus $SO(n)$ is indeed the rotation group. (Note that if n is odd, every matrix in $SO(n)$ fixes at least one direction.) The remaining matrices of $O(n)$ can be interpreted as *inversions*, transformations which change the 'handedness' of any set of n linearly independent vectors. This is shown in the next exercise. ('Handedness' of a basis will be discussed in detail in chapter 4.)

Exercise 3.15

Show that the canonical form of an element of $O(n)$ not in $SO(n)$ is the product of a matrix diag $(1, \ldots, 1, -1, 1, \ldots, 1)$ (having only one -1 on its diagonal) with the canonical form of a matrix in $SO(n)$. From this prove that it is an inversion.

Exercise 3.16

Prove that any matrix in $SO(n)$ is in a one-parameter subgroup. Prove that any matrix in $SO(3)$ is equivalent to a single rotation through a finite angle θ about some axis.

Before leaving the rotation group we need to study its Lie algebra, at least for $SO(3)$. The vector space T_e is that of all antisymmetric matrices, which has dimension three (see exercise 3.14(d)). A basis consists of the matrices

$$L_1 = \begin{pmatrix} 0 & 0 & 0 \\ 0 & 0 & -1 \\ 0 & 1 & 0 \end{pmatrix}, \quad L_2 = \begin{pmatrix} 0 & 0 & 1 \\ 0 & 0 & 0 \\ -1 & 0 & 0 \end{pmatrix},$$
$$L_3 = \begin{pmatrix} 0 & -1 & 0 \\ 1 & 0 & 0 \\ 0 & 0 & 0 \end{pmatrix}. \tag{3.63}$$

Exercise 3.17

Show that this Lie algebra basis has the brackets

$$[L_1, L_2] = L_3, \quad [L_2, L_3] = L_1, \quad [L_3, L_1] = L_2. \tag{3.64}$$

We will come back to this algebra shortly.

(iv) Another matrix group of interest in physics is $SU(n)$, which stands for Special Unitary group in n dimensions. This is a subgroup of $GL(n, C)$, the group of all complex $n \times n$ matrices of nonvanishing determinant (the General Linear group in n Complex dimensions). Since each entry may be complex and each complex number is defined by two real ones, $GL(n, C)$ has $2n^2$ (real) dimensions. Its subgroup $U(n)$ is the Unitary group, each element U obeying $U^{-1} = U^*$, where * denotes the complex-conjugate transpose (Hermitian conjugate). By analogy with $O(n)$, its Lie algebra consists of all $n \times n$ anti-Hermitian matrices. (A matrix A is *anti-Hermitian* if $A^* = -A$.) This has n^2 real dimensions, since such a matrix can have $\frac{1}{2}n(n-1)$ arbitrary complex off-diagonal elements (given by $n(n-1)$ real numbers) and n arbitrary pure-imaginary diagonal elements (contributing n real dimensions, making n^2 in all). Its subgroup $SU(n)$ is the set of all matrices in $U(n)$ with *unit determinant*. Since the determinant of any element of $U(n)$ is real, this is one extra condition, so $SU(n)$ has dimension $n^2 - 1$. Its Lie algebra is the set of all anti-Hermitian matrices with zero trace. (The trace of A is the sum $a^i{}_i$: see §1.6.)

Exercise 3.18

Show that the Lie algebra of $SU(n)$ is that of all anti-Hermitian traceless matrices. You may use the fact that any element of $U(n)$ has canonical form diag $(e^{i\phi_1}, e^{i\phi_2}, \ldots, e^{i\phi_n})$ where the numbers $\{\phi_j, j = 1, \ldots, n\}$ are real.

Exercise 3.19

(a) Show that the following matrices are a basis for T_e of $SU(2)$:

$$J_1 = \frac{1}{2}\begin{pmatrix} 0 & i \\ i & 0 \end{pmatrix}, \quad J_2 = \frac{1}{2}\begin{pmatrix} 0 & -1 \\ 1 & 0 \end{pmatrix}, \quad J_3 = \frac{1}{2}\begin{pmatrix} i & 0 \\ 0 & -i \end{pmatrix}. \tag{3.65}$$

Show that this is a three-dimensional *real* vector space: even though the matrices may contain imaginary numbers, only linear combinations of them with *real* coefficients remain in the vector space T_e of $SU(2)$.

(b) Show that this Lie algebra basis has the brackets

$$[J_1, J_2] = J_3, \quad [J_2, J_3] = J_1, \quad [J_3, J_1] = J_2. \tag{3.66}$$

These are formally identical to (3.64), and we shall see in the next section that this implies an intimate relation between $SU(2)$ and $SO(3)$.

Exercise 3.20

(a) Let tr $(A) = a^i{}_i$ denote the trace of any matrix A. Prove that tr $(B^{-1}AB)$ = tr (A).

(b) Use this and (i) the fact that the determinants of matrices obey the rule det (AB) = det (A) det (B), (ii) the result (3.56), and (iii) the canonical forms (3.55–59) to prove that for any matrix A

$$\det(\exp(A)) = \exp(\operatorname{tr}(A)). \tag{3.67}$$

(c) Use this to give an easier proof of exercise 3.18.

3.16 Lie algebras and their groups

Every Lie group G has its Lie algebra ($\mathfrak{G}$). Since every element g of G is the image of e under the left-translation g generates, and since very vector in T_e corresponds to a unique vector field in the Lie algebra, it follows that every point g of G is on one curve of each of the left-invariant congruences. Is it possible, then, to construct the group G entirely from a knowledge of its Lie algebra? The answer is a partial yes, but to phrase it we must first give a better definition of a Lie algebra than we have so far been working with.

A *Lie algebra* is a real vector space V upon which is defined a bilinear multiplication rule called [,] which produces from any two vectors $\bar{A}$ and $\bar{B}$ another vector $[\bar{A}, \bar{B}]$ satisfying:

(i) $[\bar{A},\bar{B}] = -[\bar{B},\bar{A}]$, (3.68)

(ii) $[\bar{A},[\bar{B},\bar{C}]] + [\bar{B},[\bar{C},\bar{A}]] + [\bar{C},[\bar{A},\bar{B}]] = 0.$ (3.69)

The crucial difference between this definition and the one in §2.14 is that the Lie bracket is defined *formally*, i.e. by its properties (i) and (ii), so that *any* rule for combining vectors in this manner is acceptable. The commutator of vector fields provides one such rule, and this was the only one used until now. But clearly another example is the vector space R^3 with the usual cross-product:

$$[\bar{a},\bar{b}] \equiv \bar{a} \times \bar{b}. \tag{3.70}$$

Exercise 3.21

(a) Verify that (3.70) satisfies the Jacobi identity (3.69).

(b) Show that the basis $\bar{e}_1 = (1,0,0)$, $\bar{e}_2 = (0,1,0)$, $\bar{e}_3 = (0,0,1)$ has the brackets

$$[\bar{e}_1,\bar{e}_2] = \bar{e}_3, [\bar{e}_2,\bar{e}_3] = \bar{e}_1, [\bar{e}_3,\bar{e}_1] = \bar{e}_2. \tag{3.71}$$

Compare these to (3.64) and (3.66).

We can now state but not prove a theorem which is of fundamental importance to physics, that behind every Lie algebra there is a group. Precisely, every Lie algebra is the Lie algebra of one and only one *simply-connected* Lie group. (A manifold is simply connected if every closed curve can be smoothly shrunk to a point. See Spivak (1970) or Warner (1971) for discussions and partial proofs of this theorem.) Moreover, any other Lie group with the same Lie algebra but not simply connected is *covered* by the simply-connected one. (A connected manifold M *covers* another N if there is a map π of M *onto* N such that the inverse image of some neighborhood V of any point P of N is a disjoint union of open neighborhoods of the points in $\pi^{-1}(P)$ in M. An example is given in figure 3.12.) The covering must be a homomorphism of the two groups. (See §1.4 for a definition of a homomorphism.)

The groups $SO(3)$ and $SU(2)$ illustrate this theorem nicely. First we shall show that $SU(2)$ is simply connected. We do this by considering the set H of matrices of the form

$$\begin{pmatrix} a & b \\ -\bar{b} & \bar{a} \end{pmatrix} \tag{3.72}$$

for arbitrary complex a and b, bars denoting complex conjugation.

Exercise 3.22

(a) Show that $H - \{\left(\begin{smallmatrix} 0 & 0 \\ 0 & 0 \end{smallmatrix}\right)\}$, the subset of H with non-zero determinant, is a group under multiplication, hence a Lie subgroup of $GL(2, C)$.

(b) Show that H is a real vector space (using matrix addition), has dimension 4, and has a basis consisting of J_1, J_2, and J_3 of exercise 3.19, plus the matrix $I = \begin{pmatrix} 1 & 0 \\ 0 & 1 \end{pmatrix}$.

(c) Let A be any matrix in H:

$$A = 2\alpha_1 J_1 + 2\alpha_2 J_2 + 2\alpha_3 J_3 + \alpha_4 I,$$

with $\{\alpha_j\}$ real. Show that A is in $SU(2)$ if and only if

$$\alpha_1^2 + \alpha_2^2 + \alpha_3^2 + \alpha_4^2 = 1. \tag{3.73}$$

(d) Show from this that the group $SU(2)$ has a 1–1 mapping onto the three-sphere S^3, which is a simply connected manifold. (We say that S^3 and $SU(2)$ are diffeomorphic.)

We must next find a mapping $\pi: SU(2) \to SO(3)$ which is a multiple covering. We can construct it easily by exponentiating the elements of the Lie algebra. In $SU(2)$ the element J_1 has exponential

$$\begin{aligned}\exp(tJ_1) &= \begin{pmatrix} 1 & 0 \\ 0 & 1 \end{pmatrix} + \frac{t}{2}\begin{pmatrix} 0 & \mathrm{i} \\ \mathrm{i} & 0 \end{pmatrix} + \frac{1}{2!}\left(\frac{t}{2}\right)^2 \begin{pmatrix} -1 & 0 \\ 0 & -1 \end{pmatrix} \\ &\quad + \frac{1}{3!}\left(\frac{t}{2}\right)^3 \begin{pmatrix} 0 & -\mathrm{i} \\ -\mathrm{i} & 0 \end{pmatrix} + \dots \\ &= \begin{pmatrix} \cos(t/2) & \mathrm{i}\sin(t/2) \\ \mathrm{i}\sin(t/2) & \cos(t/2) \end{pmatrix}.\end{aligned} \tag{3.74}$$

The element L_1 of $SO(3)$ has the exponential

$$\begin{aligned}\exp(sL_1) &= \begin{pmatrix} 1 & 0 & 0 \\ 0 & 1 & 0 \\ 0 & 0 & 1 \end{pmatrix} + s\begin{pmatrix} 0 & 0 & 0 \\ 0 & 0 & -1 \\ 0 & 1 & 0 \end{pmatrix} \\ &\quad + \frac{1}{2!}s^2 \begin{pmatrix} 0 & 0 & 0 \\ 0 & -1 & 0 \\ 0 & 0 & -1 \end{pmatrix} + \frac{1}{3!}s^3 \begin{pmatrix} 0 & 0 & 0 \\ 0 & 0 & 1 \\ 0 & -1 & 0 \end{pmatrix} + \dots \\ &= \begin{pmatrix} 1 & 0 & 0 \\ 0 & \cos s & -\sin s \\ 0 & \sin s & \cos s \end{pmatrix}.\end{aligned} \tag{3.75}$$

If we simply establish the natural correspondence suggested by the algebra,

$$\pi: SU(2) \to SO(3),\ \pi: \begin{pmatrix} \cos\frac{1}{2}t & \mathrm{i}\sin\frac{1}{2}t \\ \mathrm{i}\sin\frac{1}{2}t & \cos\frac{1}{2}t \end{pmatrix} \mapsto \begin{pmatrix} 1 & 0 & 0 \\ 0 & \cos t & -\sin t \\ 0 & \sin t & \cos t \end{pmatrix}, \tag{3.76}$$

then it is clear that this is a homomorphism of the two one-parameter subgroups, and it is also clear that the two elements t and $t + 2\pi$ of $SU(2)$ have the *same* image in $SO(3)$. Moreover, $t + 4n\pi$ for any integer n is the same point of $SU(2)$ as t, so we have proved that $\exp(tJ_1)$ is a *double covering* of $\exp(sL_1)$. We can generalize this to the whole group: the map

$$t\colon \exp(t_1J_1 + t_2J_2 + t_3J_3) \mapsto \exp(t_1L_1 + t_2L_2 + t_3L_3) \tag{3.77}$$

is a double covering of $SO(3)$ by $SU(2)$.

Since we know that $SU(2)$ has the global topology of the three-sphere, this double covering enables us to discover the topology of $SO(3)$. The one-parameter subgroup $\exp(tJ_1)$ of $SU(2)$ begins at e with $t = 0$ and returns to e at $t = 4\pi$. In figure 3.13 this is shown as a great circle around S^3. (But bear in mind that we have not put a metric on $SU(2)$. Only the global topology is relevant, not the actual distance relations.) The points labelled t and $t + 2\pi$ are diametrically opposite one another. In order to make them the same point of $SO(3)$, we simply

Fig. 3.12. The unit circle S^1 is covered by the real line R^1 an infinite number of times by the map $\pi\colon R^1 \to S^1$ which takes x to the point on S^1 whose coordinates in the plane R^2 are $\pi(x) = (\cos x, \sin x)$. The set $\pi^{-1}(V)$ is the union of all the open intervals shown on R.

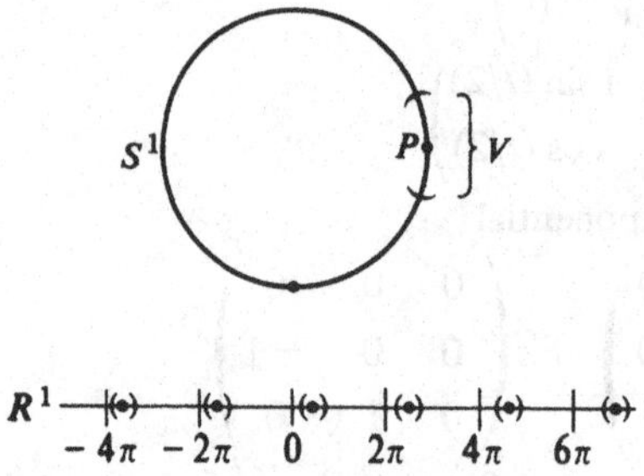

Fig. 3.13. A two-dimensional slice of S^3 containing the one-parameter subgroup $\exp(tJ_1)$ of $SU(2)$. The group $SO(3)$ is the top half of the sphere, with points on opposite ends of diameters identified with each other.

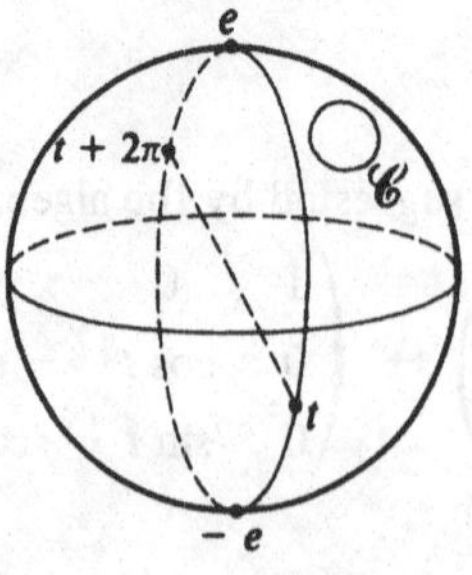

identify $SO(3)$ as the *top half* of S^3, with points on opposite ends of a diameter through the equator (e.g. $t = \pi$ and $t = 3\pi$) *identified*. This half of S^3 with these identifications is no longer simply connected. A curve such as $\mathscr{C}$ can be shrunk smoothly to a point, but the curve of the subgroup $\exp(tL_1)$ cannot be, since the two ends of the diameter of the equator cannot be brought together: they are always diametrically opposite one another. This construction also makes clear the fact that $SO(3)$ and $SU(2)$ are identical in some neighborhood of e. It is for this reason that their Lie algebras are the same. This happens for any two groups with the same Lie algebras.

To what group does the Lie algebra of equation (3.70) correspond? This is entirely a matter of interpretation. As an abstract algebra it corresponds to both groups. As a relation among vectors in R^3 it is most common to associate it with $SO(3)$ by saying that to the subgroup $\exp(\theta L_1)$ (rotation by an angle θ about the x-axis) there corresponds the 'curve' in R^3, $\exp(\theta \bar{e}_1)$ (a vector along the x-axis of length θ). This association of a rotation with a vector is very familiar to physicists, even more so in its time-differentiated version associating a rate of rotation with an angular velocity vector. This convenient identification is an accident of three dimensions: the group $SO(4)$ has dimension 6 while the vector space R^4 in which it acts has dimension 4, so no such identification is possible. But to return to R^3, we can equally well identify R^3 with $SU(2)$ in a similar fashion. In §3.18 we will see that this permits us to associate the spin of a particle with a vector in R^3 even though the spin is not an element of T_P for any P in R^3.

Before leaving Lie algebras, we must remark that we can now show that an Abelian Lie algebra is the Lie algebra of an Abelian group. An n-dimensional Abelian algebra is simply a vector space, and it is the algebra of the Lie group R^n, as discussed in §3.15. Since R^n is simply connected, any other Lie group having this algebra must be covered by R^n and must be identical to it in a neighborhood of the origin e. Since R^n is Abelian ($\bar{V} + \bar{W} = \bar{W} + \bar{V}$), so is any other Lie group with an Abelian Lie algebra.

3.17 Realizations and representations

It is usually best to regard any group as an *abstract group*, defined entirely by the group operation and, for Lie groups, by the manifold structure. Thus, $SO(3)$ as an abstract group is simply a certain three-dimensional manifold with a rule associating a product point gh with any two points g and h, the rule obeying the usual group axioms. To a physicist this abstract structure is not the aspect of group theory that is of most interest. More important is what the group *acts upon* and how it affects it. Again, $SO(3)$ is important because we associate with each point of it a rotation of our three-dimensional space. Such an association is called a realization. A *realization* of a group G is an association (map) between any element g of G and a transformation $T(g)$ of some space M in such a way that the group properties are preserved: (i) $T(e) = I$, the identity transformation (no change of M); (ii) $T(g^{-1}) = [T(g)]^{-1}$; (iii) $T(g) \circ T(h) = T(gh)$. The realization is *faithful* if the association is 1–1: $T(g) \neq T(h)$ if $g \neq h$. If M is a vector space and every $T(g)$ is a linear transformation (a $\binom{1}{1}$ tensor on that vector space) then the realization is called a *representation*. A few examples may help to make these ideas clear.

(i) Consider the effect of a rotation on the unit sphere S^2 given by the equation $x^2 + y^2 + z^2 = 1$ in R^3. Suppose we rotate by an angle θ about the x-axis. This consists of mapping any point on the sphere whose coordinates are (x, y, z) to one whose coordinates are (x', y', z') as follows

$$\begin{aligned} x' &= x, \\ y' &= y \cos\theta - z \sin\theta, \\ z' &= y \sin\theta + z \cos\theta, \end{aligned} \tag{3.78}$$

which is still on the sphere since $(x')^2 + (y')^2 + (z')^2 = 1$. This transformation is associated with the group element $\exp(\theta L_1)$ of $SO(3)$, in the notation of (3.63). To any element of the group there corresponds some transformation of S^2 into itself. Since S^2 is a manifold but not a vector space, this is a realization of $SO(3)$. On the other hand, the same transformation (3.78) can be regarded as a map of R^3 into itself, not just of S^2 into itself. Since R^3 is a vector space, this is a representation of $SO(3)$ in terms of matrices which transform vectors of R^3 into other vectors. These matrices are nothing more than the matrices we used to define the group $SO(3)$ in the first place. This illustrates a subtle but useful point of view. It is typical for a group to be defined in the first place by a (faithful) realization or representation, because this enables one to study all its properties concretely. Afterwards, however, it is more useful to regard the group as abstract because there may be *other* useful representations or realizations that one had not been aware of at first. We will illustrate these for the rotation group separately in the next section.

(ii) Every group has at least two faithful realizations: the left and right translations of itself. Any group element g defines a transformation of G which maps any h to gh (the *progressive* or *principal* realization) and one which maps h to hg^{-1} (the *retrograde* realization).

(iii)† The matrix groups that we have studied – $GL(n, \mathrm{R})$, $O(n)$, $SO(n)$, $GL(n, \mathrm{C})$, $U(n)$, $SU(n)$ – have all been studied through their faithful representations as $n \times n$ matrix transformations of n-dimensional real or complex vector spaces. But

† This example may be regarded as supplementary material.

each Lie group G has another representation as linear transformations on its own Lie algebra. This is called the *adjoint representation*, and is defined as follows. Consider first the map of G into itself given by $I_g : h \mapsto ghg^{-1}$. This is the *group adjoint realization* of G consisting of left-translation by g and right-translation by g^{-1}. (It is not necessarily faithful: if G is Abelian then I_g is the identity map $h \mapsto h$ for all g.) This realization is called the *inner automorphisms* of G. Notice that each I_g maps the identity e into itself, so that every curve through e is mapped into a (possibly different) curve through e, as shown in figure 3.14. Therefore I_g induces a map of any tangent vector in T_e to another one in T_e. This map is called Ad_g, the adjoint transformation of T_e induced by g. Now, if the solid curve in figure 3.14 is a one-parameter subgroup, say $\exp(t\bar{X})$ where $\bar{X}$ is in T_e, then so is its image under I_g, since $g(fh)g^{-1} = (gfg^{-1})(ghg^{-1})$. It follows that the dashed curve in figure 3.14 is the one-parameter subgroup generated by $Ad_g(\bar{X})$,

$$I_g[\exp(t\bar{X})] = \exp[tAd_g(\bar{X})]. \tag{3.79}$$

Now if g itself is a member of a one-parameter subgroup $g(s) = \exp(s\bar{Y})$ there should be a natural expression for $Ad_g(\bar{X})$ in terms of $\bar{Y}$. This is provided by the next exercise.

Exercise 3.23

Show that

$$Ad_{g(s)}(\bar{X}) = \exp(s\pounds_{\bar{Y}})\bar{X}. \tag{3.80}$$

Fig. 3.14. What happens to curves through e under the map $h \mapsto ghg^{-1}$, shown first as the map $h \mapsto gh$ followed by the map $gh \mapsto ghg^{-1}$. The identity e is mapped into itself but points h and f near it are generally changed, so that a tangent vector at e is mapped into another one.

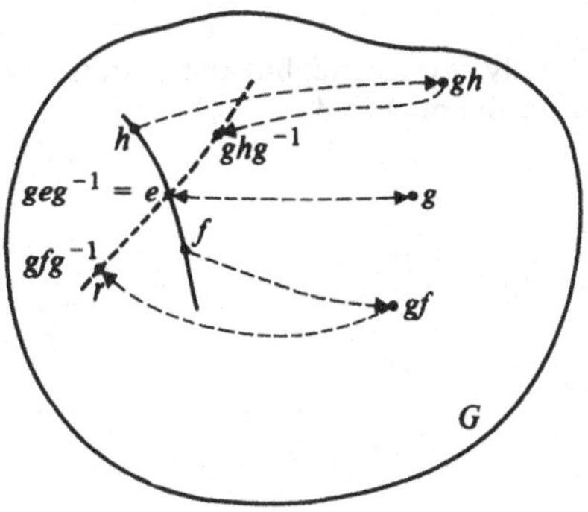

3.18 Spherical symmetry, spherical harmonics and representations of the rotation group

We have discussed Killing vectors and their relation to symmetries of Euclidean space. We can now make all of these notions precise by concentrating on the example of spherical symmetry. A manifold M with a metric tensor g is said to be *spherically symmetric* if the Lie algebra of its Killing vector fields has a subalgebra (i.e. a subspace whose brackets remain in the subspace) which is the Lie algebra of $SO(3)$. We have to speak of a subalgebra because g might have more symmetries, and here we will only consider those having to do with its spherical nature. The reader should note that it may be wrong to say M is spherical 'about some point', because the 'centers' of the spheres may not be in M (see figure 3.15). Our definition is intrinsic: the Lie subalgebra concerns vector fields of M itself. In §3.9 we saw what the Lie algebra of the vector fields $\{\bar{l}_x, \bar{l}_y, \bar{l}_z\}$ was, equation (3.30). By defining $\bar{V}_1 = -\bar{l}_x$, $\bar{V}_2 = -\bar{l}_y$, and $\bar{V}_3 = -\bar{l}_z$ we see that the Lie algebra of the vector fields $\{\bar{V}_i\}$ is identical to that of $SO(3)$, equation (3.64). This shows that our present definition of spherical symmetry implies the existence of a foliation of M into surfaces with the geometry of spheres. (Foliations were defined in §3.7.)

Suppose we concentrate now on functions defined on the two-sphere S^2. Any function on M defines such a function on any of its spheres of symmetry. We define the space of functions $L^2(S^2)$ to be the Hilbert space of all complex-valued functions on S^2 which are square-integrable: the norm

$$||f|| = \left[\int_{S^2} |f|^2 \sin\theta \, d\theta \, d\phi\right]^{1/2} \tag{3.81}$$

exists, where the integral is over the usual area element of the sphere. (Our definition of this space is a little sloppy, but accurate enough for our purposes here.) The space $L^2(S^2)$ is a vector space of infinite dimension. Its elements are functions, linear combinations of which are made with constants, and no finite number of functions is a basis. The realization of an element g of $SO(3)$ as a

Fig. 3.15. The cylinder is axially symmetric, but the centers of its circles of symmetry are not in the manifold.

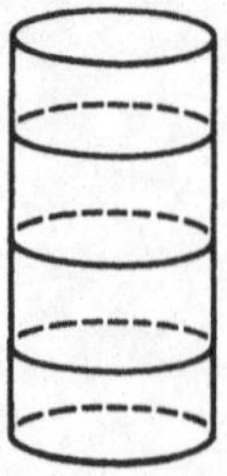

mapping $R(g)$ of S^2 into itself causes any function $f(x^i)$ on the sphere to be mapped into another one, simply by being carried along by the mapping. Therefore $R(g)$ can also be identified as a representation of $SO(3)$ in the vector space $L^2(S^2)$, an infinite-dimensional representation since $L^2(S^2)$ is of infinite dimension. The question arises whether there are finite-dimensional subspaces of $L^2(S^2)$ which provide representations of $SO(3)$. Such a subspace would have to be *invariant* under $SO(3)$, in the sense that $R(g)[f]$ for any g in $SO(3)$ and for any f in the subspace must also be in the subspace. Suppose such a subspace exists and $\{f_i, i = 1, \ldots, N\}$ is a basis for it. Then it is invariant if and only if for any numbers $\{a^i\}$ there exist $\{b^i\}$ such that

$$R(g)[a^j f_j] = b^i f_i. \tag{3.82}$$

Because the map is linear there is a relation

$$b^i = g^i{}_j a^j, \tag{3.83}$$

which defines a matrix $g^i{}_j$ corresponding to the element g of $SO(3)$. This matrix is called the representation of g in the subspace. A representation of $SO(3)$ in any vector space V is said to be *irreducible* if V contains no finite-dimensional subspaces invariant under $SO(3)$.

The construction of the irreducible representation of $SO(3)$ in $L^2(S^2)$ is treated in many books (see Gel'Fand, Minlos & Shapiro, 1963). All physicists know the basis functions of the irreducible subspaces as the *spherical harmonics*, Y_{lm}. Rather than go through their construction, let us simply try to understand them in terms of the present discussion. The claim is this. Every irreducible subspace of $L^2(S^2)$ is characterized by an integer $l \geqslant 0$ and has dimension $2l + 1$. The functions $\{Y_{lm}, m = -l, \ldots, l\}$ are basis functions for this subspace, called V_l. Moreover, the union of all these bases for all l is a basis for $L^2(S^2)$ itself, which means that the spherical harmonics are *complete*. Since any map $R(g)$ of S^2 into itself is the exponentiation of a linear combination of the vectors $\{\bar{l}_x, \bar{l}_y, \bar{l}_z\}$, V_l is invariant under $SO(3)$ if and only if it is invariant under $\bar{l}_x$, $\bar{l}_y$, and $\bar{l}_z$. A trivial example is $l = 0$, where the basis function $Y_{00} = 1$ has Lie derivatives

$$\bar{l}_x(Y_{00}) = \bar{l}_y(Y_{00}) = \bar{l}_z(Y_{00}) = 0,$$

all of which are certainly linearly dependent on Y_{00}. A better example is $l = 1$, where the three basis functions are

$$Y_{1-1} = \left(\frac{3}{8\pi}\right)^{1/2} \sin\theta\, e^{-i\phi},\ Y_{10} = \left(\frac{3}{4\pi}\right)^{1/2} \cos\theta,\ Y_{11} = \left(\frac{3}{8\pi}\right)^{1/2} \sin\theta\, e^{i\phi}. \tag{3.84}$$

Exercise 3.24

(a) Show that if x, y, z are Cartesian coordinates of R^3, then on the sphere S^2 given by $x^2 + y^2 + z^2 = 1$ the following hold

$$Y_{1\,-1} = \left(\frac{3}{8\pi}\right)^{1/2}(x - \mathrm{i}y),\ Y_{10} = \left(\frac{3}{4\pi}\right)^{1/2} z,\ Y_{11} = \left(\frac{3}{8\pi}\right)^{1/2}(x + \mathrm{i}y). \tag{3.85}$$

(b) Construct all the derivatives $\bar{l}_j Y_{1k}$, e.g.

$$\bar{l}_x(Y_{1\,-1}) = -\mathrm{i}Y_{10}/(2)^{1/2},\ \bar{l}_z(Y_{11}) = \mathrm{i}Y_{11}, \tag{3.86}$$

and show that the space V_1 is invariant under $SO(3)$.

Why is this particular basis for V_1 chosen? This is largely a matter of convenience. It is convenient that the basis should consist of eigenfunctions of relevant operators, i.e. functions which satisfy

$$Af = \alpha f \tag{3.87}$$

for some operator A and constant α. The spherical harmonics are chosen because they are eigenfunctions of both $\bar{l}_z$ and $L^2 = (\pounds_{\bar{l}_x})^2 + (\pounds_{\bar{l}_y})^2 + (\pounds_{\bar{l}_z})^2$, which was defined in exercise 3.7. The following exercise shows that this is the best one can hope to do: one cannot find nontrivial eigenfunctions of any two of $\{\bar{l}_x, \bar{l}_y, \bar{l}_z\}$.

Exercise 3.25

Assume that a function f has the properties

$$\bar{l}_x(f) = \alpha f,\ \bar{l}_y(f) = \beta f$$

for constants α and β. Show from the Lie bracket relations (3.30) that $\alpha = \beta = \bar{l}_z(f) = 0$.

Incidentally, the completeness of the basis functions comes from the fact that $\mathrm{i}\bar{l}_z$ and L^2 are commuting operators (cf. exercise 3.7) which are (or extend to) self-adjoint operators on $L^2(S^2)$. The spectral theorem of functional analysis (cf. Riesz & Sz.-Nagy, 1955) guarantees completeness of their eigenfunctions.

Actually, the representations of $SO(3)$ may be studied much more abstractly than is apparent in the above discussion. In particular one does not need to say what the vector space V is in order to develop most of the algebra. For example, our original representation of $SO(3)$ as matrices transforming vectors of R^3 is certainly irreducible, since no subspace of R^3 except the trivial one $\{0\}$ is left invariant by all rotations. It turns out to be formally *identical* to the representation $l = 1$ of the spherical harmonics, which also has dimension three $(= 2l + 1)$. In fact, equation (3.85) is simply a coordinate transformation of R^3 from (x, y, z) to $(Y_{1\,-1}, Y_{10}, Y_{11})$. The transformation involves complex numbers, but if these are just treated algebraically then the matrices $g^i{}_j$ of (3.83) expressed on the spherical-harmonic basis may be transformed into matrices expressed on the usual Cartesian basis, and these matrices turn out to be nothing more than the matrices we used to define $SO(3)$ in the first place.

Exercise 3.26

Let $\{y^{j'}, j = 1, 2, 3\}$ stand for the functions $(Y_{1\,-1}, Y_{1\,0}, Y_{1\,1})$, and let $\{x^j\}$ stand for $\{x, y, z\}$. Find the transformation matrix $\Lambda^{j'}{}_k = \partial y^{j'}/\partial x^k$ and its inverse $\Lambda^k{}_{j'}$. Find the matrix $X^{j'}{}_{k'}$ of the operator $\bar{l}_x$ on the spherical harmonic basis

$$\bar{l}_x(y^{j'}) = X^{j'}{}_{k'} y^{k'},$$

by the methods of exercise 3.24(b). Transform $X^{j'}{}_{k'}$ to the Cartesian basis

$$X^j{}_k = \Lambda^j{}_{l'} \Lambda^{r'}{}_k X^{l'}{}_{r'},$$

and show that it is just L_1 of equation (3.63).

Notice that $l = 1$ is the smallest faithful representation of $SO(3)$: $l = 0$ is not faithful. This is usually called the *fundamental representation* of $SO(3)$. We will encounter another set of irreducible representations of $SO(3)$ when we study vector spherical harmonics in §4.28. There the representation space will not be that of functions on the sphere but of vector fields on the sphere.

Finally, we need to remark on the relation between representations of $SO(3)$ and of its covering group $SU(2)$. (This passage may be skipped by readers who have not studied §3.16.) Since there is a unique element of $SO(3)$ associated with any one of $SU(2)$, any reprensentation $R(g)$ for elements g of $SO(3)$ automatically defines a representation S of $SU(2)$: for any u in $SU(2)$ the transformation $S(u)$ is $R(\pi(u))$. If u and u' both correspond to the same element of $SO(3)$, then $S(u) = S(u')$ for such representations. But $SU(2)$ will also have other representations, say T, for which $T(u) \neq T(u')$ even when $\pi(u) = \pi(u')$. These are sometimes called double-valued representations of $SO(3)$. Again we shall merely quote the result: the irreducible representations of $SU(2)$ are characterized by an index $k \geqslant 0$ which is either an integer or half an odd integer. Those for which k is an integer are representations of $SO(3)$ for the same index (i.e. $k = l$). The others are only double-valued representations of $SO(3)$. An example of the latter is provided by the matrix representation we used to define $SU(2)$, which is a representation in two complex dimensions. It has $k = \frac{1}{2}$ and is called the spin-$\frac{1}{2}$ representation. As with the $l = 1$ $SO(3)$ representation, it is the smallest faithful one for $SU(2)$. If we take any basis vector of the space (called a spinor) and operate on it by $\exp(tJ_1)$ as t goes from 0 to 4π, we see that the corresponding path in $SO(3)$, $\exp(tL_1)$, goes from 0 to 2π *twice*. When the sequence of transformations reaches $t = 2\pi$ we are back at the origin of $SO(3)$, but are at $-e$ in $SU(2)$. For this reason it is said that the spinor changes sign ($e \to -e$) if it is rotated once through an angle 2π.

It is indeed remarkable that this correspondence between representations is not simply a mathematical game. The wave-function of a spin-$\frac{1}{2}$ elementary particle is described by an element of an irreducible vector space of $SU(2)$ for nonintegral k. This is one example of what a physicist might regard as the beautiful simplicity of nature. We begin with the Lie algebra of spherical symmetry and we find that the group $SU(2)$, not $SO(3)$, is the simplest one having that algebra, in that it has the simplest global topology. We then find that, despite the difficulty of 'visualizing' the action of $SU(2)$ in R^3, nature has made the group more fundamental than $SO(3)$ by providing particles which belong to those of its representations which are not representations of $SO(3)$!

3.19 Bibliography

An old-fashioned but very complete book on Lie dervatives is K. Yano, *The Theory of Lie Derivatives and its Applications* (North-Holland, Amsterdam, 1955).

The completeness theorems for self-adjoint operators may be found in F. Riesz & B. Sz.-Nagy, *Functional Analysis* (Ungar, New York, 1955).

The close relation between Lie groups and Lie derivatives is explored in F. W. Warner, *Foundations of Differentiable Manifolds and Lie Groups* (Scott, Foresman, Glenview, Ill., 1971); in M. Spivak, *A Comprehensive Introduction to Differential Geometry* (Publish or Perish, Boston, 1970) vol. 1; and in L. Auslander & R. E. MacKenzie, *Introduction to Differentiable Manifolds* (McGraw-Hill, New York, 1963).

For more on Lie groups, see: R. Hermann, *Lie Groups for Physicists* (Benjamin, Reading, Mass., 1966); or H. Weyl, *The Theory of Groups and Quantum Mechanics* (Dover, New York, 1950). Representation theory for the most important groups is discussed in most quantum mechanics textbooks, in Weyl above, and in: H. Lipkin, *Lie Groups for Pedestrians* (North-Holland, Amsterdam, 1966); M. A. Naimark, *Linear Representations of the Lorentz Group* (Pergamon, New York, 1964); and I. M. Gel'Fand, R. A. Minlos & Z. Ya. Shapiro, *Representations of the Rotation and Lorentz Groups and Their Applications* (Pergamon, New York, 1963).

A helpful reference for the matrix algebra we have used is M. W. Hirsch & S. Smale, *Differential Equations, Dynamical Systems, and Linear Algebra* (Academic Press, New York, 1974).

4 DIFFERENTIAL FORMS

The calculus of differential forms, developed in the early part of this century by E. Cartan, is one of the most useful and fruitful analytic techniques in differential geometry. The catalogue of concepts that are unified and simplified by forms is astonishing: the theory of integration on manifolds, the cross-product, divergence, and curl of three-dimensional Euclidean geometry, determinants of matrices, orientability of manifolds, integrability conditions for systems of partial differential equations, Stokes' theorem, Gauss' theorem, and much more. As with most mathematical and physical ideas which are truly fundamental, the mathematics of forms is very simple. In this chapter, we introduce forms in the geometrical context in which they arise most naturally, and we then systematically develop their power.

A The algebra and integral calculus of forms

4.1 Definition of volume – the geometrical role of differential forms

Until now we have avoided giving our manifolds any shape or rigidity. We have mentioned the possibility of defining metric tensors, but we have concentrated on those analytic tools which are definable without reference to any particular metric. Now we will turn to the study of a particularly useful class of tensors: those which can serve to define volume elements on manifolds.

Consider the notion of volume in two dimensions, where it is called area. Any pair of (infinitesimal) vectors in Euclidean space defines an (infinitesimal) area, as in figure 4.1: the area enclosed by the parallelogram they define. Now, a given area is defined by many different pairs of vectors, which may differ from one another in length and enclosed angle, as in figure 4.2. The notion of area is, therefore, less restrictive than the notion of a metric: the Euclidean metric defines the lengths of vectors and their enclosed angle, while the specification of area gives only one number associated with the two vectors. Naturally, if a metric exists it should uniquely define the area, and we shall show how this comes about later. But it is possible to define an area for a two-manifold (or a volume on an arbitrary manifold) without having to define a metric on the manifold. Indeed, many different metrics could define the same volume.

Suppose that in a two-dimensional manifold we have at a point two linearly independent infinitesimal vectors, forming a two-dimensional parallelogram. We wish to define for this figure a (small) area, i.e. to associate with the two vectors a single number. This number ought to double if we double the length of one vector; moreover, we should require it to be additive under addition of vectors, i.e.

$$\text{area}(\bar{a}, \bar{b}) + \text{area}(\bar{a}, \bar{c}) = \text{area}(\bar{a}, \bar{b} + \bar{c}).$$

That this is true in Euclidean space is proved geometrically in figure 4.3. In the second-to-last step we have used the fact that the area of a parallelogram is

Fig. 4.1. Two pairs of vectors and the area they define.

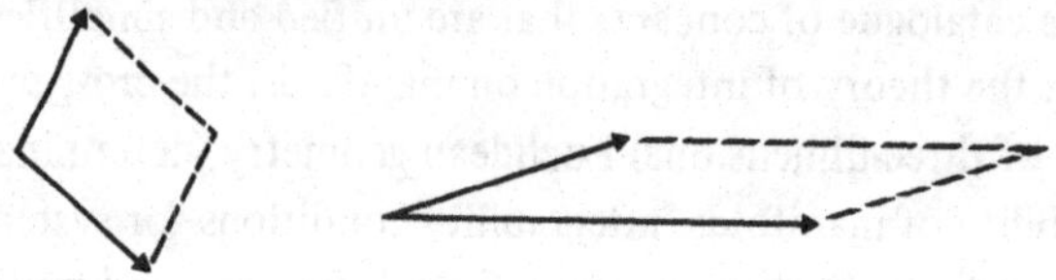

Fig. 4.2. Three pairs of vectors defining equal areas.

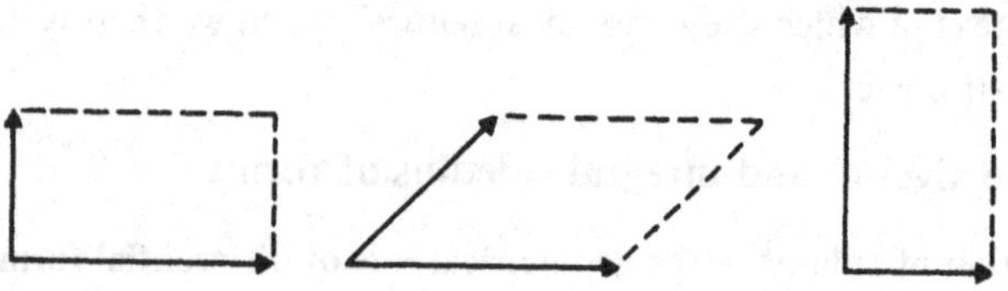

Fig. 4.3. Geometrical proof that the area of a parallelogram is the value of a tensor.

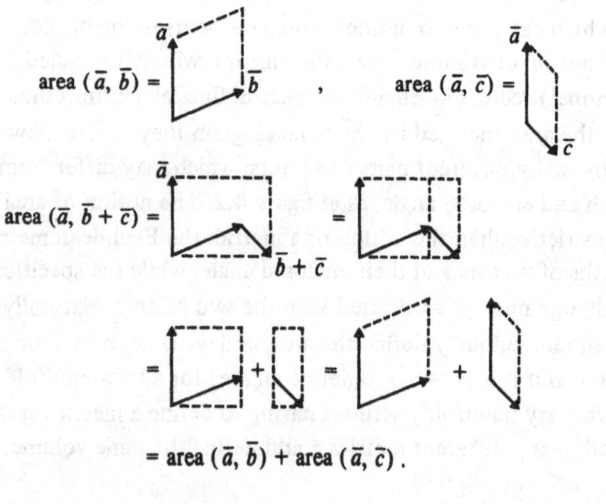

unchanged if one of its sides is displaced an arbitrary amount along the straight line it defines. So we have proved that area (,) is in fact a *tensor*, bilinear in its arguments. Since the area is a number, this is a $\binom{0}{2}$ tensor. Moreover, if $\bar{a}$ and $\bar{b}$ are parallel, the area must vanish. The following exercise shows that as a consequence the tensor must change sign if $\bar{a}$ and $\bar{b}$ are exchanged.

Exercise 4.1
Prove that if $\mathbf{B}$ is a $\binom{0}{2}$ tensor with the property that $\mathbf{B}(\bar{V}, \bar{V}) = 0$ for all $\bar{V}$, then $\mathbf{B}(\bar{U}, \bar{W}) = -\mathbf{B}(\bar{W}, \bar{U})$ for all $\bar{U}, \bar{W}$. (Hint: take $\bar{V} = \bar{U} + \bar{W}$.) We say that $\mathbf{B}$ is *antisymmetric* in its arguments.

Consider this more closely. In figure 4.4 two vectors are drawn defining a parallelogram of a certain area. In terms of components, the area is (to within a sign) the determinant

$$\text{area} = \begin{vmatrix} V^x & V^y \\ W^x & W^y \end{vmatrix}.$$

Its antisymmetry under interchange of V and W is manifest.

In ordinary uses one forgets the sign and calls the area the absolute value of that determinant. It will be convenient for us to keep the sign, since it contains information about the left- or right-handedness of the pair of vectors. We shall discuss this in more detail below. We shall also develop in more detail the evident relation between volume-tensors and determinants of matrices. But first we must develop the algebra of antisymmetric tensors. At first we will concentrate on their properties at any point, generalizing to fields later on.

4.2 Notation and definitions for antisymmetric tensors

As in exercise 4.1 above, a $\binom{0}{2}$ tensor is said to be *antisymmetric* if its value changes sign on interchange of its arguments:

$$\tilde{\omega}(\bar{U}, \bar{V}) = -\tilde{\omega}(\bar{V}, \bar{U}) \text{ for all } \bar{U}, \bar{V} \Leftrightarrow \tilde{\omega} \text{ antisymmetric.} \qquad (4.1)$$

A tensor of type $\binom{0}{p}$, $p \geqslant 3$, is said to be *completely antisymmetric* if it changes sign on interchange of *any* two of its arguments. Antisymmetric tensors can always be constructed from arbitrary ones. For example, if $\tilde{\omega}$ is a $\binom{0}{2}$ tensor and

Fig. 4.4. The area defined by $\bar{V}$ and $\bar{W}$.

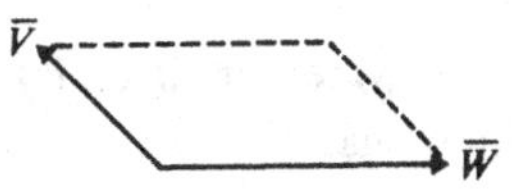

$\tilde{p}$ a $\binom{0}{3}$ tensor then their *totally antisymmetric parts* are the tensors whose values on arbitrary arguments are given by:

$$\tilde{\omega}_A(\bar{U}, \bar{V}) = \frac{1}{2!}[\tilde{\omega}(\bar{U}, \bar{V}) - \tilde{\omega}(\bar{V}, \bar{U})], \tag{4.2}$$

$$\tilde{p}_A(\bar{U}, \bar{V}, \bar{W}) = \frac{1}{3!}[\tilde{p}(\bar{U}, \bar{V}, \bar{W}) + \tilde{p}(\bar{V}, \bar{W}, \bar{U}) + \tilde{p}(\bar{W}, \bar{U}, \bar{V})$$
$$- \tilde{p}(\bar{V}, \bar{U}, \bar{W}) - \tilde{p}(\bar{W}, \bar{V}, \bar{U}) - \tilde{p}(\bar{U}, \bar{W}, \bar{V})]. \tag{4.3}$$

The rule is to take every permutation of the arguments; odd permutations contribute minus signs and even ones plus signs. The factors 1/2! and 1/3! are the conventional normalization, which is appropriate to calling $\tilde{\omega}_A$ the antisymmetric *part* of $\tilde{\omega}$. These considerations all have counterparts in index notation, obtained by letting the arbitrary vectors be basis vectors:

$$(\tilde{\omega}_A)_{ij} = \frac{1}{2!}(\omega_{ij} - \omega_{ji}) \equiv \omega_{[ij]}, \tag{4.4}$$

$$(\tilde{p}_A)_{ijk} = \frac{1}{3!}(p_{ijk} + p_{jki} + p_{kij} - p_{jik} - p_{kji} - p_{ikj}) \equiv p_{[ijk]}. \tag{4.5}$$

Here we have introduced the square bracket notation $[i \ldots k]$ to denote a completely antisymmetric set of indices, including the corresponding normalization factor. In what follows we will use a notation introduced above: a tilde (~) over a tensor's name, e.g. $\tilde{p}$, denotes a completely antisymmetric tensor. The one-form, for which we use the same notation, is a 'degenerate' case of this, because it has only one argument.

Exercise 4.2

(a) Prove that if the components of a $\binom{0}{N}$ tensor $\tilde{p}$ are antisymmetric under interchange of any two indices, then $\tilde{p}$ is a completely antisymmetric tensor.

(b) Suppose $\{A_{ijk}\}$ are the components of a completely antisymmetric tensor. Show that

$$A_{ijk} = A_{[ijk]}.$$

(c) Suppose that **A** is an antisymmetric $\binom{0}{2}$ tensor and **B** an arbitrary $\binom{2}{0}$ tensor. Show that

$$A_{ij}B^{ij} = A_{ij}B^{[ij]},$$

i.e. that the contraction of **A** with **B** involves only the antisymmetric part of **B**.

(d) Suppose **A** is as in (c) and **B** is a *symmetric* $\binom{2}{0}$ tensor: $\mathbf{B}(\tilde{\omega}, \tilde{\sigma}) = \mathbf{B}(\tilde{\sigma}, \tilde{\omega})$ for all one-forms $\tilde{\omega}$ and $\tilde{\sigma}$. Show that

$$A_{ij}B^{ij} = 0. \tag{4.6}$$

An important property of completely antisymmetric tensors is the following: on an n-dimensional vector space, a completely antisymmetric $\binom{0}{p}$ tensor ($p \leqslant n$) has at most

$$C_p^n = \frac{n!}{p!(n-p)!} \tag{4.7}$$

independent components. To see this, note that any component is defined by choosing p different numbers from the set $(1, \ldots, n)$. (They must be different, because the component vanishes if any two indices are equal, just as in exercise 4.1.) The order in which the p numbers are chosen – their order as indices on the tensor – can at most affect the sign of the component, so all components whose indices are simply rearrangements of a given set of p numbers are known if any one of them is known. The number of *independent* components is therefore the number of different sets of p numbers chosen from n numbers, which is the binomial coefficient given above.

Exercise 4.3
Prove that if $p > n$ all the components of a completely antisymmetric $\binom{0}{p}$ tensor on an n-dimensional vector space vanish.

4.3 Differential forms

A p-form ($p \geqslant 2$) is defined to be a completely antisymmetric tensor of type $\binom{0}{p}$. As before, a one-form is a $\binom{0}{1}$ tensor. A scalar function is a *zero-form*. The number p is the *degree* of the form.

Exercise 4.4
Show that the set of all p-forms for fixed p is a vector space itself, under the addition operation defined in exercise 2.4. This is then a subspace of all $\binom{0}{p}$ tensors. What is its dimension?

Just as $\binom{0}{2}$ tensors could be made from $\binom{0}{1}$ tensors using the operation $\otimes$, we define an operation $\wedge$ (called 'wedge product') for constructing two-forms from one-forms: If $\tilde{p}$ and $\tilde{q}$ are one-forms then

♦ $$\tilde{p} \wedge \tilde{q} \equiv \tilde{p} \otimes \tilde{q} - \tilde{q} \otimes \tilde{p} \tag{4.8}$$

is their wedge product. There is *no* factor of $1/2!$ in front, by contrast with equation (4.2)!

Exercise 4.5
Show that $\tilde{p} \wedge \tilde{q}$ is a two-form. Show that $\tilde{p} \wedge \tilde{p} = 0$.

Exercise 4.6
Let $\{\bar{e}_i, i = 1, \ldots, n\}$ be a basis for a vector space and $\{\tilde{\omega}^j\}$ its dual basis for one-forms. Show that $\{\tilde{\omega}^j \wedge \tilde{\omega}^k, j, k = 1, \ldots, n\}$ is a basis for the vector space of all two-forms. Hint: by considering explicitly the numbers $\alpha_{ij} = \tilde{\alpha}(\bar{e}_i, \bar{e}_j)$, where $\tilde{\alpha}$ is an arbitrary two-form, show that

♦ $$\tilde{\alpha} = \frac{1}{2!}\alpha_{ij}\tilde{\omega}^i \wedge \tilde{\omega}^j. \tag{4.9}$$

Note carefully the factor of $1/2!$ in (4.9), which occurs because the sum on (i, j) includes equal contributions from $\tilde{\omega}^i \otimes \tilde{\omega}^j$ and $\tilde{\omega}^j \otimes \tilde{\omega}^i$. This factor appears here because we did *not* put it into the definition of $\tilde{\omega}^i \wedge \tilde{\omega}^j$, equation (4.8), as some textbooks do. This is a matter of convention.

The rule for wedge products extends naturally to three-forms:

$$\begin{aligned}\tilde{p} \wedge (\tilde{q} \wedge \tilde{r}) &= (\tilde{p} \wedge \tilde{q}) \wedge \tilde{r} \\ &= \tilde{p} \wedge \tilde{q} \wedge \tilde{r} \equiv \tilde{p} \otimes \tilde{q} \otimes \tilde{r} + \tilde{q} \otimes \tilde{r} \otimes \tilde{p} + - \ldots \end{aligned} \tag{4.10}$$

using the same permutations and signs as in the previous paragraph. Notice that this expression and its generalization to higher numbers of one-forms permits one to define wedge products for arbitrary p- and q-forms, since by exercise 4.6 any p-form can be written as a linear combination of wedge products of p one-forms (the basis one-forms).

The set of all forms of arbitrary degree, equipped with the anticommutative multiplication $\wedge$, is called a *Grassmann algebra*.

Exercise 4.7
Show that the sum of the dimensions of all the N-form spaces for $p \leqslant n$ is 2^n. (Hint: use the binomial theorem.) This is the dimension of the space which has the Grassman algebra.

Exercise 4.8
Show that if $\tilde{p}$ is a one-form and $\tilde{q}$ a two-form, then

$$\begin{aligned}(\tilde{p} \wedge \tilde{q})_{ijk} &= p_i q_{jk} + p_j q_{ki} + p_k q_{ij} \\ &= 3 p_{[i} q_{jk]}\end{aligned}$$

More generally, show that if $\tilde{p}$ is a p-form and $\tilde{q}$ a q-form,

♦ $$(\tilde{p} \wedge \tilde{q})_{i\ldots jk\ldots l} = C_p^{p+q} p_{[i\ldots j} q_{k\ldots l]}\,. \tag{4.11}$$

4.4 Manipulating differential forms

The algebra of forms is fairly simple, but it can lead to difficulties keeping track of signs and factorials. In this and later sections the student should find that careful, patient reasoning is the best approach to proving any result. For instance, let us prove the commutation rule for forms. If $\tilde{p}$ is a p-form and $\tilde{q}$ a q-form, then

$$\blacklozenge \quad \tilde{p} \wedge \tilde{q} = (-1)^{pq} \tilde{q} \wedge \tilde{p}. \tag{4.12}$$

To see this, first express $\tilde{p}$ and $\tilde{q}$ as sums over their components times wedge products of the form $\tilde{\omega}^i \wedge \ldots \wedge \tilde{\omega}^j$ and $\tilde{\omega}^k \wedge \ldots \wedge \tilde{\omega}^l$ (p-factors and q-factors, respectively, in each wedge product). Now we shall show that (4.4) applies to each of the simple products

$$(\tilde{\omega}^i \wedge \ldots \wedge \tilde{\omega}^j) \wedge (\tilde{\omega}^k \wedge \ldots \wedge \tilde{\omega}^l).$$

By the associativity of the wedge product, the parentheses in this expression are unnecessary. Now, if any two factors are exchanged (e.g. $\tilde{\omega}^j$ with $\tilde{\omega}^k$) the expression changes sign. To move $\tilde{\omega}^j$ through the q-factors $\tilde{\omega}^k \wedge \ldots \wedge \tilde{\omega}^l$ requires q such exchanges, so that

$$\omega^i \wedge \ldots \wedge \tilde{\omega}^j \wedge \tilde{\omega}^k \wedge \ldots \wedge \tilde{\omega}^l = (-1)^q \tilde{\omega}^i \wedge \ldots \wedge \tilde{\omega}^k \wedge \ldots \wedge \tilde{\omega}^l \wedge \tilde{\omega}^j.$$

Doing this for each of the p-factors in $\tilde{\omega}^i \wedge \ldots \wedge \tilde{\omega}^j$ gives $[(-1)^q]^p$ times the original, which proves (4.12).

An operation we will find useful later is the *contraction* of a vector with a form. A p-form requires p vector arguments to give a real number. If it is supplied with one argument then it becomes a $(p-1)$-form. To be definite we define

$$\blacklozenge \quad \tilde{\alpha}(\bar{\xi}) \equiv \tilde{\alpha}(\bar{\xi}, \underbrace{\ ,\ ,\ ,\ldots,}_{p-1 \text{ empty slots}}), \quad [\tilde{\alpha}(\bar{\xi})]_{j\ldots k} = \alpha_{ij\ldots k}\xi^i, \tag{4.13}$$

as the $(p-1)$-form obtained by contracting $\tilde{\alpha}$ with $\bar{\xi}$. Note that putting $\bar{\xi}$ into any slot other than the first would only affect the sign of $\tilde{\alpha}(\bar{\xi})$. To get a feeling for what this means, consider $\tilde{\alpha} = \tilde{p} \wedge \tilde{q}$, where $\tilde{p}$ and $\tilde{q}$ are one-forms:

$$\begin{aligned}(\tilde{p} \wedge \tilde{q})(\bar{\xi}) &= (\tilde{p} \otimes \tilde{q} - \tilde{q} \otimes \tilde{p})(\bar{\xi}) \\ &= \tilde{p}(\bar{\xi})\, \tilde{q} - \tilde{q}(\bar{\xi}) \tilde{p}.\end{aligned}$$

Thus, although $\bar{\xi}$ is contracted with the first slot of $\tilde{p} \wedge \tilde{q}$, the permutations implicit in the $\wedge$-operation ensure that $\bar{\xi}$ is contracted with each one-form in the wedge product. Similarly, for a product of p one-forms we find

$$\begin{aligned}(\tilde{\omega}^i \wedge \tilde{\omega}^j \wedge \ldots \wedge \tilde{\omega}^k)(\bar{\xi}) &= \xi^i \tilde{\omega}^j \wedge \ldots \wedge \tilde{\omega}^k - \xi^j \tilde{\omega}^i \wedge \ldots \wedge \tilde{\omega}^k \\ &\quad + \ldots \pm \xi^k \tilde{\omega}^i \wedge \tilde{\omega}^j \wedge \ldots \\ &= p\, \xi^{[i} \tilde{\omega}^j \wedge \ldots \wedge \tilde{\omega}^{k]}.\end{aligned} \tag{4.14}$$

From this and the generalization of (4.2) it follows that if $\tilde{\alpha}$ is a p-form,

$$\tilde{\alpha}(\bar{\xi}) = \frac{1}{(p-1)!}\xi^i \alpha_{ij\ldots k}\,\tilde{\omega}^j \wedge \ldots \wedge \tilde{\omega}^k. \tag{4.15}$$

This is, of course, implied directly by (4.13) and (4.9). Similarly, if $\tilde{\alpha}$ is any form and $\tilde{\beta}$ is a p-form, then

◆ $$(\tilde{\beta} \wedge \tilde{\alpha})(\bar{\xi}) = \tilde{\beta}(\bar{\xi}) \wedge \tilde{\alpha} + (-1)^p \tilde{\beta} \wedge \tilde{\alpha}(\bar{\xi}). \tag{4.16}$$

Again this can be proved by looking at each component of $\tilde{\beta} \wedge \tilde{\alpha}$.

Exercise 4.9
Prove (4.16).

Widely used alternative notations for $\tilde{\alpha}(\bar{\xi})$ are $\bar{\xi}\rfloor\tilde{\alpha}$ and $i_{\bar{\xi}}\tilde{\alpha}$.

4.5 Restriction of forms

An elementary but important concept is that of restricting a form to a subspace of the original vector space V. Since a p-form $\tilde{\alpha}$ is a $\binom{0}{p}$ tensor, its domain is the set of all vectors in V (strictly, the domain is the *product space* $V \times V \times \ldots \times V$, p 'copies' of V). The *restriction* of $\tilde{\alpha}$ to a subspace W of V is the same p-form $\tilde{\alpha}$ whose domain is now restricted to vectors in W. We call this $\tilde{\alpha}|_W$:

$$\tilde{\alpha}|_W(\bar{X}, \ldots, \bar{Y}) = \tilde{\alpha}(\bar{X}, \ldots, \bar{Y}),$$

where all of $\bar{X}, \ldots, \bar{Y}$ are in W. Thus, $\tilde{\alpha}|_W$ is defined only on W. Note that if the dimension m of W is less than p, the restriction $\tilde{\alpha}|_W$ is necessarily zero (any p-form is zero on an m-dimensional space if $p > m$), and if $p = m$ then $\tilde{\alpha}|_W$ has only one independent component. The operation of restricting a form is often called *sectioning* it, because the picture one has is of the vector subspace W being a plane passing through (sectioning) the series of surfaces that represent a form. A form is said to be *annulled* by a vector subspace if its restriction to it vanishes.

4.6 Fields of forms

As with any tensor, a *field* of p-forms on a manifold M is a rule (with appropriate differentiability conditions) giving a p-form at each point of M. Then all our remarks up to now apply to forms as functions on the space T_P at any point P of M. Only one point needs to be made: since a submanifold S of M picks out a subspace V_P of the manifold's tangent space T_P at every point P of S, we define the restriction of a p-form field $\tilde{\alpha}$ to S to be that field formed by restricting $\tilde{\alpha}$ at P to V_P. We have seen an example of this for a one-form in §3.6.

4.7 Handedness and orientability

In an n-dimensional manifold there is only a one-dimensional space of n-forms at any point (equation 4.7). Choose some n-form field, and call it $\tilde{\omega}$. Consider a vector basis $\{\bar{e}_1, \ldots, \bar{e}_n\}$ at a point P. Since these are linearly independent vectors, the number $\tilde{\omega}(\bar{e}_1, \ldots, \bar{e}_n)$ is nonzero if and only if $\tilde{\omega} \neq 0$ at P. Therefore $\tilde{\omega}$ separates the set of *all* vector bases at P into two classes, those for which $\tilde{\omega}(\bar{e}_1, \ldots, \bar{e}_n)$ is positive and those for which it is negative. These classes are in fact independent of $\tilde{\omega}$. For, if $\tilde{\omega}'$ is any other n-form nonzero at P, then there exists a number $f \neq 0$ such that $\tilde{\omega}' = f\tilde{\omega}$. Any two bases which made $\tilde{\omega}$ positive will give $\tilde{\omega}'$ the same sign (positive if $f > 0$, negative if $f < 0$) and so will again be in the same class. So all bases at a point can be put into one of two classes: *right-handed* and *left-handed*. (Which class has which name is, of course, a convention; what is important is that the classes themselves are distinct.) A manifold is said to be (internally) *orientable* if it is possible to define handedness consistently (i.e. continuously) over the entire manifold, in the sense that it is possible to define a continuous vector basis $\{\bar{e}_1(P), \ldots, \bar{e}_n(P)\}$ whose handedness is the same everywhere. Clearly this is equivalent to being able to define an n-form which is continuous and nonzero everywhere. Euclidean space is orientable; the Möbius band is not.

4.8 Volumes and integration on oriented manifolds

We return now to our view that forms are related to volume-elements. In an n-dimensional manifold, a set of n linearly independent ('infinitesimal') vectors define a region of nonzero volume, an n-dimensional parallelepiped. The volume of this region is then the value of an n-form. One is free to choose any n-form as the volume n-form; which one chooses will be determined by the particular problem one is solving.

Now, integration of a function on a manifold involves essentially multiplying the value of the function by the volume of a small coordinate element and then adding up all such values. Following our discussion of volume-forms, we shall introduce a useful notation for this. Suppose $\tilde{\omega}$ is an n-form on a region U of an n-dimensional manifold M whose coordinates are $\{x^1, \ldots, x^n\}$. Then because all n-forms at a point form a one-dimensional vector space, there exists some $f(x^1, \ldots, x^n)$ such that

$$\tilde{\omega} = f\,\tilde{d}x^1 \wedge \ldots \wedge \tilde{d}x^n.$$

To integrate over the U, we divide it up into tiny regions ('cells') spanned by n-tuples of vectors $\{\Delta x^1\, \partial/\partial x^1, \Delta x^2\, \partial/\partial x^2, \ldots, \Delta x^n\, \partial/\partial x^n\}$, where the $\{\Delta x^i\}$ are very small numbers. The integral of the function f over one small cell is approximately the value of f times the product

$$\Delta x^1 \Delta x^2 \ldots \Delta x^n = \tilde{d}x^1 \wedge \ldots \wedge \tilde{d}x^n(\Delta x^1\, \partial/\partial x^1, \ldots, \Delta x^n\, \partial/\partial x^n).$$

Thus, we have

$$\int_{\text{cell}} f(x^1, \ldots, x^n)\, \mathrm{d}^n x \cong \tilde{\omega}(\text{cell}). \tag{4.17}$$

Adding up all the contributions from the different cells and taking the limit as the size of each cell goes to zero gives what we call the integral of $\tilde{\omega}$ over U:

$$\blacklozenge \quad \int \tilde{\omega} \equiv \int f(x^1, \ldots, x^n)\, \mathrm{d}x^1 \ldots \mathrm{d}x^n, \tag{4.18}$$

where the integral on the right is the ordinary integral of calculus and the integral on the left is our new notation. Since the version on the left does not mention coordinates, we must prove that it really does not depend on the coordinates chosen for U. We shall restrict our proof to two dimensions, since the generalization will be obvious. Consider first coordinates λ and μ. Then we have

$$\int \tilde{\omega} \equiv \int f(\lambda, \mu)\, \tilde{\mathrm{d}}\lambda \wedge \tilde{\mathrm{d}}\mu = \int f(\lambda, \mu)\, \mathrm{d}\lambda\, \mathrm{d}\mu.$$

When we change to coordinates x and y, the chain rule gives

$$\tilde{\mathrm{d}}\lambda = \tilde{\mathrm{d}}\lambda(x, y) = \frac{\partial \lambda}{\partial x}\tilde{\mathrm{d}}x + \frac{\partial \lambda}{\partial y}\tilde{\mathrm{d}}y,$$

$$\tilde{\mathrm{d}}\mu = \frac{\partial \mu}{\partial x}\tilde{\mathrm{d}}x + \frac{\partial \mu}{\partial y}\tilde{\mathrm{d}}y,$$

which follows from the definition of $\tilde{\mathrm{d}}\lambda$ as a gradient. So we get (remember $\tilde{\mathrm{d}}x \wedge \tilde{\mathrm{d}}x \equiv 0$ since it is antisymmetric)

$$\begin{aligned}
\tilde{\mathrm{d}}\lambda \wedge \tilde{\mathrm{d}}\mu &= \left(\frac{\partial \lambda}{\partial x}\tilde{\mathrm{d}}x + \frac{\partial \lambda}{\partial y}\tilde{\mathrm{d}}y\right) \wedge \left(\frac{\partial \mu}{\partial x}\tilde{\mathrm{d}}x + \frac{\partial \mu}{\partial y}\tilde{\mathrm{d}}y\right) \\
&= \frac{\partial \lambda}{\partial x}\frac{\partial \mu}{\partial y}\tilde{\mathrm{d}}x \wedge \tilde{\mathrm{d}}y + \frac{\partial \lambda}{\partial y}\frac{\partial \mu}{\partial x}\tilde{\mathrm{d}}y \wedge \tilde{\mathrm{d}}x \\
&= \left(\frac{\partial \lambda}{\partial x}\frac{\partial \mu}{\partial y} - \frac{\partial \lambda}{\partial y}\frac{\partial \mu}{\partial x}\right)\tilde{\mathrm{d}}x \wedge \tilde{\mathrm{d}}y.
\end{aligned} \tag{4.19}$$

The factor in front of $\tilde{\mathrm{d}}x \wedge \tilde{\mathrm{d}}y$ is the Jacobian of the coordinate transformation, $\partial(\lambda, \mu)/\partial(x, y)$. From ordinary calculus we know that is how volume elements *do* in fact transform. Therefore, the (λ, μ)-integral of f is related to the (x, y)-integral of f in exactly the right way.

But the value of $\int \tilde{\omega}$ is not quite independent of the coordinates originally chosen. What we have shown is that a coordinate transformation does not change its value, but there is an ambiguity of sign in equation (4.17). This equation provided the original definition of $\int \tilde{\omega}$, and this definition would have given us the opposite sign had our original coordinate system had a basis of the opposite handedness to the one we chose. The right-hand side of (4.17) would have been the same – the form is basis-independent – but on the left-hand side f would have changed sign. (This change is not the sort of coordinate transformation we discussed above, in which $\mathrm{d}^n x$ would be multiplied by a negative Jacobian and all would be well. It is a change in the original identification of the symbol $\int \tilde{\omega}$ with an integral from the calculus.) This ambiguity cannot be avoided. It is conventional to choose an orientation in U – i.e. define which of the two sets of bases is right-handed – and to use a right-handed coordinate system in the definition (4.17). We therefore find that the integral of $\tilde{\omega}$ over the region U is independent of everything except orientation.

It was important in this argument that U be covered by a single coordinate system. Can we extend this integral to all of M, which may not have a global coordinate system? It is clear that if two coordinate patches have a single connected overlap region, the orientation chosen on one induces a unique orientation on the other, and the integral over the union of the two regions is well-defined. Clearly this can be extended to M as a whole if and only if M is orientable. From now on we shall restrict ourselves to integration on orientable manifolds, but it must be mentioned that the theory has been extended to non-orientable manifolds by de Rham, and this can have interesting physical applications (see the paper by Sorkin (1977) in the bibliography).

Integration as we have defined it is always done over forms of the maximum degree: n-forms on n-dimensional manifolds. One can of course integrate a p-form over a p-dimensional submanifold, provided the submanifold is itself internally orientable. How is the internal orientation of a submanifold S related to that of M? Suppose M is orientable and P is a point of S. Given an n-form $\tilde{\omega}$ defined as 'right-handed' (or 'positively oriented') at P, is there a unique 'induced' orientation for p-forms of S at P? Unfortunately not, because $\tilde{\omega}$ on its own does not do anything for S, as its restriction to S is zero since $p < n$. What is usually done is to reduce $\tilde{\omega}$ from an n-form to a p-form by defining $n-p$ linearly independent 'normal vectors' at P not tangent to S and defining the restriction of the p-form

$$\tilde{\omega}(\bar{n}_1, \ldots, \bar{n}_{n-p})$$

to S to be right-handed. This definition clearly depends on the choice of the vectors $\{\bar{n}_i\}$, including the order in which they are numbered. Such a choice is called choosing an *external orientation* for S at P. We shall give an example of this in our proof of Stokes' theorem below. If it is possible to define the external orientation $\{\bar{n}_i, i = 1, \ldots, n-p\}$ continuously over all of S (and 'continuously' means keeping the $\bar{n}_i$ linearly independent and not tangent to S) then S is said to be *externally orientable*.

It is clear that if some open region of M containing S is orientable then either S is both internally and externally orientable or it is neither, and that if no such region of M is orientable S may be one but not both. For example, consider a Möbius strip as a two-dimensional submanifold of R^3 (figure 4.5) and a curve in the strip as a one-dimensional submanifold of the strip (figure 4.6). Set up a right-handed triad of vectors at any point P of the strip, two lying in the strip and one out of it. Carry them continuously once around the strip, keeping the two always tangent to it. The outward pointing one always returns pointing to the opposite side: the Möbius band is not externally orientable in R^3. Similarly, set up two vectors in the strip, one tangent to the curve $\mathscr{C}_1$ and the other not. Transport these continuously around and the outward pointing one returns pointing to the other side of the curve in the strip. Since we know that the curve is internally orientable (this is a property independent of any space it is embedded in) it cannot be externally orientable in a larger nonorientable manifold.

Fig. 4.5. The Möbius band in R^3. It is easiest to imagine it made of rubber, lying flat on the page except near the top of the figure, where the single twist is. A triad at P, carried clockwise around (dashed path) returns in a way which cannot be continuously deformed into its original while keeping vectors 1 and 2 in the band and all three linearly independent.

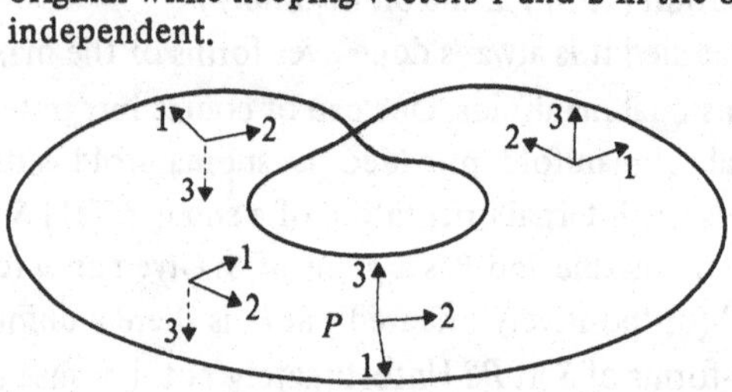

Fig. 4.6. Curves in the Möbius band. Curve $\mathscr{C}_1$ is not externally oriented: vectors 1 and 2 begin at P and are transported as in figure 4.5. They return in a way which cannot be continuously deformed into the original while keeping 1 tangent to the curve and both linearly independent. But curve $\mathscr{C}_2$ is externally orientable because it has a neighborhood (dotted line) in which a consistent choice of orientation is possible.

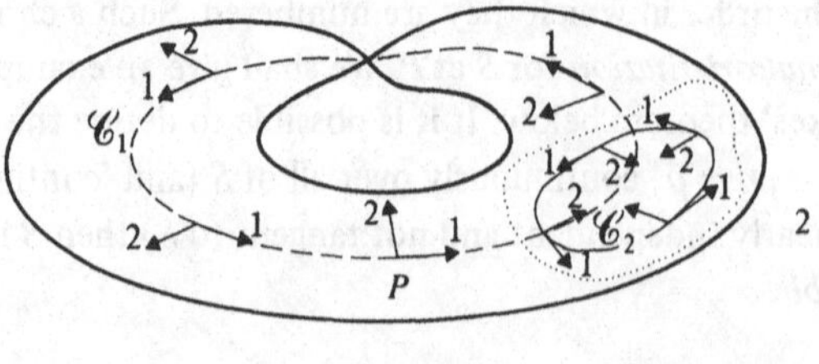

By contrast, the curve $\mathscr{C}_2$ is both internally and externally orientable in the strip because it does not 'feel' the nonorientability of the strip: it has a neighborhood in the strip which is orientable.

4.9 N-vectors, duals, and the symbol $\epsilon_{ij\ldots k}$

We have so far considered completely antisymmetric $\binom{0}{N}$ tensors, but of course the Grassmann algebra can be constructed for $\binom{N}{0}$ tensors in a parallel fashion. A completely antisymmetric $\binom{N}{0}$ tensor is called a *N*-vector. As for forms, the vector space of all *p*-vectors at any point in an *n*-dimensional manifold is of dimension C^n_p.

Notice that at a point there are four spaces which have equal dimension: the vector spaces of *p*-forms, $(n-p)$-forms, *p*-vectors, and $(n-p)$-vectors all have dimension $C^n_p = C^n_{n-p}$. Under certain circumstances one can find a 1–1 mapping between various of these spaces. We saw in §2.29 that a metric tensor gives a 1–1 map from $\binom{0}{p}$ tensors to $\binom{p}{0}$ tensors. It is not hard to see that this map preserves antisymmetry, so that it maps *p*-forms into *p*-vectors invertibly. Whether or not a metric is defined, however, a volume *n*-form $\tilde{\omega}$ (i.e. an *n*-form which is nowhere zero) provides a mapping between *p*-forms and $(n-p)$-vectors. This map is called the *dual* map, and we now show how to construct it. (Do not confuse this map, which depends on $\tilde{\omega}$ and maps a single $\binom{0}{p}$ tensor into a unique $\binom{n-p}{0}$ tensor and viceversa, with the concept of a dual basis for one-forms discussed in chapter 2, which does not involve $\tilde{\omega}$ and maps a set of n $\binom{1}{0}$ tensors into a unique *set* of n $\binom{0}{1}$ tensors and vice versa.)

A given q-vector **T** with components $T^{i\ldots k} = T^{[i\ldots k]}$ (q indices) defines a tensor $\tilde{A}$ by the equation

$$\blacklozenge \qquad A_{j\ldots l} = \frac{1}{q!}\,\omega_{i\ldots kj\ldots l}T^{i\ldots k}. \tag{4.20}$$

Symbolically, we write

$$\tilde{A} = \tilde{\omega}(\mathbf{T})$$

or simply

$$\blacklozenge \qquad \tilde{A} = {}^*\mathbf{T}. \tag{4.21}$$

We say that $\tilde{A}$ is the *dual* of **T** with respect to $\tilde{\omega}$. From (4.20) and the antisymmetry of $\omega_{i\ldots l}$ under interchange of any two indices it is clear that $\tilde{A}$ is a completely antisymmetric tensor of degree $n-q$ (which is the number of indices on $\tilde{\omega}$ left over after contracting with the q indices on **T**). That is, $\tilde{A}$ is an $(n-q)$-form. This map defines a unique $(n-q)$-form from any q-vector. We will show that it is invertible below, but first we will show that one has already become familiar with this map in the form of the *cross-product* in three-dimensional Euclidean vector algebra.

To understand this, recall that in Euclidean space one usually does not distinguish between vectors and one-forms: in Cartesian coordinates the components of a vector and of its associated one-form are equal. So consider two vectors $\bar{U}$ and $\bar{V}$ and their one-forms $\tilde{U}$ and $\tilde{V}$. The two-form $\tilde{U} \wedge \tilde{V}$ has $C_2^3 = 3$ independent components, which are $U_1V_2 - U_2V_1, U_1V_3 - U_3V_1, U_2V_3 - U_3V_2$. The vector $\bar{U} \times \bar{V}$ has the same components, and it is easy to show that

$$*(\bar{U} \times \bar{V}) = \tilde{U} \wedge \tilde{V} \quad \text{(three dimensions).} \tag{4.22}$$

Exercise 4.10

Prove (4.22) by using (4.20).

This illuminates a number of odd things about the cross-product: why it exists at all, why it does not exist in other than three-dimensions (only in three dimensions does the dual map take vectors into two-forms), and why $\bar{U} \times \bar{V}$ is an 'axial' vector. This last fact comes from the fact that it is conventional to define $\tilde{\omega}$ in Euclidean space to give a positive volume to the basis $(\bar{e}_1, \bar{e}_2, \bar{e}_3)$. If the handedness of the basis is changed, then so is the sign of $\tilde{\omega}$ and, consequently, the sign of $\bar{U} \times \bar{V}$ (which depends on the sign of $\tilde{\omega}$ in that it must be mapped by $\tilde{\omega}$ into $\tilde{U} \wedge \tilde{V}$, whose sign does not change). Under an inversion of coordinates, then, the conventional cross-product changes sign.

The map between **T** and ***T** is invertible because they each have the same number of components. (Said another way, contracting **T** with $\tilde{\omega}$ in (4.20) loses no information from **T** because **T** is already antisymmetric on all its indices.) That is, for a given p-form $\tilde{A}$ there is a *unique* $(n-p)$-vector **T** for which $\tilde{A} = *\mathbf{T}$. This can be formalized by defining an n-vector $\omega^{i\ldots k}$, the inverse of $\tilde{\omega}$, by the equation

$$\blacklozenge \qquad \omega^{i\ldots k}\omega_{i\ldots k} = n!. \tag{4.23}$$

The factor of $n!$ is used because the sum in (4.23) has $n!$ equal terms: $\omega^{123\ldots n} \times \omega_{123\ldots n} = \omega^{213\ldots n}\omega_{213\ldots n} = \ldots$. The $n!$ factor assures the normalization

$$\omega^{123\ldots n} = \frac{1}{\omega_{123\ldots n}}. \tag{4.24}$$

Then we say that **S** *is the dual of* $\tilde{B}$ *with respect to* $\tilde{\omega}$,

$$\blacklozenge \qquad \mathbf{S} = *\tilde{B}, \tag{4.25}$$

if (for $\tilde{B}$ a p-form)

$$\blacklozenge \qquad S^{i\ldots k} = \frac{1}{p!}\omega^{l\ldots m\, i\ldots k}B_{l\ldots m}. \tag{4.26}$$

To illustrate the inverse property of the two dual relations, let us look first at scalar functions. The function f, viewed as a zero-vector, has the n-form dual $f\tilde{\omega}$. This n-form has the dual zero-vector

$$^*(f\tilde{\omega}) = \frac{1}{n!}\,\omega^{l\ldots m}(f\omega_{l\ldots m}) = f.$$

Thus we have proved $^{**}f = f$.

The general relation of this sort is found as follows. Start with a p-form $\tilde{B}$ and define the $(n-p)$-vector $\mathbf{S}$ by (4.26). We take the dual of $\mathbf{S}$:

$$\begin{aligned}(^*\mathbf{S})_{j\ldots l} &= \frac{1}{(n-p)!}\,\omega_{i\ldots k\,j\ldots l}S^{i\ldots k}\\ &= \frac{1}{p!(n-p)!}\,\omega_{i\ldots k\,j\ldots l}\,\omega^{r\ldots s\,i\ldots k}B_{r\ldots s}\\ &= \frac{(-1)^{p(n-p)}}{p!(n-p)!}\,\omega_{i\ldots k\,j\ldots l}\,\omega^{i\ldots k\,r\ldots s}B_{r\ldots s}.\end{aligned}$$

To get the last line one has to move each of the $(n-p)$ indices $i\ldots k$ 'through' (by permutation) all of the p indices $r\ldots s$, giving $(n-p)$ factors of $(-1)^p$. Now, fix the indices $(j\ldots l)$ at, say, $(1\ldots p)$. (Their names clearly cannot matter.) Then in the sum (for fixed $(r\ldots s)$)

$$\omega_{i\ldots k\,1\ldots p}\,\omega^{i\ldots k\,r\ldots s}$$

the indices $i\ldots k$ must be chosen from the set $(p+1,\ldots,n)$. There will therefore be at most $(n-p)!$ nonzero terms in this sum, and each such term will be equal to every other one, just as in (4.23). So we have

$$\omega_{i\ldots k\,1\ldots p}\,\omega^{i\ldots k\,r\ldots s} = (n-p)!\;\omega_{p+1\ldots n\,1\ldots p}\,\omega^{p+1\ldots n\,r\ldots s}.$$

Moreover, this will be zero unless $(r\ldots s)$ is a permutation of $(1\ldots p)$, for otherwise the second ω will have repeated indices. In the sum over $(r\ldots s)$,

$$\omega^{p+1\ldots n\,r\ldots s}B_{r\ldots s},$$

there are thus at most $p!$ nonzero terms, and again each of them equals every other. So we have

$$\omega^{p+1\ldots n\,r\ldots s}B_{r\ldots s} = p!\;\omega^{p+1\ldots n\,1\ldots p}B_{1\ldots p}.$$

Combining all these results gives

$$(^*\mathbf{S})_{1\ldots p} = (-1)^{p(n-p)}\omega_{p+1\ldots n\,1\ldots p}\,\omega^{p+1\ldots n\,1\ldots p}B_{1\ldots p}.$$

But, from (4.24) we see that

$$\omega_{p+1\ldots n\,1\ldots p}\,\omega^{p+1\ldots n\,1\ldots p} = 1,$$

and so

$$(^*\mathbf{S})_{1\ldots p} = (-1)^{p(n-p)}B_{1\ldots p}.$$

Since the labels $1\ldots p$ could have stood for any indices, we have proved that

◆ $$**\tilde{B} = (-1)^{p(n-p)}\tilde{B}. \qquad (4.27a)$$

Similarly, had we started with a q-vector **T**, we would have found

◆ $$**\mathbf{T} = (-1)^{q(n-q)}\mathbf{T}. \qquad (4.27b)$$

Notice that if n is odd, the factor $(-1)^{p(n-p)}$ is always $+1$.

As mentioned before, a metric maps p-forms into p-vectors. Combined with the dual map this gives a map from p-forms to $(n-p)$-forms, or q-vectors to $(n-q)$-vectors. This map is usually simply called * as well. But some caution regarding signs is necessary when the metric is indefinite (when, as in relativity, some lengths are positive and some negative). This is discussed in more detail in a later section, and an example of the use of this metric dual is given in exercise 5.13.

In the algebra of forms it is often convenient to introduce the completely antisymmetric *Levi–Civita symbols*

◆ $$\epsilon_{ij\ldots k} = \epsilon^{ij\ldots k} \equiv \begin{cases} +1 \text{ if } ij\ldots k \text{ is an even permutation of } 1, 2, \ldots, n; \\ -1 \text{ if } ij\ldots k \text{ is an odd permutation of } 1, 2, \ldots, n; \\ 0 \text{ otherwise.} \end{cases} \qquad (4.28)$$

For instance, the form $\tilde{d}x^1 \wedge \tilde{d}x^2 \wedge \tilde{d}x^3$ on a three-dimensional manifold has components ϵ_{ijk} in the coordinate system (x^1, x^2, x^3), but will have components $h\epsilon_{ijk}$ in other coordinates, where h is some function. Suppose that a volume-form $\tilde{\omega}$ has components

$$\omega_{ij\ldots k} = f\epsilon_{ij\ldots k}, \qquad (4.29)$$

where f is some function. Then its inverse is

$$\omega^{ij\ldots k} = \frac{1}{f}\epsilon^{ij\ldots k}. \qquad (4.30)$$

4.10 *Tensor densities*

We have taken the point of view that any nonzero n-form on an n-dimensional manifold defines a volume element. It sometimes happens that a given problem has two or three such n-forms. (An example of this is the flow of a perfect fluid, discussed in chapter 5. In the three-dimensional manifold of Euclidean space there are three physically defined three-forms: one whose integral gives the volume of a region, another the mass, and a third a conserved quantity related to the vorticity.) This makes it more convenient on occasion to relate all such forms to the *coordinate-dependent* n-form $\tilde{d}x^1 \wedge \tilde{d}x^2 \wedge \ldots \wedge \tilde{d}x^n$, whose components are simply $\epsilon_{ij\ldots k}$. If $\tilde{\omega}$ is an n-form of interest, then the relation (4.29) rewritten as

$$\omega_{ij\ldots k} = \mathfrak{w}\,\epsilon_{ij\ldots k},$$

defines a quantity $\mathfrak{w}$ which is called a *scalar density*. Although $\mathfrak{w}$ is a function on the manifold, it is not a true scalar because it depends on the coordinates. Under a change of coordinates to $x^{i'} = f^{i}(x^{j})$, the components of $\tilde{\omega}$ are multiplied by the Jacobian J of the transformation (equation (4.19)) while $\epsilon_{ij\ldots k}$ is *by definition* unchanged. So $\mathfrak{w}$ obeys the law

$$\mathfrak{w}' = J\mathfrak{w}.$$

This is the transformation law for a scalar density of weight 1. (The term 'weight' is defined below.) It is possible to extend this to tensor densities. Suppose, for instance, that $\mathbf{T}$ is a $\binom{2}{n}$ tensor on an n-dimensional manifold which is completely antisymmetric in its vector arguments:

$$T^{ij}\underbrace{{}_{k\ldots l}}_{n \text{ indices}} = T^{ij}{}_{[k\ldots l]}.$$

Then upon contraction with two one-forms $\tilde{\alpha}$ and $\tilde{\beta}$, $\mathbf{T}$ produces a volume-form $\tilde{t}(\tilde{\alpha}, \tilde{\beta})$:

$$\tilde{t}(\tilde{\alpha}, \tilde{\beta}) = \mathbf{T}(\tilde{\alpha}, \tilde{\beta};\ ,\ldots,),$$

$$t_{k\ldots l} = T^{ij}{}_{k\ldots l}\alpha_i\beta_j.$$

(Such tensors may arise in physics. For instance, the stress tensor mentioned in chapter 2 gives the density of stress when given two one-forms; the *total* stress is the integral of this over the volume, the integral of the contraction of the $\binom{2}{n}$ tensor obtained by multiplying the stress tensor by the volume-form.) It is possible to write the components of $\mathbf{T}$ as

$$T^{ij}{}_{k\ldots l} = \mathfrak{T}^{ij}\epsilon_{k\ldots l},$$

which defines the numbers $\{\mathfrak{T}^{ij}\}$, which are components of a $\binom{2}{0}$ tensor density. (It is conventional to use German letters to denote densities.) The transformation law for such a density is

$$\mathfrak{T}^{i'j'} = J\Lambda^{i'}{}_{k}\Lambda^{j'}{}_{l}\mathfrak{T}^{kl}, \tag{4.31}$$

where again J is the Jacobian (determinant of $\Lambda^{i'}{}_{j}$). This is the transformation law for a $\binom{2}{0}$ tensor density of weight 1.

The term *weight* refers to the number of factors of J in the transformation law. For instance, a number $\mathfrak{w}$ which transforms by

$$\mathfrak{w}' = J^2\mathfrak{w}$$

is a scalar density of weight two. The generalization to tensor densities and to other weights is obvious. (An ordinary tensor is a density of weight zero.) The interpretation of densities of weights other than zero or one is more complicated, but such quantities do on occasion prove useful. In this book we shall not deal with densities, preferring to use the n-forms themselves.

4.11 Generalized Kronecker deltas

The Levi–Civita symbol has many useful and interesting properties, some of which we explore in this and the next section. As we saw earlier, one often encounters products of ϵs, such as $\epsilon^{ij\ldots k}\epsilon_{il\ldots m}$. It is possible to develop a systematic and convenient method of handling them.

First we note that, in two dimensions, for any nonzero two-form $\tilde{\omega}$ we have

$$\omega_{ij}\omega^{kl} = \epsilon_{ij}\epsilon^{kl} = \delta^k{}_i\delta^l{}_j - \delta^k{}_j\delta^l{}_i. \tag{4.32}$$

The first equality follows from (4.29) and (4.30). To establish the second it is easiest simply to note that both sides are antisymmetric in (k, l) and (i, j), so it suffices to consider the case $i \neq j, k \neq l$. There is, up to a sign, only one such term, $\epsilon_{12}\epsilon^{12} = 1$. Clearly the right-hand side of (4.32) also gives one, which proves the result. A nearly identical chain of reasoning leads to the general result for n dimensions:

$$\begin{aligned}\epsilon_{ij\ldots k}\epsilon^{lm\ldots r} &= \delta^l{}_i\delta^m{}_j\ldots\delta^r{}_k - \delta^l{}_j\delta^m{}_i\ldots\delta^r{}_k + \ldots \\ &= n!\,\delta^l{}_{[i}\delta^m{}_j\ldots\delta^r{}_{k]}.\end{aligned} \tag{4.33}$$

There is an abbreviated notation for this. We define the p-delta symbol by

♦ $$\delta^{i\ldots j}_{k\ldots l} = p!\,\delta^i{}_{[k}\ldots\delta^j{}_{l]}, \tag{4.34}$$

where the sets $(i\ldots j)$ and $(k\ldots l)$ each contain p indices. Then we have as a special case

♦ $$\epsilon_{ij\ldots k}\epsilon^{lm\ldots r} = \delta^{lm\ldots r}_{ij\ldots k}. \tag{4.35}$$

The p-delta symbol can be obtained from the $(p+1)$-delta symbol by contraction, conventionally on the first indices. We begin with

$$\delta^{ijk\ldots l}_{imr\ldots s} = (p+1)!\,\delta^i{}_{[i}\delta^j{}_m\delta^k{}_r\ldots\delta^l{}_{s]}.$$

The terms can be arranged as:

$$\begin{aligned} &= p!\,\delta^i{}_i\delta^j{}_{[m}\delta^k{}_r\ldots\delta^l{}_{s]} - p!\,\delta^i{}_m\delta^j{}_{[i}\delta^k{}_r\ldots\delta^l{}_{s]} \\ &\quad - p!\,\delta^i{}_r\delta^j{}_{[m}\delta^k{}_i\ldots\delta^l{}_{s]} - \ldots - p!\,\delta^i{}_s\delta^j{}_{[m}\delta^k{}_r\ldots\delta^l{}_{i]} \\ &= p!\,\{n\,\delta^j{}_{[m}\delta^k{}_r\ldots\delta^l{}_{s]} - \delta^j{}_{[m}\delta^k{}_r\ldots\delta^l{}_{s]} \\ &\quad - \delta^j{}_{[m}\delta^k{}_r\ldots\delta^l{}_{s]} - \ldots - \delta^j{}_{[m}\delta^k{}_r\ldots\delta^l{}_{s]}\},\end{aligned}$$

which gives

♦ $$\delta^{ij\ldots l}_{im\ldots s} = (n-p)\,\delta^{j\ldots l}_{m\ldots s} \tag{4.36}$$

for the single contraction of a $(p+1)$-delta in an n-dimensional space.

Exercise 4.11

(a) Justify each of the steps in the derivation of (4.36).

(b) Obtain the p-delta from the n-delta by $n-p$ contractions:

$$\underbrace{\delta^{r\ldots s}_{r\ldots s}}_{n-p}\underbrace{{}^{i\ldots j}_{k\ldots l}}_{p} = (n-p)!\,\delta^{i\ldots j}_{k\ldots l}. \tag{4.37}$$

As an example of the utility of this algebra, we shall calculate the *triple cross-product* in three-dimensional Euclidean space. In Cartesian coordinates the * operator uses ϵ, so

$$(\bar{U}\times\bar{V})_i = \epsilon_{ijk}U^jV^k,$$

and therefore

$$\begin{aligned}[\bar{W}\times(\bar{U}\times\bar{V})]_i &= \epsilon_{ijk}W^j\epsilon^k{}_{lm}U^lV^m\\ &= \epsilon_{kij}\epsilon^{klm}W^jU_lV_m.\end{aligned}$$

Using (4.34) and (4.36) we have

$$\begin{aligned}[\bar{W}\times(\bar{U}\times\bar{V})]_i &= (\delta^l{}_i\delta^m{}_j-\delta^l{}_j\delta^m{}_i)W^jU_lV_m\\ &= U_i(\bar{W}\cdot\bar{V})-V_i(\bar{W}\cdot\bar{U}).\end{aligned}$$

This derivation is so quick that it should make memorization of the triple cross-product formula completely unnecessary!

4.12 Determinants and $\epsilon_{ij\ldots k}$

Consider a 2×2 matrix with elements A^{ij}. We shall show that

$$\det(A) = \epsilon_{ij}A^{1i}A^{2j}. \tag{4.38}$$

To show this, write the sum on the right-hand side out explicitly:

$$\epsilon_{ij}A^{1i}A^{2j} = \epsilon_{12}A^{11}A^{22}+\epsilon_{21}A^{12}A^{21},$$

where we have used the fact that $\epsilon_{11}=\epsilon_{22}=0$. Now, we also have that $\epsilon_{12}=-\epsilon_{21}=1$, so we get

$$A^{11}A^{22}-A^{12}A^{21},$$

which is the definition of the determinant of the matrix. The next exercise generalizes this to $n\times n$ matrixes.

Exercise 4.12

(a) Show that the determinant of an $n\times n$ matrix with elements A^{ij} $(i,j=1,\ldots,n)$ is

◆ $$\det(A) = \epsilon_{ij\ldots k}A^{1i}A^{2j}\ldots A^{nk}. \tag{4.39}$$

(Hint: the determinant of an $n\times n$ matrix is defined in terms of $(n-1)\times(n-1)$ determinants by the cofactor rule. Use that rule to prove (4.39) by induction from the 2×2 case.)

(b) Show that

$$\det(A) = \frac{1}{n!}\epsilon_{ab\ldots c}\epsilon_{ij\ldots k}A^{ai}A^{bj}\ldots A^{ck}.$$

Exercise 4.13

If a manifold has a metric, let $\{\tilde{\omega}^i\}$ be an orthonormal basis for one-forms, and define $\tilde{\omega}$ to be the preferred volume-form

$$\tilde{\omega} = \tilde{\omega}^1 \wedge \tilde{\omega}^2 \wedge \ldots \wedge \tilde{\omega}^n.$$

Show that, if $\{x^{k'}\}$ is an arbitrary coordinate system,

◆ $$\tilde{\omega} = |g|^{1/2}\tilde{\mathrm{d}}x^{1'} \wedge \tilde{\mathrm{d}}x^{2'} \wedge \ldots \wedge \tilde{\mathrm{d}}x^{n'}, \tag{4.40}$$

where g is the determinant of the matrix of components $g_{i'j'}$ of the metric tensor in these coordinates.

Again it is interesting to look explicitly at the three-dimensional Euclidean case. The volume of a parallelepiped formed by the three vectors $\bar{a}$, $\bar{b}$, and $\bar{c}$ is the determinant of the matrix whose rows are the components of those vectors. From (4.39), therefore,

$$\begin{aligned}\text{volume} &= \epsilon_{ijk}a^ib^jc^k = a^i(\epsilon_{ijk}b^jc^k)\\ &= a^i(\bar{b}\times\bar{c})_i = \bar{a}\cdot(\bar{b}\cdot\bar{c}),\end{aligned}$$

another well-known expression for the volume.

4.13 Metric volume elements

In exercise 4.13 we used the metric of a manifold to define a certain orthonormal basis $\{\tilde{\omega}^i\}$, from which we constructed an n-form $\tilde{\omega}$ (equation (4.40)) which we called 'the preferred volume-form'. Does this form deserve the name 'preferred': is it unique, or does it depend upon the particular orthonormal basis (which certainly is not unique) used to define it? The answer is that it *is* unique, apart from a sign. To see this, note that the components of $\tilde{\omega}$ on the original basis are, by definition, $\epsilon_{ij\ldots k}$. If $\{\tilde{\omega}^{j'}\}$ is any other orthonormal basis, then the components of $\tilde{\omega}$ on this basis are $J\epsilon_{i'j'\ldots k'}$, where J is the Jacobian of the transformation from $\{\tilde{\omega}^j\}$ to $\{\tilde{\omega}^{j'}\}$. But, because the two bases are orthonormal, this Jacobian is ± 1 (proved below). Therefore, the form $\tilde{\omega}$ differs from the 'preferred' form defined by $\{\tilde{\omega}^{j'}\}$ by at most a sign. If we adopt a convention for handedness, we can define $\tilde{\omega}$ by right-handed orthonormal bases, and it is unique. So a metric defines a unique volume-form for an oriented manifold. On intuitive grounds, of course, this is not at all surprising.

To prove this result we used the fact that the Jacobian of a transformation from one orthonormal basis to another – which is just the determinant of the transformation matrix $\Lambda^{i'}{}_j$ – has absolute value one. This is not hard to establish. We start from the general transformation law for the metric tensor's components

$$g_{i'j'} = \Lambda^k{}_{i'}\Lambda^l{}_{j'}g_{kl},$$

which can be written in matrix language as (cf. §2.29)

$$(g') = (\Lambda)^{\mathrm{T}}(g)(\Lambda).$$

The determinant of this transformation law for g_{ij} gives

$$\det(g') = \det(g)\,[\det(\Lambda)]^2.$$

But in an orthonormal basis, g_{ij} is a matrix which has ± 1 on the diagonal and zero elsewhere. (Recall that if g_{ij} is an *indefinite* metric not all diagonal elements will have the same sign.) So the determinant of g_{ij} is ± 1, and has the *same* sign in all orthonormal bases. Therefore we have for the Jacobian

$$\det(\Lambda) = J = \pm 1.$$

For indefinite metrics, the dual operation * can be defined in either of two ways, which arise because $\omega^{ij\ldots k}$, the inverse of the volume-form, has two 'natural' definitions which may differ by a sign. The point of view we took earlier was that

$$\omega^{ij\ldots k}\omega_{ij\ldots k} = n!\,,$$

i.e. that

$$\omega^{12\ldots n} = (\omega_{12\ldots n})^{-1}.$$

But if there is a metric one might like to define an n-vector $\tilde{\omega}'$ by raising the indices of $\tilde{\omega}$:

$$(\tilde{\omega}')^{ij\ldots k} = g^{il}g^{jm}\ldots g^{kr}\omega_{lm\ldots r}.$$

From equations (4.39) and (4.40) it follows that

$$(\tilde{\omega}')^{ij\ldots k} = |g|^{1/2}\det(g^{lm})\,\epsilon^{ij\ldots k}.$$

Now, since (g^{lm}) is the matrix inverse to g_{ij}, its determinant is g^{-1}, and we have

$$(\tilde{\omega}')^{12\ldots n} = \frac{|g|^{1/2}}{g}, \tag{4.41}$$

whereas we had

$$(\tilde{\omega})^{12\ldots n} = \frac{1}{(\tilde{\omega})_{12\ldots n}} = \frac{1}{|g|^{1/2}}. \tag{4.42}$$

If g is negative, these differ by a sign. It is conventional in relativity, where g is negative, to use $\tilde{\omega}'$ in the inverse-dual relations. This introduces an extra minus sign into equations like (4.27).

B The differential calculus of forms and its applications

Where there is an integral calculus there is also a differential calculus, and so we shall introduce the exterior derivative, which operates on forms and produces forms which are their derivatives. The exact sense in which exterior differentiation is the inverse of integration is shown in Stokes' theorem, proved below, which is the generalization of the fundamental theorem of calculus,

$$\int_a^b \mathrm{d}f = f(b) - f(a). \tag{4.43}$$

We will then go on to show the close relationship between differential forms and partial differential equations.

4.14 The exterior derivative

We want to define a derivative operator on forms which preserves their character as forms and which is inverse to integration, in the sense of (4.43) above. Note that if M is a one-dimensional manifold, the operator $\tilde{\mathrm{d}}$ which takes a zero-form f to a one-form $\tilde{\mathrm{d}}f$ does indeed satisfy (4.43) above. So what we want is to extend $\tilde{\mathrm{d}}$ to forms of higher degree. By analogy with the operation of $\tilde{\mathrm{d}}$ on zero-forms, it must raise the degree of a form. Thus, if $\tilde{\alpha}$ is a p-form, then $\tilde{\mathrm{d}}\tilde{\alpha}$ is to be a $(p+1)$-form. The appropriate way to extend $\tilde{\mathrm{d}}$ is as follows (where $\tilde{\alpha}$ is a p-form and $\tilde{\beta}$, $\tilde{\gamma}$ are q-forms):

- ♦ (i) $\tilde{\mathrm{d}}(\tilde{\beta} + \tilde{\gamma}) = (\tilde{\mathrm{d}}\tilde{\beta}) + (\tilde{\mathrm{d}}\tilde{\gamma})$
- ♦ (ii) $\tilde{\mathrm{d}}(\tilde{\alpha} \wedge \tilde{\beta}) = (\tilde{\mathrm{d}}\tilde{\alpha}) \wedge \tilde{\beta} + (-1)^p \tilde{\alpha} \wedge \tilde{\mathrm{d}}\tilde{\beta}$,
- ♦ (iii) $\tilde{\mathrm{d}}(\tilde{\mathrm{d}}\tilde{\alpha}) = 0$.

Property (ii) is just the Leibniz rule apart from the $(-1)^p$, which comes about because one has to bring the operator $\tilde{\mathrm{d}}$ 'through' the p-form $\tilde{\alpha}$ in order to get at $\tilde{\beta}$, and this involves 'exchanging' it with p one-forms, each exchange contributing a factor of -1. This property guarantees that $\tilde{\mathrm{d}}$ will preserve the rule (4.12). (A derivative with the property (ii) is called *antiderivation*.) Property (iii) is at first sight surprising, but on examination for the case where $\tilde{\alpha}$ is a function f proves sensible: the one-form $\tilde{\mathrm{d}}f$ has components $\partial f/\partial x^i$. A second derivative would have components that were linear combinations of $\partial^2 f/\partial x^j \partial x^i$. But to be a two-form this second derivative would have to be antisymmetric in i and j, whereas $\partial^2 f/\partial x^i \partial x^j$ is symmetric (partial derivatives commute). Therefore it is sensible that it vanishes. The properties (i)–(iii) plus the definition of $\tilde{\mathrm{d}}$ on functions uniquely determine $\tilde{\mathrm{d}}$. (This is a theorem whose rather long proof may be found in any of the standard references.)

Exercise 4.14

(a) Show that

$$\tilde{\mathrm{d}}(f\tilde{\mathrm{d}}g) = \tilde{\mathrm{d}}f \wedge \tilde{\mathrm{d}}g. \tag{4.44}$$

(b) Use (a) to show that if

$$\tilde{\alpha} = \frac{1}{p!}\alpha_{i\ldots j}\tilde{\mathrm{d}}x^i \wedge \ldots \wedge \tilde{\mathrm{d}}x^j$$

is the expression for the p-form $\tilde{\alpha}$ in a coordinate basis, then

$$\tilde{\mathrm{d}}\tilde{\alpha} = \frac{1}{p!}\frac{\partial}{\partial x^k}(\alpha_{i\ldots j})\,\tilde{\mathrm{d}}x^k \wedge \tilde{\mathrm{d}}x^i \wedge \ldots \wedge \tilde{\mathrm{d}}x^j,$$

and hence that

◆ $$(\tilde{\mathrm{d}}\tilde{\alpha})_{ki\ldots j} = (p+1)\frac{\partial}{\partial x^{[k}}\alpha_{i\ldots j]}\,. \tag{4.45}$$

4.15 Notation for derivatives

We shall have frequent occasion to use partial derivatives from now on. There is a standard and convient notation that for any function f on the manifold

$$\frac{\partial f}{\partial x^i} \equiv f_{,i}. \tag{4.46}$$

Notice that f might itself be the component of a tensor, in which case the comma follows all other indices:

◆ $$\frac{\partial V^i{}_j}{\partial x^k} \equiv V^i{}_{j,k}. \tag{4.47}$$

Second derivatives are denoted by more indices after the comma, but conventionally no extra commas are used:

$$\frac{\partial^2 f}{\partial x^k \partial x^i} = f_{,ik}. \tag{4.48}$$

The indices are to be read left-to-right to find the order in which the derivatives are applied (the *opposite* to the $\partial/\partial x^k$ convention). Note carefully that partial differentiation is *not* an allowed tensor operation on components as discussed in §2.27. That is, the functions $\{V^i{}_{j,k}\}$ do not in general equal the functions

$$\Lambda^i{}_{a'}\Lambda^{b'}{}_j\Lambda^{c'}{}_k V^{a'}{}_{b',c'},$$

which are obtained by transforming the partial derivatives from another set of coordinates. (Recall the discussion in §3.4 of the problems involved in defining differentiation of tensors on a manifold.) An exception to this rule is differentiation of a *scalar* function, where we have seen that $f_{,i}$ is the component of the one-form $\tilde{d}f$. (Here it is worth recalling the distinction between *scalar* and *function* drawn in §2.28.) An example of our new notation is afforded by the Lie bracket:

$$[\bar{U}, \bar{V}]^i = U^j V^i{}_{,j} - V^j U^i{}_{,j}.$$

Although each term on the right separately does not transform as a tensor, together they do. Similarly, the partial derivatives in (4.45) appear in a combination which also transforms as a tensor.

Exercise 4.15
Show that $V^i{}_{,k}$ does not transform as a tensor under a general coordinate transformation, and then show that $[\bar{U}, \bar{V}]^i$ does transform as a vector.

With the convention that the derivative index is placed after all the others, (4.45) becomes

$$(\tilde{d}\tilde{\alpha})_{i\ldots jk} = (-1)^p(p+1)\alpha_{[i\ldots j,k]}. \tag{4.49}$$

4.16 Familiar examples of exterior differentiation

Just as the wedge product gave us the cross-product in three dimensions, so the exterior derivative ('wedge-derivative') gives us the curl. Consider a vector $\bar{a}$. The exterior derivative of its associated one-form is

$$\begin{aligned}\tilde{d}\tilde{a} &= \tilde{d}(a_1\tilde{d}x^1 + a_2\tilde{d}x^2 + a_3\tilde{d}x^3)\\ &= a_{1,j}\tilde{d}x^j \wedge \tilde{d}x^1 + a_{2,j}\tilde{d}x^j \wedge \tilde{d}x^2 + a_{3,j}\tilde{d}x^j \wedge \tilde{d}x^3.\end{aligned}$$

Since $\tilde{d}x^1 \wedge \tilde{d}x^1 = 0$ and similarly for indices 2 and 3, this becomes

$$\begin{aligned}\tilde{d}\tilde{a} &= (a_{1,2} - a_{2,1})\,\tilde{d}x^2 \wedge \tilde{d}x^1 + (a_{2,3} - a_{3,2})\,\tilde{d}x^3 \wedge \tilde{d}x^2\\ &\quad + (a_{3,1} - a_{1,3})\,\tilde{d}x^1 \wedge \tilde{d}x^3.\end{aligned}$$

The curl is clearly involved here. To isolate it as a vector, we take the dual:

$$\begin{aligned}{}^*\tilde{d}\tilde{a} &= (a_{1,2} - a_{2,1})\,{}^*(\tilde{d}x^2 \wedge \tilde{d}x^1) + \ldots\\ &= (a_{1,2} - a_{2,1})\,\epsilon^{213}\frac{\partial}{\partial x^3} + \ldots\\ &= (a_{2,1} - a_{1,2})\frac{\partial}{\partial x^3} + \ldots\end{aligned}$$

$${}^*\tilde{d}\tilde{a} = \bar{\nabla} \times \bar{a}. \tag{4.50}$$

So the curl operator in three dimensions is ${}^*\tilde{d}$.

Not only the curl, but also the divergence comes from exterior differentiation. In this case the appropriate operator is $\tilde{d}^*$. That is, start with a vector $\bar{a}$ and take its dual:

$$\begin{aligned} {}^*(\bar{a}) &= {}^*\left(a^1 \frac{\partial}{\partial x^1} + a^2 \frac{\partial}{\partial x^2} + a^3 \frac{\partial}{\partial x^3}\right) \\ &= a^1\,{}^*\left(\frac{\partial}{\partial x^1}\right) + \ldots \\ &= \tfrac{1}{2} a^1 \epsilon_{1jk}\, \tilde{d}x^j \wedge \tilde{d}x^k + \ldots \\ &= a^1 (\tilde{d}x^2 \wedge \tilde{d}x^3) + \ldots . \end{aligned}$$

Then the exterior derivative of this is

$$\begin{aligned} \tilde{d}^*\bar{a} &= a^1{}_{,j} \tilde{d}x^j \wedge \tilde{d}x^2 \wedge \tilde{d}x^3 + \ldots \\ &= a^1{}_{,1} \tilde{d}x^1 \wedge \tilde{d}x^2 \wedge \tilde{d}x^3 + \ldots \\ &= (a^i{}_{,i})\, \tilde{d}x^1 \wedge \tilde{d}x^2 \wedge \tilde{d}x^3. \end{aligned} \tag{4.51}$$

(In going from the first line to the second only $j = 1$ survives in the wedge product.) We have therefore shown that

♦ $$\tilde{d}^*\bar{a} = (\bar{\nabla}\cdot\bar{a})\tilde{\omega}, \tag{4.52}$$

where $\tilde{\omega} = \tilde{d}x^1 \wedge \tilde{d}x^2 \wedge \tilde{d}x^3$ is the Euclidean volume-element in Cartesian coordinates. We shall generalize this divergence formula to arbitrary manifolds and arbitrary p-vectors in §4.23.

Exercise 4.16

Use (4.50), (4.52), and property (iii) of §4.14 to show that (in three-dimensional Euclidean vector calculus) the divergence of a curl and the curl of a gradient both vanish.

4.17 Integrability conditions for partial differential equations

Exterior differentiation, like forms themselves, is closely related to familiar concepts from calculus. As an example, consider the system of partial differential equations

$$\frac{\partial f}{\partial x} = g(x, y), \quad \frac{\partial f}{\partial y} = h(x, y). \tag{4.53}$$

By letting (x, y) be coordinates of a manifold, this can be written as

$$f_{,i} = a_i,$$

where $a_x = g$ and $a_y = h$. This equation, in turn, has the coordinate-independent form

$$\tilde{d}f = \tilde{a}, \tag{4.54}$$

where $\tilde{a}$ is a one-form with components g and h. Now, if f is a solution to this equation then we get a valid equation by operating with $\tilde{d}$ upon it:

$$\tilde{d}(\tilde{d}f) = \tilde{d}\tilde{a}.$$

But the left-hand side vanishes by property (iii) of the definition of $\tilde{d}$, so we have that a *necessary* condition for the solution to exist is that

$$\tilde{d}\tilde{a} = 0.$$

In component language this is

$$a_{[i,j]} = 0,$$

which is really only one equation (a two-form on a two-dimensional manifold):

$$\frac{\partial a_x}{\partial y} - \frac{\partial a_y}{\partial x} = 0 \Rightarrow \frac{\partial g}{\partial y} - \frac{\partial h}{\partial x} = 0. \tag{4.55}$$

These are, of course, the integrability conditions for the equations. Thus, the exterior calculus gives a geometric derivation of these conditions, and it is usually the easiest way to derive them because of the conciseness of its notation. The fact that the integrability conditions are *sufficient* conditions for the existence of a solution is assured by Frobenius' theorem, in the version described in §4.26.

4.18 Exact forms

By definition of the exterior derivative $\tilde{d}$, the statement $\tilde{\alpha} = \tilde{d}\tilde{\beta}$ implies $\tilde{d}\tilde{\alpha} = 0$. It is natural to ask for the converse: if $\tilde{d}\tilde{\alpha} = 0$, do we know there exists a $\tilde{\beta}$ such that $\tilde{\alpha} = \tilde{d}\tilde{\beta}$? A form $\tilde{\alpha}$ for which $\tilde{d}\tilde{\alpha} = 0$ is said to be *closed*; a form $\tilde{\alpha}$ for which $\tilde{\alpha} = \tilde{d}\tilde{\beta}$ is said to be *exact*. Is a closed form exact? In the next section we will prove that the answer is yes in the following sense. Consider a neighborhood $\mathscr{D}$ of a point P, in which $\tilde{\alpha}$ is everywhere defined and in which $\tilde{d}\tilde{\alpha} = 0$. Then there exists a sufficiently small neighborhood of P in which a form $\tilde{\beta}$ is everywhere defined and for which $\tilde{\alpha} = \tilde{d}\tilde{\beta}$. Clearly, $\tilde{\beta}$ is not unique: $\tilde{\beta} + \tilde{d}\tilde{\gamma}$ for any $\tilde{\gamma}$ (of the right degree) also works.

We only claim that a closed form is exact *locally*, because the statement is not always true globally. Given an arbitrary region $\mathscr{D}$ of a manifold in which $\tilde{\alpha}$ is defined and closed, it may not be possible to find a single $\tilde{\beta}$ defined everywhere in $\mathscr{D}$ for which $\tilde{\alpha} = \tilde{d}\tilde{\beta}$.

We give the following example in R^2. In figure 4.7, consider the annulus enclosed between the curves $\mathscr{C}_1$ and $\mathscr{C}_2$, and consider Cartesian coordinates x and y, whose origin P is inside $\mathscr{C}_2$. The one-form

$$\tilde{\alpha} = \frac{x\tilde{d}y - y\tilde{d}x}{x^2 + y^2}$$

is defined everywhere between the curves and has the property $\tilde{\mathrm{d}}\tilde{\alpha} = 0$, as one can easily verify. Is there a function f such that $\tilde{\alpha} = \tilde{\mathrm{d}}f$? If we introduce the usual polar coordinates r and θ, then it is easy to see that $\tilde{\alpha} = \tilde{\mathrm{d}}\theta$, so we apparently have the answer 'yes'. But there is a problem: θ is not a single-valued continuous function everywhere in the region of interest, the region between $\mathscr{C}_1$ and $\mathscr{C}_2$. Therefore, although $\tilde{\alpha}$ is well-defined everywhere in this region, there is no function f such that $\tilde{\alpha} = \tilde{\mathrm{d}}f$ everywhere. The answer is 'yes' locally, but 'no' globally. This problem would go away if we ignored $\mathscr{C}_2$ and considered the whole interior of $\mathscr{C}_1$, since $\tilde{\alpha}$ is not defined at $x = y = 0$. Similarly, if we considered the region shown in figure 4.8, then again the problem goes away: in this case $\tilde{\alpha}$ is defined everywhere inside $\mathscr{C}$, and θ can be chosen single-valued and continuous inside $\mathscr{C}$ as well. So in this simple example we have found that, whereas *locally* $\tilde{\mathrm{d}}\tilde{\alpha} = 0 \Rightarrow \tilde{\alpha} = \tilde{\mathrm{d}}f$, the *global* question (whether f is defined everywhere) depends on the region being considered.

It is clear that we are dealing with one aspect of the topology of a region or a manifold. The study of those topological properties which determine the relation between closed and exact forms is called *cohomology theory*. After we have proved Stokes' theorem we will have enough mathematical machinery to take at least a brief look at cohomology theory in §4.24.

Fig. 4.7. An annular region of R^2. The region does not include its boundaries.

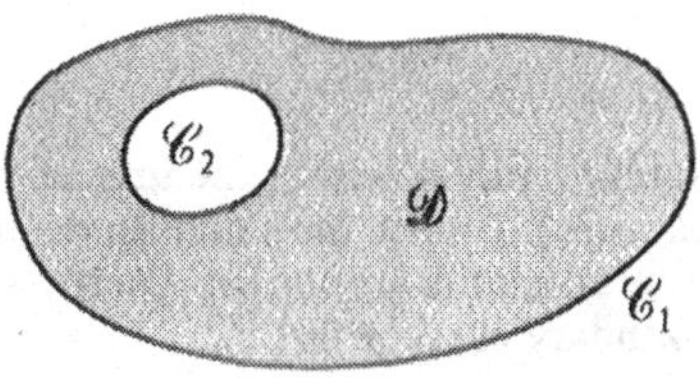

Fig. 4.8. A region of R^2 similar to that in figure 4.7 but whose boundary is a single connected curve. The discontinuity in θ (where θ jumps from 2π down to 0) on any circle $r =$ const about P can be made to take place outside $\mathscr{C}$.

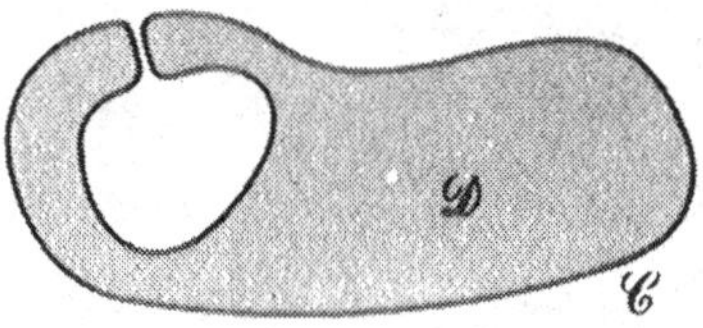

4.19 *Proof of the local exactness of closed forms*

We shall prove the following theorem, known as the Poincaré lemma. Let $\tilde{\alpha}$ be a closed p-form ($\tilde{d}\tilde{\alpha} = 0$) defined everywhere in a region U of M, and let U have a 1–1 differentiable map onto the unit open ball of R^n, i.e. the interior of the sphere S^{n-1} defined by $(x^1)^2 + (x^2)^2 + \ldots + (x^n)^2 = 1$. Then in U there is $(p-1)$-form $\tilde{\beta}$ for which $\tilde{\alpha} = \tilde{d}\tilde{\beta}$.

Before proving this let us see what this map is. Clearly it means that U is covered by a single topologically Cartesian coordinate system. This is really a topological condition on U: the region shown in figure 4.7 does not have such a coordinate system while that in figure 4.8 does, as illustrated in figure 4.9. Other kinds of regions also have such a map. For instance R^n itself can be mapped onto its unit open ball by the equations

$$x^i \to \frac{2}{\pi} x^i \frac{\arctan r}{r}, \tag{4.56}$$

$$r = [(x^1)^2 + (x^2)^2 + \ldots + (x^n)^2]^{1/2}, \tag{4.57}$$

because these imply

$$r \to \frac{2}{\pi} \arctan r. \tag{4.58}$$

This is a C^∞ map even at the origin, as one can see by expanding arctan r in its Taylor series

$$\arctan r = r - \tfrac{1}{3} r^3 + \tfrac{1}{5} r^5 - + \ldots.$$

To prove the theorem we use the coordinates x^i in U and construct the form $\tilde{\beta}$ we seek. Suppose $\tilde{\alpha}$ is

Fig. 4.9. A map from the region in figure 4.8 onto the unit open ball of R^2 (the interior of the unit circle). Dotted lines map to dotted lines, dashed to dashed, and a few typical points are shown. Clearly such a map can be made C^∞ if the boundary curve $\mathscr{C}$ is C^∞.

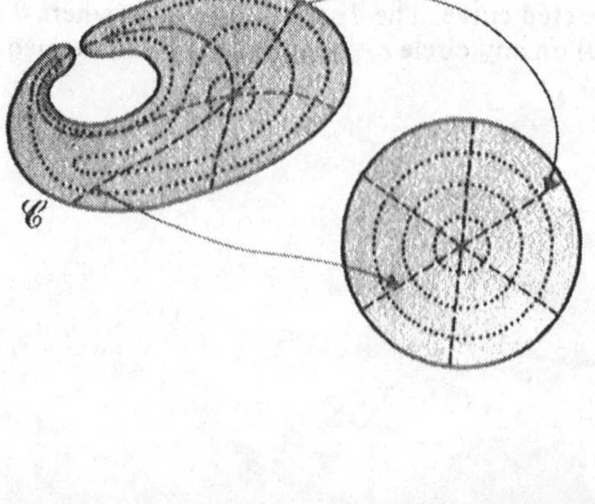

$$\tilde{\alpha} = \alpha_{i\dots k}(x^1, \dots, x^n)\,\tilde{d}x^i \wedge \dots \wedge \tilde{d}x^k, \qquad (4.59)$$

where each component $\alpha_{i\dots k}$ has p indices. Contract $\tilde{\alpha}$ with the 'radial vector' $\bar{r}$ whose components at any point on the coordinate basis are $(x^1, \dots, x^n)$, and call this $(p-1)$-form $\tilde{\mu}$. It is, by equation (4.13),

$$\tilde{\mu} = \tilde{\alpha}(\bar{r}) = \alpha_{ij\dots k}x^i \mathrm{d}x^j \wedge \dots \wedge \mathrm{d}x^k. \qquad (4.60)$$

Now we define the functions

$$\beta_{j\dots k}(x^1, \dots, x^n) = \int_0^1 t^{p-1}\alpha_{ij\dots k}(tx^1, tx^2, \dots, tx^n)x^i \mathrm{d}t. \qquad (4.61)$$

This integral is along the radial line in the coordinate system which $\{x^i\}$ happens to lie on. The functions define a $(p-1)$-form $\tilde{\beta}$

$$\tilde{\beta} = \beta_{j\dots k}\tilde{d}x^j \wedge \dots \wedge \tilde{d}x^k,$$

and the claim is that $\tilde{\alpha} = \tilde{d}\tilde{\beta}$.

The proof of this claim is straightforward algebra. From (4.45) we have

$$(\tilde{d}\tilde{\beta})_{ij\dots k} = p\frac{\partial}{\partial x^{[i}}\beta_{j\dots k]}. \qquad (4.62)$$

The derivative is easy:

$$\frac{\partial}{\partial x^i}\beta_{j\dots k} = \int_0^1 t^{p-1}\alpha_{ij\dots k}(tx^1, \dots, tx^n)\,\mathrm{d}t + \int_0^1 t^p x^l \alpha_{lj\dots k,\,i}(tx^1, \dots, tx^n)\,\mathrm{d}t. \qquad (4.63)$$

In order to antisymmetrize this on $[ij \dots k]$ we invoke (for the first time) the closure of $\tilde{\alpha}$:

$$0 = \alpha_{[lj\dots k,\,i]} = \alpha_{l[j\dots k,\,i]} - \alpha_{[j|l|\dots k,\,i]} - \alpha_{[kj\dots|l|,\,i]} - \alpha_{[ij\dots k],\,l}, \qquad (4.64)$$

where vertical bars separate out an index which is not included in the antisymmetrization implied by [ ]. But the components of $\tilde{\alpha}$ are already antisymmetric on all indices, so the first p terms are equal, and we have

$$0 = p\alpha_{l[j\dots k,\,i]} - \alpha_{[ij\dots k],\,l}. \qquad (4.65)$$

Putting this into the second integral of the antisymmetrized version of (4.63) and inserting this into (4.62) gives

$$\begin{aligned}(\tilde{d}\tilde{\beta})_{ij\dots k} &= \int_0^1 [pt^{p-1}\alpha_{ij\dots k}(tx^1, \dots, tx^n) + t^p x^l \alpha_{ij\dots k,\,l}(tx^1, \dots, tx^n)]\,\mathrm{d}t \\ &= \int_0^1 \frac{\mathrm{d}}{\mathrm{d}t}[t^p\alpha_{ij\dots k}(tx^1, \dots, tx^n)]\,\mathrm{d}t \\ &= \alpha_{ij\dots k}(x^1, \dots, x^n). \end{aligned} \qquad (4.66)$$

This proves the theorem.

Exercise 4.17

Prove equations (4.64) and (4.65).

Exercise 4.18

Use the local exactness theorem to show that locally (in three-dimensional Euclidean vector calculus) a curl-free vector field is a gradient and a divergence-free vector field is a curl.

There are two cautionary observations to be made here. The first is, as noted in §4.18, the $(p-1)$-form $\tilde{\beta}$ we constructed is not the only one for which $\tilde{d}\tilde{\beta} = \tilde{\alpha}$. The second is that we have merely given a sufficient condition for a closed form to be exact. Cohomology theory reveals many more complicated manifolds on which a closed form is still exact. (See §4.24.)

4.20 Lie derivatives of forms

We shall prove the following useful expression for the Lie derivative of a p-form $\tilde{\omega}$ with respect to a vector field $\bar{V}$:

$$\text{♦} \qquad £_{\bar{V}}\tilde{\omega} = \tilde{d}[\tilde{\omega}(\bar{V})] + (\tilde{d}\tilde{\omega})(\bar{V}). \tag{4.67}$$

That is, the p-form $£_{\bar{V}}\tilde{\omega}$ is the sum of two p-forms; the first is the exterior derivative of $\tilde{\omega}(\bar{V})$, the contraction of $\tilde{\omega}$ on $\bar{V}$; the second is the contraction of $\tilde{d}\tilde{\omega}$ on $\bar{V}$. The proof is rather long and may be omitted on a first reading. The result, (4.67), has a nice naturalness: $£_{\bar{V}}\tilde{\omega}$ is a p-form involving $\bar{V}$ and $\tilde{\omega}$; if it can be constructed using $\tilde{d}$ at all (which we should expect, since both derivatives involve only the differential structure of the manifold), then it must involve the only two p-forms which one can construct from $\bar{V}$, $\tilde{\omega}$, and $\tilde{d}$. In fact, it is just their sum.

The proof proceeds by induction. We shall drop tildes over the symbols in the rest of this section, for the sake of clarity.

The first part of the proof is the case where ω is a zero-form, a function f. Then its contraction on $\bar{V}$ is by definition zero, while its exterior derivative is df. If $\bar{V} = d/d\lambda$, then we know that $df(\bar{V}) = df/d\lambda$, but this is also equal to $£_{\bar{V}}f = \bar{V}(f) = df/d\lambda$. This proves the expression in the simplest case.

The next case is ω, a one-form. Then we use component notation:

$$\begin{aligned} \omega(\bar{V}) &= \omega_i V^i \Rightarrow d[\omega(\bar{V})] = (\omega_i V^i)_{,j} dx^j \\ d\omega &= d(\omega_i dx^i) = (d\omega_i) \wedge dx^i \\ &= \omega_{i,j} dx^j \wedge dx^i = \omega_{i,j}(dx^j \otimes dx^i - dx^i \otimes dx^j) \end{aligned}$$

$$\Rightarrow (\mathrm{d}\omega)(\bar{V}) = \omega_{i,j}[\mathrm{d}x^j(\bar{V})\mathrm{d}x^i - \mathrm{d}x^i(\bar{V})\mathrm{d}x^j]$$
$$= \omega_{i,j}V^j\,\mathrm{d}x^i - \omega_{i,j}V^i\,\mathrm{d}x^j.$$

These expressions combine to give

$$\mathrm{d}[\omega(\bar{V})] + \mathrm{d}\omega(\bar{V}) = [\omega_{i,j}V^j + \omega_j V^j{}_{,i}]\ \mathrm{d}x^i,$$

which is the same as $\pounds_{\bar{V}}\omega$ from equation (3.14).

The rest of the proof proceeds by induction. Since a general p-form can be represented as a sum of functions times wedge products of p one-forms, as

$$\omega = \frac{1}{p!}\,\omega_{i\ldots k}\mathrm{d}x^i \wedge \ldots . \wedge \mathrm{d}x^k,$$

it suffices to prove the theorem for a form which can be written as

$$\omega = fa \wedge b, \tag{4.68}$$

where we assume the theorem has been established for a and b. Then we have

$$\pounds_{\bar{V}}\omega = (\pounds_{\bar{V}}f)a \wedge b + f(\pounds_{\bar{V}}a) \wedge b + fa \wedge (\pounds_{\bar{V}}b)$$
$$= \mathrm{d}f(\bar{V})a \wedge b + f\{\mathrm{d}[a(\bar{V})] + (\mathrm{d}a)(\bar{V})\} \wedge b$$
$$+ fa \wedge \{\mathrm{d}[b(\bar{V})] + (\mathrm{d}b)(\bar{V})\}.$$

But we also know that (if a is a p-form)

$$\mathrm{d}[\omega(\bar{V})] = \mathrm{d}[fa(\bar{V}) \wedge b + (-1)^p fa \wedge b(\bar{V})]$$
$$= \mathrm{d}f[(a \wedge b)(\bar{V})] + f\{d[a(\bar{V})] \wedge b + (-1)^{p-1}a(\bar{V}) \wedge \mathrm{d}b$$
$$+ (-1)^p \mathrm{d}a \wedge b(\bar{V}) + a \wedge [\mathrm{d}b(\bar{V})]\},$$

and

$$(\mathrm{d}\omega)(\bar{V}) = [\mathrm{d}f \wedge a \wedge b + f\mathrm{d}a \wedge b + (-1)^p fa \wedge \mathrm{d}b]\,(\bar{V})$$
$$= \mathrm{d}f(\bar{V})a \wedge b - \mathrm{d}f \wedge [(a \wedge b)(\bar{V})] + f[\mathrm{d}a(\bar{V}) \wedge b$$
$$+ (-1)^{p+1}\mathrm{d}a \wedge b(\bar{V}) + (-1)^p a(\bar{V}) \wedge \mathrm{d}b + a \wedge \mathrm{d}b(\bar{V})].$$

Thus, adding these gives the same expression as for $\pounds_{\bar{V}}\omega$ above. This establishes the expression's validity for general forms.

4.21 Lie derivatives and exterior derivatives commute

A very important consequence of (4.67) is the fact that *Lie and exterior differentiation commute*: (4.69) below. To prove this, note that for any form ω (again omitting tildes for clarity),

$$\pounds_{\bar{V}}\mathrm{d}\omega = \mathrm{d}[(\mathrm{d}\omega)(\bar{V})],$$

since $\mathrm{dd}\omega = 0$. But, by using the Lie derivative formula once more we have

$$(\mathrm{d}\omega)(\bar{V}) = \pounds_{\bar{V}}\omega - \mathrm{d}[\omega(\bar{V})].$$

So that (again because $\mathrm{dd} = 0$) we get

♦ $$\pounds_{\bar{V}}(\mathrm{d}\omega) = \mathrm{d}(\pounds_{\bar{V}}\omega). \tag{4.69}$$

Lie differentiation and exterior differentiation commute! This is actually a special case of a more fundamental property of d which is established in more complete treatments of the subject, namely that there is a sense in which d commutes with any differentiable mapping of the manifold. (This commutation property may make it easier for you to prove the second-to-last part of exercise 3.9!)

4.22 Stokes' theorem

We are now in a position to show that exterior differentiation and integration are inverse to one another. Since integration of forms on an n-dimensional manifold is defined only for n-forms, the inverse property applies only to exterior derivatives of $(n-1)$-forms. Moreover, since only *definite* integration of n-forms is defined (i.e. the integral produces a number, not another form), the inverse relation analogous to equation (4.43) at the beginning of part B of this chapter will have to relate the integral of the n-form $\tilde{d}\tilde{\omega}$ to another integral, that of $\tilde{\omega}$. But the $(n-1)$-form $\tilde{\omega}$ can be integrated only over $(n-1)$-dimensional hypersurfaces, so we are led naturally to look for a theorem relating the integral of $\tilde{d}\tilde{\omega}$ over a finite region to the integral of $\tilde{\omega}$ over the region's *boundary*, which is $(n-1)$-dimensional. Our approach to this theorem (equation (4.75) below) will, however, be somewhat indirect, in order to avoid some of the lengthy calculations the usual proofs employ. We shall begin by looking at the change (to first order) in the value of an integral when the region of integration is slightly changed.

Accordingly, let us consider the integral of an n-form $\tilde{\omega}$ over a region U of an n-dimensional manifold M. Let U have a smooth orientable boundary called ∂U, by which we mean an orientable submanifold of M of dimension $n-1$ which divides M–∂U into disjoint sets U and CU (the complement of U) in such a way that any continuous curve joining a point of U to a point of CU must contain a point of ∂U. For simplicity we will assume ∂U is connected, although this is not necessary. Examples are given in figure 4.10. Now let $\bar{\xi}$ be any vector field on M, and consider a change in the region of integration generated by Lie dragging the region (*but not the form* $\tilde{\omega}$) along $\bar{\xi}$. Thus there is a family of regions $U(\epsilon)$ and boundaries $\partial U(\epsilon)$ obtained by moving along $\bar{\xi}$ a parameter distance ϵ from the original ones, $U = U(0)$ and $\partial U = \partial U(0)$. This is illustrated in figure 4.11.

The change in the integral of $\tilde{\omega}$ is simply the integral over $\delta U(\epsilon)$, the region between the boundaries:

$$\int_{U(\epsilon)} \tilde{\omega} - \int_{U(0)} \tilde{\omega} = \int_{\delta U(\epsilon)} \tilde{\omega}. \tag{4.70}$$

We will calculate this. Let V be a patch of ∂U covered by coordinates we shall call

$\{x^2, x^3, \ldots, x^n\}$. By Lie dragging along $\bar{\xi}$, we construct coordinates $\{x^1 = \epsilon, x^2, x^3, \ldots, x^n\}$ for a neighborhood in M of any such patch V in which $\bar{\xi}$ is not tangent to ∂U (see figure 4.12). This defines and provides a coordinate system for the region $\delta V(\epsilon)$ between $\partial U(0)$ and $\partial U(\epsilon)$ 'above' $V = V(0)$. We will first calculate the integral of $\tilde{\omega}$ over this region and then extend it to all of $\delta U(\epsilon)$.

In our coordinates we write

$$\tilde{\omega} = f(x^1, \ldots, x^n)\, \tilde{d}x^1 \wedge \ldots \wedge \tilde{d}x^n.$$

If ϵ is small, its integral is†

Fig. 4.10. A manifold with a 'handle'. Curves $\mathscr{C}_1$ and $\mathscr{C}_2$ are not boundaries since they do not divide M into an 'inside' and 'outside'. The union $\mathscr{C}_1 \cup \mathscr{C}_2$ is a boundary consisting of disconnected submanifolds. By contrast, $\mathscr{C}_3$ is a connected boundary.

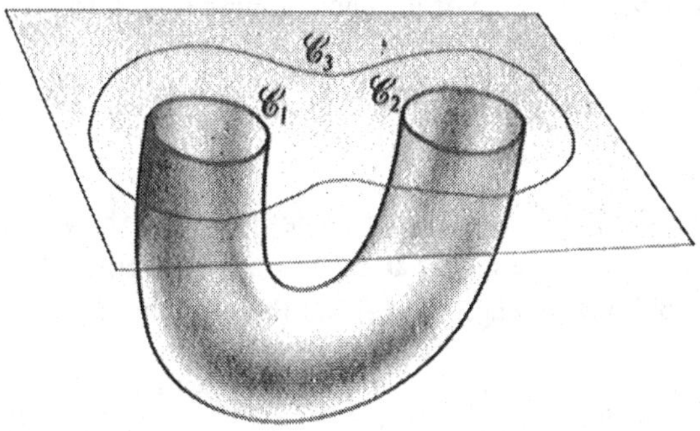

Fig. 4.11. The deformation of $U = U(0)$ into $U(\epsilon)$ by displacement a parameter distance ϵ along the integral curves of $\bar{\xi}$. Arrows represent the vector $\epsilon\bar{\xi}$ (for small ϵ). The region between ∂U and $\partial U(\epsilon)$ is $\delta U(\epsilon)$.

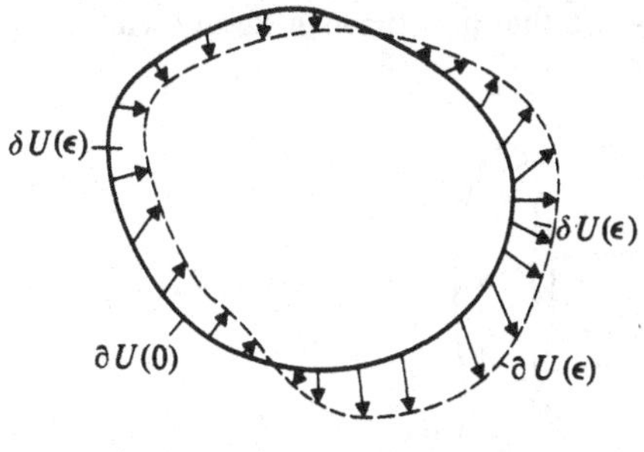

† The symbol $o(\epsilon)$ stands for any function $g(\epsilon)$ for which $g(\epsilon)/\epsilon \to 0$ as $\epsilon \to 0$.

$$\int_{\delta V(\epsilon)} \tilde{\omega} = \int_{V(0)} \left[\int_0^\epsilon f \, \mathrm{d}x^1 \right] \mathrm{d}x^2 \ldots \mathrm{d}x^n$$

$$= \epsilon \int_{V(0)} f(0, x^2, \ldots, x^n) \, \mathrm{d}x^2 \ldots \mathrm{d}x^n + o(\epsilon)$$

$$= \epsilon \int_V \tilde{\omega}(\bar{\xi}) \Big|_{\partial U} + o(\epsilon). \tag{4.71}$$

The last line follows from (4.13) and the fact that $\partial/\partial x^1 = \bar{\xi}$.

Equation (4.71) is independent of the coordinate we constructed, but it does require that $\bar{\xi}$ should not be tangent to ∂U in V. So it obviously applies to any region of ∂U bounded by points where $\bar{V}$ is tangent to ∂U. If these points form submanifolds of ∂U of lower dimensionality (as in figure 4.11), then they will not cause a problem. They will just divide ∂U into different regions $V_{(i)}$ in each of which (4.71) holds. If on the other hand $\bar{\xi}$ is tangent to ∂U in an open region in ∂U then the Lie dragging simply maps that region into itself and does not change the integral of $\tilde{\omega}$ at all, so (4.71) still holds, both sides being zero. We can therefore apply (4.71) over all of ∂U and combine it with (4.70) to get

◆ $$\frac{\mathrm{d}}{\mathrm{d}\epsilon} \int_{U(\epsilon)} \tilde{\omega} = \lim_{\epsilon \to 0} \frac{1}{\epsilon} \left[\int_{U(\epsilon)} \tilde{\omega} - \int_{U(0)} \tilde{\omega} \right] = \int_{\partial U} \tilde{\omega}(\bar{\xi}) \Big|_{\partial U}. \tag{4.72}$$

Now, we can obtain another expression for $(\mathrm{d}/\mathrm{d}\epsilon) \int \tilde{\omega}$ from the very construction of the Lie dragging of the region along $\bar{\xi}$; at any new point the integrand differs from the old one by $\epsilon \pounds_{\bar{\xi}} \tilde{\omega} + o(\epsilon)$. We therefore have

◆ $$\frac{\mathrm{d}}{\mathrm{d}\epsilon} \int_{U(\epsilon)} \tilde{\omega} = \int_{U(0)} \pounds_{\bar{\xi}} \tilde{\omega}. \tag{4.73}$$

But the expression (4.30) for $\pounds_{\bar{\xi}} \tilde{\omega}$ is particularly simple in this case, since $\mathrm{d}\tilde{\omega}$ is an $(n+1)$-form and so vanishes identically:

Fig. 4.12. A coordinate system for the neighborhood of a patch V of ∂U in which $\bar{\xi}$ is never tangent to ∂U. The region $\delta V(\epsilon)$ is all the points on those integral curves of $\bar{\xi}$ that pass through V, and which are a parameter distance $\leqslant \epsilon$ from V.

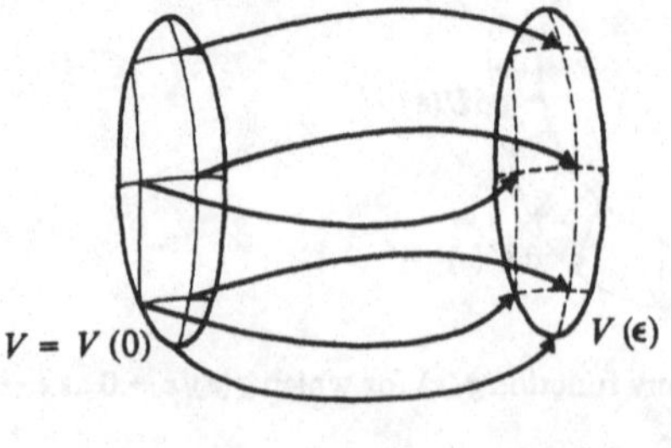

$$\int_U \pounds_{\bar{\xi}}\tilde{\omega} = \int_U \tilde{d}[\tilde{\omega}(\bar{\xi})].$$

Combining this with our previous expression for $(d/d\epsilon)\int_U \tilde{\omega}$, we get the *divergence theorem* (the reason for whose name will become clear in the next section):

$$\blacklozenge \qquad \int_U \tilde{d}[\tilde{\omega}(\bar{\xi})] = \int_{\partial U} \tilde{\omega}(\bar{\xi})\Big|_{\partial U}. \tag{4.74}$$

Now, since $\tilde{\omega}$ and $\bar{\xi}$ are arbitrary, $\tilde{\omega}(\bar{\xi})$ is an arbitrary $(n-1)$-form. So we can rewrite this as *Stokes' theorem* for an arbitrary $(n-1)$-form $\tilde{\alpha}$ defined on M:

$$\blacklozenge \qquad \int_U \tilde{d}\tilde{\alpha} = \int_{\partial U} \tilde{\alpha}, \tag{4.75}$$

where on the right-hand side we must of course restrict $\tilde{\alpha}$ to ∂U.

That this is what one knows as Stokes' theorem in Euclidean vector calculus is easily seen by letting M be two-dimensional as in figure 4.11. Then let $\tilde{\alpha}$ be a one-form, $\tilde{\alpha} = \alpha_i \tilde{d}x^i$, $\tilde{d}\tilde{\alpha} = (\alpha_{i,j} - \alpha_{j,i})\tilde{d}x^j \otimes \tilde{d}x^i$. The restriction of $\tilde{\alpha}$ to ∂U means allowing it to operate only on $\bar{l} = d/d\lambda$, a vector tangent to the curve ∂U. Then we get

$$\int (\alpha_{1,2} - \alpha_{2,1})\, dx^1 dx^2 = \oint_{\partial U} \alpha_i \frac{dx^i}{d\lambda} d\lambda = \oint_{\partial U} \alpha_i dx^i.$$

This is the usual Stokes' theorem.

4.23 Gauss' theorem and the definition of divergence

Stokes' theorem also embodies what is usually known as Gauss' theorem in vector calculus. For example, return to (4.74) and consider coordinates in which $\tilde{\omega} = \tilde{d}x^1 \wedge \ldots \wedge \tilde{d}x^n$, in some region W of M. Then its contraction with $\bar{\xi}$ is

$$\tilde{\omega}(\bar{\xi}) = \xi^1 \tilde{d}x^2 \wedge \ldots \wedge \tilde{d}x^n - \xi^2 \tilde{d}x^1 \wedge \tilde{d}x^3 \ldots \tilde{d}x^n \pm \ldots, \tag{4.76}$$

and

$$\begin{aligned} \tilde{d}[\tilde{\omega}(\bar{\xi})] &= \xi^1{}_{,1} \tilde{d}x^1 \wedge \tilde{d}x^2 \wedge \ldots \wedge \tilde{d}x^n \\ &\quad + \xi^2{}_{,2} \tilde{d}x^1 \wedge \tilde{d}x^2 \wedge \tilde{d}x^3 \wedge \ldots \tilde{d}x^n + \ldots \\ &= \xi^i{}_{,i} \tilde{\omega}. \end{aligned}$$

By analogy with Euclidean geometry we define the '*$\tilde{\omega}$-divergence*' of a vector field $\bar{\xi}$:

$$\blacklozenge \qquad (\text{div}_{\tilde{\omega}} \bar{\xi})\tilde{\omega} \equiv \tilde{d}[\tilde{\omega}(\bar{\xi})]. \tag{4.77}$$

If in the patch V of ∂U defined in §4.22 we again use coordinates such that ∂U is a surface of constant x^1, then the *restriction* of $\tilde{\omega}(\bar{\xi})$ to ∂U is again

$$\tilde{\omega}(\bar{\xi})|_{\partial U} = \xi^1 \tilde{d}x^2 \wedge \ldots \wedge \tilde{d}x^n = \tilde{d}x^1(\bar{\xi})\tilde{d}x^2 \wedge \ldots \wedge \tilde{d}x^n.$$

More generally, if $\tilde{n}$ is a one-form normal to ∂U (i.e. $\tilde{n}(\bar{\eta}) = 0$ on any vector $\bar{\eta}$ tangent to ∂U), and if $\tilde{\alpha}$ is any $(n-1)$-form such that

$$\tilde{\omega} = \tilde{n} \wedge \tilde{\alpha},$$

then we get $\tilde{\omega}(\vec{\xi})|_{\partial U} = \tilde{n}(\vec{\xi})\tilde{\alpha}|_{\partial U}$. This gives (4.74) in the form

$$\blacklozenge \quad \int_U (\mathrm{div}_{\tilde{\omega}}\vec{\xi})\tilde{\omega} = \int_{\partial U} \tilde{n}(\vec{\xi})\tilde{\alpha}, \tag{4.78}$$

where $\tilde{\alpha}$ is restricted to ∂U and $\tilde{n} \wedge \tilde{\alpha} = \tilde{\omega}$. If the coordinate system for (4.76) covers all of U, then this is

$$\int_U \xi^i{}_{,i}\mathrm{d}^n x = \oint_{\partial U} \xi^i n_i \mathrm{d}^{n-1}x, \tag{4.79}$$

which is the usual version of Gauss' theorem in R^n.

Exercise 4.19

Show that, although $\tilde{\alpha}$ as defined in (4.78) is not unique, the restriction $\tilde{\alpha}|_{\partial U}$ *is* unique once $\tilde{n}$ is given. Show that $\tilde{n}$ is fixed up to the scale transformation $\tilde{n} \to f\tilde{n}$, where f is any nowhere-zero function, and so show that $\tilde{\alpha}|_{\partial U}$ is unique up to $\tilde{\alpha}|_{\partial U} \to f^{-1}\tilde{\alpha}|_{\partial U}$. In this way conclude that $\tilde{n}(\vec{\xi})\tilde{\alpha}|_{\partial U}$ is *unique*.

The arbitrariness of $\tilde{\omega}$ in the definition we have used for the divergence of $\vec{\xi}$ can be eliminated if there is a metric, by using the metric volume element (§4.13). This is ambiguous up to a sign but (4.34) shows that $\mathrm{div}_{\tilde{\omega}}\vec{\xi}$ is in fact independent of this sign. Equation (4.79) shows that the usual divergence in R^n uses the form $\tilde{\mathrm{d}}x^1 \wedge \ldots \wedge \tilde{\mathrm{d}}x^n$, which is the metric volume element of the Euclidean metric.

Exercise 4.20

From (4.77) show that, if coordinates are chosen in which $\tilde{\omega} = f\tilde{\mathrm{d}}x^1 \wedge \ldots \wedge \tilde{\mathrm{d}}x^n$, then

$$\mathrm{div}_{\tilde{\omega}}\vec{\xi} = \frac{1}{f}(f\xi^i)_{,i}. \tag{4.80}$$

Exercise 4.21

In Euclidean three-space the preferred volume three-form is $\tilde{\omega} = \tilde{\mathrm{d}}x \wedge \tilde{\mathrm{d}}y \wedge \tilde{\mathrm{d}}z$. Show that in spherical polar coordinates this is $\tilde{\omega} = r^2 \sin\theta\, \tilde{\mathrm{d}}r \wedge \tilde{\mathrm{d}}\theta \wedge \tilde{\mathrm{d}}\phi$. Use (4.80) to show that the divergence of a vector $\vec{\xi} = \xi^r\, \partial/\partial r + \xi^\theta\, \partial/\partial\theta + \xi^\phi\, \partial/\partial\phi$ is,

$$\mathrm{div}\,\vec{\xi} = \frac{1}{r^2}\frac{\partial}{\partial r}(r^2\xi^r) + \frac{1}{\sin\theta}\frac{\partial}{\partial\theta}(\sin\theta\,\xi^\theta) + \frac{\partial\xi^\phi}{\partial\phi}.$$

Exercise 4.22

In fluid dynamics (and in many other branches of physics) one deals with the equation of continuity, which is written in ordinary tensor-calculus form as

$$\frac{\partial \rho}{\partial t} + \operatorname{div}(\rho \bar{V}) = 0.$$

Here ρ is the density of mass (or other conserved quantity) and $\bar{V}$ is its rate of flow. Defining $\tilde{\omega} = \tilde{d}x \wedge \tilde{d}y \wedge \tilde{d}z$ as in exercise 4.21 above, and using the *comoving* time-derivative operator $(\partial/\partial t + £_{\bar{V}})$ (which we will discuss in detail in chapter 5), show that the equation of continuity is

$$\left(\frac{\partial}{\partial t} + £_{\bar{V}}\right)(\rho\tilde{\omega}) = 0.$$

This permits one to regard $\rho\tilde{\omega}$ as a dynamically conserved volume three-form on the fluid. The 'volume' it assigns to any fluid element is that element's mass.

Exercise 4.23

(a) Show from (4.77) that another expression for the divergence of a vector $\bar{\xi}$ is

$$\operatorname{div}_{\tilde{\omega}}\bar{\xi} = {}^*\tilde{d}{}^*\bar{\xi}, \tag{4.81}$$

where the *-operation is the dual with respect to $\tilde{\omega}$ introduced earlier.

(b) For any p-vector $\mathbf{F}$ define

$$\operatorname{div}_{\omega}\mathbf{F} = (-1)^{n(p-1)}\,{}^*\tilde{d}{}^*\mathbf{F}. \tag{4.82}$$

Show that $\operatorname{div}_{\tilde{\omega}}\mathbf{F}$ is a $(p-1)$-vector. Show that if $\tilde{\omega}$ has components $\epsilon_{i\ldots j}$ in some coordinate system, then

$$(\operatorname{div}_{\tilde{\omega}}\mathbf{F})^{i\ldots j} = F^{ki\ldots j}{}_{,k} \tag{4.83}$$

in those coordinates.

(c) Generalize (4.80) to p-vectors.

Exercise 4.24

(a) On the sphere S^2 use Stokes' theorem to prove that a two-form $\tilde{\omega}$ is exact (is the exterior derivative of another form) only if

$$\int_{S^2} \tilde{\omega} = 0.$$

(Hint: S^2 has no boundary.)

(b) Show that the two-form

$$\tilde{\omega} = x^1\,\tilde{d}x^2 \wedge \tilde{d}x^3$$

defined on R^3 has the following value when integrated over the unit sphere S^2 as a submanifold of R^3:

$$\int_{S^2} \tilde{\omega}\Big|_{S^2} = \frac{4}{3}\pi.$$

(Hint: what is $\tilde{d}\tilde{\omega}$ in R^3?) Since any two-form on S^2 is closed (why?), this proves that not every closed two-form on S^2 is exact.

(c) Show that every closed one-form $\tilde{\beta}$ on S^2 is exact. (Hint: integrate $\tilde{d}\tilde{\beta}$ over a part of S^2.)

4.24 *A glance at cohomology theory*

Exercise 4.24 above illustrates how Stokes' theorem may be used to study those global properties of a manifold which determine the relation between closed and exact forms. Let $Z^p(M)$ be the set of all closed p-forms on M (all $\tilde{\alpha}$ such that $\tilde{d}\tilde{\alpha} = 0$) and let $B^p(M)$ be the set of all exact p-forms on M (all $\tilde{\alpha}$ such that $\tilde{\alpha} = \tilde{d}\tilde{\beta}$). Both sets are vector spaces over the real numbers: for example, if $\tilde{\alpha}$ and $\tilde{\beta}$ are closed p-forms then $a\tilde{\alpha} + b\tilde{\beta}$ is also closed for any real numbers a and b. In fact, B^p is a subspace of Z^p, since $\tilde{d}\tilde{d}\tilde{\beta} \equiv 0$. We now show how $Z^p(M)$ can be split up into equivalence classes modulo the addition of elements of $B^p(M)$. Closed forms $\tilde{\alpha}_1$ and $\tilde{\alpha}_2$ are said to be equivalent ($\tilde{\alpha}_1 \approx \tilde{\alpha}_2$) if their difference is an element of $B^p(M)$:

$$\tilde{\alpha}_1 \approx \tilde{\alpha}_2 \Leftrightarrow \tilde{\alpha}_1 - \tilde{\alpha}_2 = \tilde{d}\tilde{\beta}. \tag{4.84}$$

The equivalence class of $\tilde{\alpha}_1$ is the set of all closed forms equivalent to it. The set of all equivalence classes is called the *pth de Rham cohomology vector space* of M, $H^p(M)$.

Exercise 4.25

(a) A relation $\approx$ is called an *equivalence relation* if it has the following properties: (i) for any $\tilde{\alpha}$, $\tilde{\alpha} \approx \tilde{\alpha}$; (ii) if $\tilde{\alpha} \approx \tilde{\beta}$ then $\tilde{\beta} \approx \tilde{\alpha}$; and (iii) if $\tilde{\alpha} \approx \tilde{\beta}$ and $\tilde{\beta} \approx \tilde{\gamma}$ then $\tilde{\alpha} \approx \tilde{\gamma}$. Show that (4.84) does define an equivalence relation.

(b) If Z^p and B^p were, respectively, *any* vector space and a subspace then the set of equivalence classes we have defined is called the *quotient space* of Z^p by B^p, denoted by Z^p/B^p. Show that this is a vector space. (You must define addition of equivalence classes. Prove and then use the following result: if $\tilde{\alpha}_1$ and $\tilde{\alpha}_2$ are in equivalence classes A_1 and A_2 respectively, then the sum of any element A_1 and any element of A_2 is in the equivalence class of $\tilde{\alpha}_1 + \tilde{\alpha}_2$.)

(c) Consider the vector space R^2 and its subspace $R^1{}_x$ consisting of all vectors of the form $(a, 0)$ for arbitrary real a. Show that $R^2/R^1{}_x$ is the congruence of straight lines parallel to the x-axis.

We can translate the result of §4.19 into the statement that for the open ball in n-dimensions or any region U diffeomorphic to it, $H^p(U) = 0$ for $p \geqslant 1$, since all closed p-forms are equivalent to one another and hence to the zero p-form. It is also easy to compute $H^0(U)$, or in fact $H^0(M)$ for *any* connected manifold M. A zero-form is just a function, so $Z^0(M)$ is the space of functions f for which $\tilde{d}f = 0$, i.e. the constant functions. This is simply R^1. Moreover, since there is no such thing as a (-1)-form, the space $B^0(M)$ is just the zero-function. The equivalence relation $\approx$ is therefore just the usual algebraic equality: constants f and g are equivalent ($f \approx g$) if and only if they are equal ($f = g$). Therefore $H^0(M) = Z^0(M) = R^1$. If M is not a connected manifold then a function in $Z^0(M)$ need be constant only on each connected component of M, but may have different values on different components. Then $H^0(M) = Z^0(M) = R^m$, where m is the number of components of M.

Exercise 4.24 clearly can be generalized to any number of dimensions and shows that $H^n(S^n) \neq 0$ (part (b)) and $H^{n-1}(S^n) = 0$ (part (c)). These are special cases of

$$\begin{aligned} H^n(S^n) &= R^1, \\ H^p(S^n) &= 0, 0 < p < n, \\ H^0(S^n) &= R^1. \end{aligned} \tag{4.85}$$

The proof of this plus many other interesting results can be found in Spivak (1970, volume 1). Among the many applications of cohomology theory in Spivak is the *fixed-point theorem*: for even n the sphere S^n does not possess a nowhere-zero vector field.

Exercise 4.26

For odd n, find a nowhere-zero vector field on S^n. (Hint: regard S^{2m+1} as a submanifold of R^{2m+2} and consider the effect on S^{2m+1} of the rotation corresponding to the following matrix of $SO(2m + 2)$: $T = \mathrm{diag}(A_1, A_2, \ldots, A_{m+1})$, where each A_j is the 2×2 matrix

$$A = \begin{pmatrix} \cos\theta & -\sin\theta \\ \sin\theta & \cos\theta \end{pmatrix}$$

independently of j. Show that the vector field $\mathrm{d}/\mathrm{d}\theta$ on S^{2m+1} which is tangent to the congruence generated by this one-parameter subgroup of mappings does not vanish anywhere.)

Exercise 4.27

(a) Generalize exercise 4.24(b) to show that the $(n-1)$-form defined on R^n

$$\tilde{\omega} = \epsilon_{ij\ldots k} x^i \tilde{d}x^j \wedge \ldots \wedge \tilde{d}x^k \tag{4.86}$$

is nowhere zero when restricted to the sphere S^{n-1} defined by $(x^1)^2 + (x^2)^2 + \ldots + (x^n)^2 = 1$.

(b) Show that $H^{n-1}(S^{n-1}) = R^1$ implies that if $\tilde{\alpha}$ is any $(n-1)$-form on S^{n-1} then $\tilde{\alpha} - a\tilde{\omega}$ is exact, where $a = \int_{S^{n-1}} \tilde{\alpha} / \int_{S^{n-1}} \tilde{\omega}$.

(c) By taking the dual of this relation show that if f is any function on S^{n-1}, it can always be represented in the form $f = c + \operatorname{div}_{\tilde{\omega}} \bar{V}$ for some constant c and vector field $\bar{V}$ on S^{n-1}.

(d) For the circle S^1 prove that $H^1(S^1) = R^1$ by constructing a function f for which $\tilde{d}f = \tilde{\alpha} - a\tilde{\omega}$, as in (b) above.

Exercise 4.28

(a) Suppose a one-form $\tilde{\alpha}$ on M has the property $\int_{\mathscr{C}} \tilde{\alpha} = 0$ for *any* closed curve $\mathscr{C}$ in M. Show that $\tilde{\alpha}$ is exact, i.e. that there is a function f such that $\tilde{\alpha} = \tilde{d}f$.

(b) A connected manifold M is simply connected if every closed curve can be smoothly contracted to a single point. Show that M is simply connected if and only if $H^1(M) = 0$.

Before leaving the subject of cohomology we must take two short remarks. First, the *dimension* of $H^p(M)$ is called the pth-Betti number b^p of M. Second, although our definition of $H^p(M)$ relied on the differential structure of the manifold, it is one of the most fundamental theorems of cohomology (the de Rham theorem) that the cohomology groups depend only on the topological structure of M and not its differentiability. See Warner (1971) for further discussion.

4.25 Differential forms and differential equations

The example mentioned in §4.17 of the way in which exterior differentiation has a natural relation to integrability conditions also illustrates that, at least for first-order partial differential equations, there is a natural way to write the equations as relations among forms. This is so important that we shall expand upon it here.

Consider the equation

$$\frac{dy}{dx} = f(x, y),$$

rewritten as

$$\mathrm{d}y = f(x,y)\,\mathrm{d}x. \tag{4.87}$$

On a two-dimensional manifold M whose coordinates are x and y, we are tempted to write the one-form equation

$$\tilde{\mathrm{d}}y - f\tilde{\mathrm{d}}x = 0, \tag{4.88}$$

where f is now a function on M. What meaning does this equation have? Surely on such a manifold the one-forms $\tilde{\mathrm{d}}y$ and $\tilde{\mathrm{d}}x$ are linearly independent, so (4.88) cannot really be true: it is not an identity. But we do not expect it to be an identity. It comes from (4.87), which is a relation between 'increments' $\mathrm{d}y$ and $\mathrm{d}x$ *for solutions only*. A solution of (4.87) is a relation of the form $y = g(x)$, which defines a curve (or a path, at any rate) in M: a one-dimensional submanifold of M. Vectors tangent to this submanifold have slope $\mathrm{d}y/\mathrm{d}x$ equal to $f(x,y)$. Consider one such vector $\bar{V}$ at some point P, with components $(1, f(P))$. For such a vector, $\tilde{\mathrm{d}}y(\bar{V}) = f(P)$ and $\tilde{\mathrm{d}}x(\bar{V}) = 1$. Therefore the one-form in (4.88) is *zero* on $\bar{V}$:

♦ $$(\tilde{\mathrm{d}}y - f\tilde{\mathrm{d}}x)(\bar{V}) = 0.$$

This is the meaning of (4.88): solutions to the original differential equation define submanifolds of M whose tangent vectors *annul* the form (4.88). Equation (4.88) is true when *restricted* to this submanifold. Conversely, if there exist submanifolds whose tangent vectors annul (4.88), then these submanifolds are solutions of (4.87). Naturally there is not just one solution submanifold but a whole family of them, distinguished from one another by, say, the 'initial value' of the solution y at some fixed $x = x_0$ (or equivalently by the arbitrary constant of integration in the solution to (4.87)).

One can of course generalize this picture. Any given set of forms (not necessarily one-forms) $\{\gamma_i, i = 1, \ldots, N\}$ defines at any point P a subspace of T_P which annuls them. A solution to the forms (or to their associated differential equations) is the submanifold formed by the meshing together of these little tangent subspaces. The question of whether this meshing together is possible is clearly related to the theorem of Frobenius, proved in chapter 3. We reformulate this theorem in the language of forms in the next section.

The first question for the physicist, however, is usually to find the set (or *a* set) of forms which is equivalent to a given set of differential equations. An example of this is given in exercise 4.32, where the equations are first-order. A more complicated example is provided by the second-order harmonic oscillator equation

$$\frac{\mathrm{d}^2x}{\mathrm{d}t^2} + \omega^2 x = 0, \tag{4.89}$$

where ω is, for convenience, taken to be a constant. To put this in the language of forms, we write it as two first-order equations:

$$\frac{dx}{dt} = y, \quad \frac{dy}{dt} = -\omega^2 x.$$

Then it is clear that finding a submanifold that annulls the forms

$$\tilde{\alpha} \equiv \tilde{d}x - y\tilde{d}t, \quad \tilde{\beta} \equiv \tilde{d}y + \omega^2 x \tilde{d}t,$$

is equivalent to solving (4.89). The whole manifold is three-dimensional, with coordinates (x, y, t). A solution submanifold is one-dimensional, since annulling $\tilde{\alpha}$ and $\tilde{\beta}$ amounts to two restrictions on the vectors at any point of the manifold. Further instructive examples may be found in the papers by Estabrook (1976) and by Harrison & Estabrook (1971) in the bibliography. We now turn to the problem of the *existence* of solutions to these equations.

4.26 Frobenius' theorem (differential forms version)

We now return to one of the most important theorems of differential calculus on manifolds, whose Lie derivative version we gave in §3.7. In order to re-cast it in terms of differential forms we first need some definitions. A set of forms $\{\tilde{\beta}_i\}$ of any degree defines at each point P a subspace of vectors X_P of T_P, each of which annuls each $\tilde{\beta}_i$. This is called the *annihilator* of the set of forms at P. The *complete ideal* of the set at P is all the forms at P whose restriction to X_P vanishes. (Notice that if $\tilde{\gamma}$ is *any* form at P, $\tilde{\gamma} \wedge \tilde{\beta}_i$ is zero when restricted to the annihilator of $\tilde{\beta}_i$, and so is in the complete ideal.) Any such complete ideal has a set of linearly independent one-forms $\{\tilde{\alpha}_j\}$ which *generates* it, in the sense that the complete ideal of $\{\tilde{\alpha}_j\}$ is the same as that of $\{\tilde{\beta}_i\}$. Exercise 4.29 constructs such a set of generators.

Exercise 4.29

Let $\{\bar{e}_1, \ldots, \bar{e}_m\}$ be a basis for X_P and augment it with any other set of vectors $\{\bar{e}_{m+1}, \ldots, \bar{e}_n\}$ to form a basis for T_P. Show that the dual basis one-forms $\{\tilde{\omega}^{m+1}, \ldots, \tilde{\omega}^n\}$ generate the complete ideal. Show that any form in this ideal can be written as $\Sigma_{i=m+1}^{n} \tilde{\gamma}^i \wedge \tilde{\omega}^i$ for some $\{\tilde{\gamma}^i\}$.

Exercise 4.30

Let $\{\tilde{\alpha}_j, j = 1, \ldots, m\}$ be a set of linearly independent one-forms. Show that any form $\tilde{\gamma}$ is in their complete ideal if and only if

$$\tilde{\gamma} \wedge \tilde{\alpha}_1 \wedge \tilde{\alpha}_2 \wedge \ldots \wedge \tilde{\alpha}_m = 0. \tag{4.90}$$

The above algebra extends naturally to fields of forms. The complete ideal of a set of fields $\{\tilde{\beta}_i\}$ is the set of fields which are annulled by the annihilator X_P of $\{\tilde{\beta}_i\}$ at every point P. An ideal is said to be a *differential ideal* if, for every $\tilde{\gamma}$ in the ideal, $\tilde{d}\tilde{\gamma}$ is also in it. A set of one-forms $\{\tilde{\alpha}_j\}$ is said to be *closed* if each form $\tilde{d}\tilde{\alpha}_j$ is in the complete ideal generated by the $\tilde{\alpha}_j$s.

Exercise 4.31

(a) Show that a closed set of one-forms generates a differential ideal.
(b) On an n-dimensional manifold, show that any linearly independent set of n or $n-1$ one-forms is closed.

The *Frobenius theorem* can now be stated: suppose $\{\tilde{\alpha}_i, i = 1, \ldots, m\}$ are a linearly independent set of one-form fields in an open region U of an n-dimensional manifold M. If and only if they are closed, there exist functions $\{P_{ij}, Q_j, i, j = 1, \ldots, m\}$ such that

$$\blacklozenge \quad \tilde{\alpha}_i = \sum_{j=1}^{m} P_{ij}\tilde{d}Q_j. \tag{4.91}$$

Before proving this in the next section, let us see what it means. We are looking for solutions of the differential equations $\{\tilde{\alpha}_i = 0\}$, which are shown by (4.91) to be equivalent to $\{\tilde{d}Q_j = 0\}$. But this latter set is easy to solve: $\{Q_j = \text{const}\}$. So *the functions* $\{Q_j\}$ *are the solutions to the equations* $\{\tilde{\alpha}_i = 0\}$. Each set of values $\{Q_j\}$ defines an m-dimensional submanifold of M. Its tangent vectors annul $\{\tilde{d}Q_j\}$ by definition and therefore annul $\{\tilde{\alpha}_i\}$. This is the link with our previous version of Frobenius' theorem. The requirement that the set of one-forms be closed is the dual of the requirement that the set of vector fields annulling them be a Lie algebra, as discussed in more detail below.

Forms $\{\tilde{\alpha}_i\}$ satisfying (4.91) are said to be *surface-forming*. We can now establish the sufficiency of the integrability conditions discussed in §4.17. In that case the manifold had dimension two and the solution submanifold dimension one. The equation

$$\tilde{\alpha} = \tilde{d}f$$

is of the form (4.91), so a function f exists if and only if $\tilde{d}\tilde{\alpha} = 0$. A more complicated example follows.

Exercise 4.32

Consider the set of coupled linear inhomogeneous differential equations for the functions f and g in the independent variables x and y

$$\frac{\partial f}{\partial x} + A_1 f + B_1 g = C_1,$$
$$\frac{\partial f}{\partial y} + A_2 f + B_2 g = C_2,$$
$$\frac{\partial g}{\partial x} + D_1 g + E_1 f = F_1, \tag{4.92}$$
$$\frac{\partial g}{\partial y} + D_2 g + E_2 f = F_2,$$

where $A_i, B_i, C_i, D_i, E_i, F_i$ $(i = 1, 2)$ are functions of x and y. We wish to establish the integrability conditions for these equations.

(a) In the four-dimensional manifold M whose coordinates are (x, y, f, g) we define two one-forms

$$\tilde{\alpha} = \tilde{d}f + f\tilde{A} + g\tilde{B} - \tilde{C},$$
$$\tilde{\beta} = \tilde{d}g + g\tilde{D} + f\tilde{E} - \tilde{F}, \tag{4.93}$$

with the one-form $\tilde{A}$ being defined as

$$\tilde{A} = A_1\,\tilde{d}x + A_2\,\tilde{d}y, \tag{4.94}$$

and similarly for $\tilde{B}, \tilde{C}, \ldots$. *Show* that finding a two-dimensional submanifold $\mathscr{H}$ in M on which $\tilde{\alpha}|_{\mathscr{H}} = \tilde{\beta}|_{\mathscr{H}} = 0$ is equivalent to solving (4.92).

(b) By Frobenius' theorem, if $(\tilde{\alpha}, \tilde{\beta})$ are closed, then there exist functions U, V, W, X, Y, Z of the four variables (x, y, f, g) such that

$$\tilde{\alpha} = W\,\tilde{d}U + X\,\tilde{d}V,$$
$$\tilde{\beta} = Y\,\tilde{d}U + Z\,\tilde{d}V.$$

Show that

$$U(x, y, f, g) = \text{const},$$
$$V(x, y, f, g) = \text{const},$$

defines a *solution* to (4.92).

(c) By (b), a necessary and sufficient condition for a solution to exist is that the two-forms $\tilde{d}\tilde{\alpha}$ and $\tilde{d}\tilde{\beta}$ be in the ideal of $(\tilde{\alpha}, \tilde{\beta})$. *Show* that this is true if and only if

$$\tilde{d}\tilde{A} + \tilde{B}\wedge\tilde{E} = \tilde{d}\tilde{B} + \tilde{B}\wedge\tilde{D} + \tilde{A}\wedge\tilde{B} = \tilde{d}\tilde{C} + \tilde{B}\wedge\tilde{F} + \tilde{A}\wedge\tilde{C}$$
$$= \tilde{d}\tilde{D} + \tilde{E}\wedge\tilde{B} = \tilde{d}\tilde{E} + \tilde{E}\wedge\tilde{A} + \tilde{D}\wedge\tilde{E} = \tilde{d}\tilde{F} + \tilde{E}\wedge\tilde{C} + \tilde{D}\wedge\tilde{F} = 0.$$

(Hint: the realization that by (4.94) $\tilde{d}\tilde{A}$ is proportional to $\tilde{d}x \wedge \tilde{d}y$ helps simplify the algebra enormously.)

(d) *Show* that the conditions in (c) lead to the integrability conditions for (4.92):

$$\frac{\partial A_1}{\partial y} - \frac{\partial A_2}{\partial x} + B_2E_1 - B_1E_2 = 0,$$

$$\frac{\partial B_1}{\partial y} - \frac{\partial B_2}{\partial x} + B_2D_1 + A_2B_1 - B_1D_2 - A_1B_2 = 0,$$

and so on.

What does Frobenius' theorem have to say about the existence of solutions to equation (4.89)? The answer is simple: since any two linearly independent one-forms in a three-dimensional manifold automatically have a closed ideal (cf. exercise 4.31(b)), there must exist functions f, g, h, l, m, n for which

$$\tilde{\alpha} = h\tilde{d}f + l\tilde{d}g,$$
$$\tilde{\beta} = m\tilde{d}f + n\tilde{d}g.$$

Then the one-dimensional submanifolds defined by $f = \text{const}$, $g = \text{const}$ annul the forms α and β, and so are the solution submanifolds.

Our version of Frobenius' theorem does not directly deal with systems of differential equations described by sets of forms including two-forms or forms of higher degree. This case can be handled by finding a set of one-forms which generate the same complete ideal, as in exercise 4.29. It will not always be the case that these one-forms are algebraically equivalent to the original set, i.e. they might not give differential equations equivalent to the original ones. If they do, Frobenius' theorem applies directly. If not, then a more subtle approach is needed. See Choquet-Bruhat *et al.* (1977) for a discussion.

4.27 *Proof of the equivalence of the two versions of Frobenius' theorem*

Let us recall the geometrically more transparent version given in chapter 3: a given set of q vector fields $\{\bar{V}_{(i)}, i = 1, \ldots, q\}$, which at every point form a p-dimensional vector space, will mesh to form a p-dimensional hypersurface if and only if all the Lie brackets $[\bar{V}_{(i)}, \bar{V}_{(j)}]$ $(i, j = 1, \ldots, q)$ are linear combinations of the q vector fields. The version given in this chapter involves forms and the closure of their exterior derivatives; this is a picture 'dual', or complementary, to one with vectors and the closure of their Lie brackets. The key element in the correspondence between the two pictures is that if the vector fields define an r-dimensional subspace of T_P at a point P, of an n-dimensional manifold, then they define in a natural way an $(n-r)$-dimensional subspace of T_P^*, the space of one-forms at P, by the requirement that the forms be annulled by the vectors. Conversely, the same requirement allows a set of q one-forms to define an $(n-q)$-dimensional subspace of T_P. What we have, in effect, is that a submanifold can be described either by giving at every point the r-dimensional subspace of T_P which contains the vectors tangent to it, *or* by giving the $(n-r)$-dimensional subspace of one-forms annulled by those vectors. The proof of the equivalence between the two versions of Frobenius' theorem has two steps.

(1) Consider a submanifold of dimension p in a manifold of dimension n: there are $n-p$ different functions $Q_{(k)}$ which (locally) define the hypersurface by the $n-p$ equations $Q_{(k)} = \text{const}$. The forms $\tilde{d}Q_{(k)}$ are, by hypothesis, all linearly independent, and they are all anulled by any vector $\bar{V}$ tangent to the submanifold: $\langle \tilde{d}Q_{(k)}, \bar{V}\rangle = 0$. On the other hand, the tangent space to the submanifold is a p-dimensional vector space, which therefore defines a $(n-p)$-dimensional subspace of one-forms, such that any one-form $\tilde{\beta}$ in this subspace is annulled by all the $\bar{V}_{(i)}$: $\langle \tilde{\beta}, \bar{V}_{(i)}\rangle = 0$. Let $\{\tilde{\alpha}_{(k)}, k = 1, \ldots, n-p\}$ be any basis for the subspace. It is clear that the forms $\tilde{d}Q_{(k)}$ are also a basis, so that any $\tilde{\alpha}_{(k)}$ can be written as a linear combination of all the $\tilde{d}Q_{(k)}$s, as in equation (4.40). So the equivalence proof must now show that the condition on the vector fields – closure of their Lie brackets – is equivalent to the closure condition on the forms $\{\tilde{\alpha}_{(k)}\}$.

(2) This is done by beginning with the equation

$$\langle \tilde{\alpha}_{(i)}, \bar{V}_{(j)}\rangle = 0\ (i = 1, \ldots, n-p; j = 1, \ldots, p),$$

and taking its Lie derivative with respect to any $\bar{V}_{(k)}$:

$$0 = \pounds_{\bar{V}_{(k)}}\langle \tilde{\alpha}_{(i)}, \bar{V}_{(j)}\rangle = \langle \pounds_{\bar{V}_{(k)}}\tilde{\alpha}_i, \bar{V}_{(j)}\rangle + \langle \tilde{\alpha}_{(i)}, \pounds_{\bar{V}_{(k)}}\bar{V}_{(j)}\rangle.$$

By the rules for the Lie derivatives of forms we have

$$\langle \pounds_{\bar{V}_{(k)}}\tilde{\alpha}_{(i)}, \bar{V}_{(j)}\rangle = \langle \tilde{d}\langle \tilde{\alpha}_{(i)}, \bar{V}_{(k)}\rangle, \bar{V}_{(j)}\rangle + \langle \tilde{d}\tilde{\alpha}_{(i)}(\bar{V}_{(k)}), \bar{V}_{(j)}\rangle.$$

The first term vanishes because $\tilde{\alpha}_{(i)}$ is by definition annulled by $\bar{V}_{(k)}$, while the second one is just $\tilde{d}\tilde{\alpha}_{(i)}(\bar{V}_{(k)}, \bar{V}_{(j)})$, the value of $\tilde{d}\tilde{\alpha}_{(i)}$ on two vectors in the original set. Now, if $\pounds_{\bar{V}_{(k)}}\bar{V}_{(j)}$ is a linear combination of some $\bar{V}_{(i)}$, then it annuls $\tilde{\alpha}_{(i)}$ and we have that $\tilde{d}\tilde{\alpha}_{(i)}$ is annulled by the $\{\bar{V}_{(j)}\}$ as well. Therefore, $\tilde{d}\tilde{\alpha}_{(i)}$ is in the ideal, and closure of the Lie brackets implies closure of the forms. Conversely it is easy to see that closure of the forms (which implies $\tilde{d}\tilde{\alpha}_{(i)}(\bar{V}_{(k)}, \bar{V}_{(j)}) = 0$) implies closure of the Lie brackets.

4.28 *Conservation laws*

A particularly nice approach to conservation laws for differential equations is afforded by forms. Suppose solving a system of equations is equivalent to finding surfaces that annul a certain set of forms $\{\tilde{\alpha}_i\}$. Suppose further that there exists a form $\tilde{\gamma}$, a linear combination of $\{\tilde{\alpha}_i\}$,

$$\tilde{\gamma} = A_1\tilde{\alpha}_1 + \ldots,$$

such that

$$\tilde{d}\tilde{\gamma} = 0.$$

Then there exists another form $\tilde{\sigma}$ such that

$$\tilde{\gamma} = \tilde{d}\tilde{\sigma},$$

in a suitable region U of a solution surface H, and

$$\tilde{\gamma}|_H = \tilde{d}\tilde{\sigma}|_H = 0. \tag{4.95}$$

Applying Stokes' theorem to the integral of $\tilde{d}\tilde{\sigma}$ on the region U of H gives

$$\int_U \tilde{d}\tilde{\sigma} = \oint_{\partial U} \tilde{\sigma}.$$

But by (4.95) this vanishes:

$$\oint_{\partial U} \tilde{\sigma}|_H = 0$$

on the boundary of the region of a solution surface. This is a kind of integral conservation law, as we now illustrate for the harmonic oscillator.

The solution surfaces of (4.95) are one-dimensional curves, so the form $\tilde{d}\tilde{\sigma}$ must be a one-form, and $\tilde{\sigma}$ is in fact a zero-form (a function). Since $\tilde{d}\tilde{\sigma}$ is the same as $\tilde{\gamma}$, consider the form (notation same as before (§4.25))

$$\tilde{\gamma} = \omega^2 x\tilde{\alpha} + y\tilde{\beta}.$$

It is easy to verify that

$$\tilde{d}\tilde{\gamma} \equiv 0, \tag{4.96}$$

and in fact that

$$\tilde{\gamma} = \tilde{d}(\tfrac{1}{2}y^2 + \tfrac{1}{2}\omega^2 x^2). \tag{4.97}$$

Then on a solution curve, for which $\tilde{\alpha} = \tilde{\beta} = 0$ and hence $\tilde{\gamma} = 0$, we have that

$$\tilde{d}(\tfrac{1}{2}y^2 + \tfrac{1}{2}\omega^2 x^2) = 0,$$

$$0 = \int \tilde{d}(\tfrac{1}{2}y^2 + \tfrac{1}{2}\omega^2 x^2)$$

$$= (\tfrac{1}{2}y^2 + \tfrac{1}{2}\omega^2 x^2)|_{p_1}^{p_2}$$

where p_1 and p_2 are the endpoints of the region of the curve we integrated over. This just expresses the constancy of the energy, $\frac{1}{2}y^2 + \frac{1}{2}\omega^2 x^2$, along a solution curve.

For an application of this point of view to equations having soliton solutions, the interested reader is referred to Estabrook & Wahlquist (1975) (see bibliography).

Exercise 4.33

Verify equations (4.96) and (4.97).

4.29 *Vector spherical harmonics*

We resume here our discussion of spherical harmonics in §3.18. In that section we noted that a finite-dimensional representation of $SO(3)$ in the space of functions on S^2, $L^2(S^2)$, had the basis $\{Y_{lm}, m = -l, \ldots, l\}$. How do we create a related basis for vector fields on S^2? The space of all vector fields can be given a natural norm in terms of the metric g| of S^2, whose components are $\{g_{\theta\theta} = 1, g_{\phi\phi} = \sin^2\theta, g_{\theta\phi} = 0\}$ in the usual spherical coordinates. If we let $\tilde{\omega}$ be the metric-induced volume form on S^2 (§4.13) then the space $L^2_{1,0}(S^2)$ is the vector space of all vector fields $\bar{V}$ on S^2 whose norm

$$\|\bar{V}\|^2 = \int_{S^2} \text{g|}(\bar{V}, \bar{V})\, \tilde{\omega} \tag{4.98}$$

is finite. What we want are vector fields in $L^2_{1,0}(S^2)$ which are eigenfunctions of $\bar{l}_z$ and L^2.

We use two facts: first, g| and hence $\tilde{\omega}$ are invariant under $\bar{l}_z$ and L^2; and second, exterior differentiation and Lie differentiation commute. From the function Y_{lm} we construct the one-form $\tilde{d}Y_{lm}$, and from it the vector $\bar{\nabla}Y_{lm}$ with components (indices A, B run over 1 and 2)

$$(\bar{\nabla}Y_{lm})^A = g^{AB}(Y_{lm})_{,B}. \tag{4.99}$$

Evidently this is also an eigenfunction of $\bar{l}_z$ and L^2:

$$£_{\bar{l}_z}\bar{\nabla}Y_{lm} = im\,\bar{\nabla}Y_{lm}, \tag{4.100a}$$

$$L^2(\bar{\nabla}Y_{lm}) = -l(l+1)\bar{\nabla}Y_{lm}. \tag{4.100b}$$

But we cannot stop with *one* sort of vector harmonic, since we need to span a two-dimensional vector space. Here we take advantage of the fact that there is another way (on a two-dimensional manifold) to construct a vector from a one-form: the dual operation. So we also have $^*\tilde{d}Y_{lm}$, which is of course also an eigenfunction.

Exercise 4.34

Show that $\bar{\nabla}Y_{lm}$ and $^*\tilde{d}Y_{lm}$ are in general linearly independent vectors at each point.

It follows from the completeness theorem quoted in §3.18 that the two sets of vector spherical harmonics

◆ $$\bar{Y}^+_{lm} \equiv \bar{\nabla}Y_{lm}, \tag{4.101a}$$

◆ $$\bar{Y}^-_{lm} \equiv {}^*\tilde{d}Y_{lm} \tag{4.101b}$$

form a complete set for representing vectors on the two-sphere.

It is possible to follow this procedure further and define second-rank tensor spherical harmonics. This would, however, involve us with the covariant derivative on the sphere, which we have not yet discussed (see chapter 6). Interested readers may consult the paper by Regge & Wheeler (1957) listed in the bibliography.

Note that we have discussed only scalars and vectors *on the sphere.* Most applications involve larger manifolds with spherical symmetry, in which the spheres are submanifolds. As a simple example, consider three-dimensional Euclidean space E^3. A function on E^3 can be expanded in a series $\Sigma f_{lm}(r) \times Y_{lm}$, where its r-dependence is entirely contained in $\{f_{lm}\}$. A vector field $\bar{V}$ on E^3 can be split into two fields

$$\bar{V} = \bar{V}_\perp + \bar{V}_\mathrm{T},$$

where $\bar{V}_\perp$ is perpendicular to the spheres (parallel to $\bar{e}_r$) and $\bar{V}_\mathrm{T}$ is tangent to the spheres. If we write $\bar{V}_\perp$ as $v\bar{e}_r$, where v is a function, then under a rotation v transforms as a scalar function on the sphere while $\bar{V}_\mathrm{T}$ transforms as a vector on the sphere. So $\bar{V}_\mathrm{T}$ must be expanded in terms of vector spherical harmonics while v is expanded in scalar spherical harmonics. (Many authors multiply these scalars by $\bar{e}_r$ and call the resulting set a third kind of vector spherical harmonic.) We shall employ these in our examination of cosmological models in chapter 5 part E.

There are other equivalent formulations of vector spherical harmonics which, at first sight, seem to have very little to do with the ones defined here. These are defined by the algebraic methods of group theory (cf. Edmonds, 1957). The set presented here are convenient to use in differential equations, where the derivatives we have used occur naturally.

4.30 Bibliography

E. Cartan's own point of view on differential forms is very lucidly set out in E. Cartan, *Les Systemes Differentials Exterieurs et Leurs Applications Geometriques* (Hermann, Paris, 1945). An excellent introduction with more detail than we have room for, and with many applications in physics and engineering, is H. Flanders, *Differential Forms with Applications to the Physical Sciences* (Academic Press, New York, 1963). A recent and modern discussion which presupposes very little mathematical background is M. Schreiber, *Differential Forms: a Heuristic Introduction* (Springer, Berlin, 1977). A rigorous and advanced discussion of forms can be found in Y. Choquet-Bruhat, C. Dewitt-Morette & M. Dillard-Bleick, *Analysis, Manifolds, and Physics* (North-Holland, Amsterdam, 1977).

For a discussion of Stokes' theorem on nonorientable manifolds, with an application to the problem of the apparent nonexistence of magnetic monopoles, see R. Sorkin, *J. Phys. A* **10**, 717 (1977).

For further discussion of cohomology theory, see M. Spivak, *A*

Comprehensive Introduction to Differential Geometry (Publish or Perish, Boston, 1970) vol. 1; or F. W. Warner, *Foundations of Differentiable Manifolds and Lie Groups* (Scott, Foresman, Glenview, Ill. 1971).

The usefulness of forms in exploring the structure of differential equations is illustrated by F. B. Estabrook & H. D. Wahlquist, Prolongation structure of nonlinear evolution equations, *J. Math. Phys.* **16**, 1 (1975), and The geometric approach to sets of ordinary differential equations and Hamiltonian dynamics, *SIAM Review* **17**, 201 (1975). See also B. K. Harrison & F. B. Estabrook, Geometric approach to invariance groups and solutions of partial differential systems, *J. Math. Phys.* **12**, 653 (1971); and F. B. Estabrook, Some old and new techniques for the practical use of exterior differential forms, in *Bäcklund Transformation* ed. R. N. Miura, Lecture notes in mathematics, no. 515 (Springer-Verlag, Heidelberg, 1976).

There are many formulations of vector spherical harmonics. A standard reference is A. R. Edmonds, *Angular Momentum in Quantum Mechanics* (Princeton University Press, 1957). For the extension of our methods to tensors, see T. Regge & J. A. Wheeler, *Phys. Rev.* **108**, 1063 (1957). Other definitions of tensor spherical harmonics may be found in D. A. Akyeampong, *J. Math. Phys.* **20**, 505–8 (1979); and E. T. Newman & R. Penrose, *J. Math. Phys.* **7**, 863 (1966).

5 APPLICATIONS IN PHYSICS

A Thermodynamics

5.1 Simple systems

We confine our attention at first to a one-component fluid, for which the equation of conservation of energy is

$$\delta Q = P\mathrm{d}V + \mathrm{d}U, \tag{5.1}$$

where U is the internal energy of the fluid and δQ is the heat absorbed as the fluid does work $P\mathrm{d}V$ and changes its energy. We shall interpret this equation as a relation among various one-forms in the two-dimensional manifold whose coordinates are (V, U), on which the function $P(V, U)$ is defined (called the equation of state). Then since $\tilde{\mathrm{d}}V$ and $\tilde{\mathrm{d}}U$ are one-forms, so is $\tilde{\delta}Q$. But is $\tilde{\delta}Q$ an *exact* one-form? That is, can one find a function $Q(V, U)$ such that $\tilde{\delta}Q = \tilde{\mathrm{d}}Q$? If this were true, then one would have $\widetilde{\mathrm{d}\mathrm{d}}Q = 0$, which would mean

$$\begin{aligned} 0 = \tilde{\mathrm{d}}P\wedge\tilde{\mathrm{d}}V &= \left[\left(\frac{\partial P}{\partial V}\right)_U \tilde{\mathrm{d}}V + \left(\frac{\partial P}{\partial U}\right)_V \tilde{\mathrm{d}}U\right]\wedge\tilde{\mathrm{d}}V \\ &= \left(\frac{\partial P}{\partial U}\right)_V \tilde{\mathrm{d}}U\wedge\tilde{\mathrm{d}}V. \end{aligned}$$

(Subscripts on derivatives indicate which variable is fixed during differentiation.) Thus, a function Q can exist only if $(\partial P/\partial U)_V$ vanishes everywhere: this would be a strange fluid indeed!

Since $\tilde{\delta}Q$ is a one-form in a two-space, its ideal is automatically *closed*, so by Frobenius' theorem (§4.26) there must exist functions $T(U, V)$ and $S(U, V)$ such that $\tilde{\delta}Q = T\tilde{\mathrm{d}}S$. Thus, we define the temperature and entropy functions for the single-component gas in thermodynamic equilibrium simply as a representation of the one-form in equation (5.1):

♦ $$T\tilde{\mathrm{d}}S = P\tilde{\mathrm{d}}V + \tilde{\mathrm{d}}U. \tag{5.2}$$

It is important to understand that this is a purely mathematical definition of T and S, and it has no relation to the second law of thermodynamics, which we will consider in a moment. No mathematical *identity* of this sort would hold for a multi-component fluid. (We shall see that the second law of thermodynamics is equivalent to requiring $\tilde{\delta}Q = T\tilde{d}S$ for composite systems. Because this is not an automatic identity, the second law *is* a physical law: it restricts the possible mathematical nature of physical systems.)

5.2 Maxwell and other mathematical identities

Taking the exterior derivative of (5.2) gives

$$\tilde{d}T \wedge \tilde{d}S = \tilde{d}P \wedge \tilde{d}V. \tag{5.3}$$

Suppose we write $T = T(S, V), P = P(S, V)$. Then (5.3) gives (since $\tilde{d}S \wedge \tilde{d}S \equiv 0$, $\tilde{d}V \wedge \tilde{d}V \equiv 0$):

$$\left(\frac{\partial T}{\partial V}\right)_S \tilde{d}V \wedge \tilde{d}S = \left(\frac{\partial P}{\partial S}\right)_V \tilde{d}S \wedge \tilde{d}V = -\left(\frac{\partial P}{\partial S}\right)_V \tilde{d}V \wedge \tilde{d}S.$$

From this we conclude

$$\left(\frac{\partial T}{\partial V}\right)_S = -\left(\frac{\partial P}{\partial S}\right)_V, \tag{5.4}$$

which is known as one of the Maxwell identities. Similarly, by writing $S = S(T, V), P = P(T, V)$, we can deduce

$$\left(\frac{\partial S}{\partial V}\right)_T = \left(\frac{\partial P}{\partial T}\right)_V, \tag{5.5}$$

another Maxwell identity. By dividing (5.2) by T and then taking the exterior derivative we get

$$\frac{1}{T}\tilde{d}P \wedge \tilde{d}V - \frac{P}{T^2}\tilde{d}T \wedge \tilde{d}V - \frac{1}{T^2}\tilde{d}T \wedge \tilde{d}U = 0.$$

By writing $U = U(T, V), P = P(T, V)$, we get

$$\frac{1}{T}\left(\frac{\partial P}{\partial T}\right)_V \tilde{d}T \wedge \tilde{d}V - \frac{P}{T^2}\tilde{d}T \wedge \tilde{d}V - \frac{1}{T^2}\left(\frac{\partial U}{\partial V}\right)_T \tilde{d}T \wedge \tilde{d}V = 0,$$

or

$$T\left(\frac{\partial P}{\partial T}\right)_V - P = \left(\frac{\partial U}{\partial V}\right)_T. \tag{5.6}$$

Exercise 5.1

Derive the identity

$$T\left(\frac{\partial P}{\partial T}\right)_S - P = \left(\frac{\partial P}{\partial T}\right)_S\left(\frac{\partial U}{\partial S}\right)_T - \left(\frac{\partial P}{\partial S}\right)_T\left(\frac{\partial U}{\partial T}\right)_S \tag{5.7}$$

by multiplying (5.2) by $1/P$ and differentiating.

Another important relation which follows easily from the use of forms is

$$\left(\frac{\partial T}{\partial P}\right)_S\left(\frac{\partial S}{\partial T}\right)_P\left(\frac{\partial P}{\partial S}\right)_T = -1, \tag{5.8}$$

which is equally true of any set of three of (P, V, U, T, S). We prove this by writing

$$T = T(P, S), S = S(T, P), P = P(T, S), \tag{5.9}$$

which is possible since the manifold is two-dimensional. Then we have the successive identities:

$$\begin{aligned}\tilde{\mathrm{d}}T \wedge \tilde{\mathrm{d}}S &= \left(\frac{\partial T}{\partial P}\right)_S \tilde{\mathrm{d}}P \wedge \tilde{\mathrm{d}}S \\ &= \left(\frac{\partial T}{\partial P}\right)_S\left(\frac{\partial S}{\partial T}\right)_P \tilde{\mathrm{d}}P \wedge \tilde{\mathrm{d}}T \\ &= \left(\frac{\partial T}{\partial P}\right)_S\left(\frac{\partial S}{\partial T}\right)_P\left(\frac{\partial P}{\partial S}\right)_T \tilde{\mathrm{d}}S \wedge \tilde{\mathrm{d}}T,\end{aligned}$$

from which follows (5.8). Notice that the derivation here relies only on the ability to write (5.9), so that it is really an identity among any three functions on a two-dimensional manifold.

The ease with which the Maxwell identities and (5.8) can be derived using forms is an illustration of the natural way in which they fit into thermodynamics: the one-forms $\tilde{\mathrm{d}}P$, $\tilde{\mathrm{d}}S$, etc. are the mathematically precise substitutes for the physicists' rather fuzzier concept of the infinitesimals $\mathrm{d}P$, $\mathrm{d}S$, etc.

5.3 Composite thermodynamic systems: Caratheodory's theorem

We now consider composite thermodynamic systems, the parts of which may exchange energy with each other and with the outside world. In this case the law of conservation of energy is (for a system with N parts)

$$\begin{aligned}\tilde{\delta}Q &= P_1\tilde{\mathrm{d}}V_1 + \tilde{\mathrm{d}}U_1 + P_2\tilde{\mathrm{d}}V_2 + \tilde{\mathrm{d}}U_2 + \dots \\ &= \sum_{i=1}^{N} (P_i\tilde{\mathrm{d}}V_i + \tilde{\mathrm{d}}U_i).\end{aligned} \tag{5.10}$$

We regard this as a relation among one-forms on a $2N$-dimensional manifold whose coordinates are $(V_i, U_i; i = 1, \dots, N)$, and we assume that each P_i can be expressed as a function of these coordinates. The question arises of whether one can define an entropy and temperature for the system as a whole, i.e. whether T and S exist such that

◆ $$\tilde{\delta}Q = T\tilde{\mathrm{d}}S. \tag{5.11}$$

This equation is just the statement that $\tilde{\delta}Q$ is *integrable* (in the sense of the Frobenius theorem). Now the Frobenius theorem tells us that the necessary and sufficient condition for this to be true is $\tilde{d}\tilde{\delta Q} \wedge \tilde{\delta Q} = 0$. It is easy to see from (5.6) that this will not generally be true, so we can conclude that for a *general* interacting system there is no global temperature or entropy function. But the situation can be different for an *equilibrium* system, because the conditions for mechanical and thermodynamic equilibrium among the constituent parts restrict the problem (we assume) to a submanifold of the $2N$-dimensional one. We shall from now on let the world 'manifold' refer to this *equilibrium submanifold*, and examine the possibility that $\tilde{\delta Q}$ is integrable in it from the point of view of *Caratheodory*.

If $\tilde{\delta Q}$ is integrable, then every point of the manifold is on one and only one integral submanifold; these submanifolds are defined by S = const. None of these surfaces intersect. Therefore, starting at one point, it is *not* possible to reach an arbitrary point of the manifold along a curve on which δQ is everywhere zero. In other words, if an entropy function exists it is not possible to reach every equilibrium state of the system along an adiabatic path of equilibria. The physically interesting question is whether the *converse* is true: if we know that not every state is reachable along a path for which $\delta Q = 0$, can we say that $\tilde{\delta Q}$ is integrable? This is interesting because one version of the second law of thermodynamics asserts that it is impossible in a closed system to transfer heat from a colder to a hotter body without making other changes as well. By a closed system we mean one for which $\delta Q = 0$, so that the second law tells us that not every state can be achieved with $\delta Q = 0$. So does the second law imply the existence of an entropy function? Caratheodory's theorem says it does.

What we shall prove is that if $\tilde{\delta Q}$ is not integrable then all points in the neighborhood of some initial point P are reachable from P on a curve which annuls $\tilde{\delta Q}$. Since $\tilde{\delta Q}$ is not integrable, the version of Frobenius' theorem given in §4.26 shows us that there are at least two vector fields $\bar{V}$ and $\bar{W}$ for which $\tilde{\delta Q}(\bar{V}) = \tilde{\delta Q}(\bar{W}) = 0$ in a neighborhood of any point P, but $\tilde{\delta Q}([\bar{V}, \bar{W}]) \neq 0$ at P. That is, the one-form $\tilde{\delta Q}$ defines at each point P a subspace K_P of T_P, the vectors of which annul $\tilde{\delta Q}$; the nonintegrability of $\tilde{\delta Q}$ means that vector fields everywhere in K_P do not form a hypersurface: at least one of their Lie brackets does not lie in K_P (see figure 5.1). Because annulling $\tilde{\delta Q}$ is only one equation, K_P has

Fig. 5.1. The tangent hyperplane K_P contains the vectors annulling $\tilde{\delta Q}$ but not all of their Lie brackets at P.

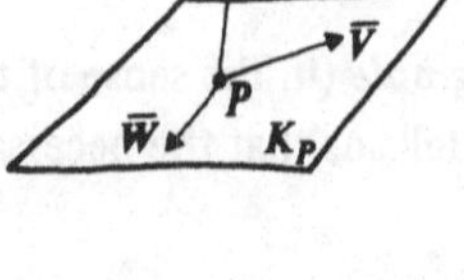

dimension $n-1$, where n is the dimension of the equilibrium manifold. Now, recall the exponentiation notation for the Taylor series introduced in §2.13. If we take any vector field U which is in K_P at all points P, and we move along it a parameter distance ϵ from P, we reach the point whose coordinates are $x^i = \exp(\epsilon \bar{U}) x^i|_P$, where we use $\bar{U}$ as a derivative operator on the function x^i along the curve. The set of all points in a small neighborhood of P reachable in this way may be called $\exp(\epsilon K_P)$: it is the representation in the manifold of the vector space K_P. This set of points is locally like a piece of an $(n-1)$-dimensional hypersurface. We shall show that, by following the curves of $\bar{V}$ and $\bar{W}$ defined above, we can reach points 'above' or 'below' this 'hypersurface' – i.e. that we can reach *all* points near P. The trip we make is the following: we move first a distance ϵ along $\bar{V}$, then ϵ along $\bar{W}$, then $-\epsilon$ along $\bar{V}$, and finally $-\epsilon$ along $\bar{W}$. This takes us to (cf. equation (2.6))

$$\begin{aligned} x^i &= \mathrm{e}^{-\epsilon\bar{W}}\mathrm{e}^{-\epsilon\bar{V}}\mathrm{e}^{\epsilon\bar{W}}\mathrm{e}^{\epsilon\bar{V}}x^i|_P \\ &= (1+\epsilon^2[\bar{W},\bar{V}]+O(\epsilon^3))x^i|_P. \end{aligned} \tag{5.12}$$

This means that we wind up almost back at P, but a parameter distance ϵ^2 away from it along $[\bar{V}, \bar{W}]$. This point is not in $\exp(\epsilon K_P)$, since $[\bar{V}, \bar{W}]$ is not in K_P. It is on one side of $\exp(\epsilon K_P)$; to finish on the other side we would have travelled first on $\bar{W}$, then on $\bar{V}$. Now, our path was along $\bar{V}$ or $\bar{W}$ everywhere, so it was adiabatic: $\delta Q = 0$ everywhere. It is clear, therefore, that if $\tilde{\delta Q}$ is not integrable, all states of the system will be reachable along adiabatic paths. This proves that the second law requires integrability of $\tilde{\delta Q}$ in the equilibrium manifold and the existence of an entropy function for composite systems in equilibrium.

B Hamiltonian mechanics

5.4 Hamiltonian vector fields

The Hamiltonian version of a dynamical system of equations begins with the Lagrangian $\mathscr{L}(q, q_{,t})$ for some dynamical variable $q(t)$. The momentum p is defined as

$$p = \partial\mathscr{L}/\partial(q_{,t}), \tag{5.13}$$

and the Hamiltonian H as

$$H = pq_{,t} - \mathscr{L} = H(p, q). \tag{5.14}$$

The dynamical equation

$$\frac{\mathrm{d}}{\mathrm{d}t}\frac{\partial\mathscr{L}}{\partial q_{,t}} - \frac{\partial\mathscr{L}}{\partial q} = 0, \tag{5.15}$$

and the definition of p can be written, respectively, as

$$\frac{\partial H}{\partial q} = -\frac{dp}{dt}, \quad \text{and} \quad \frac{\partial H}{\partial p} = \frac{dq}{dt}. \tag{5.16}$$

We now make a geometric picture of Hamiltonian dynamics by defining a manifold M called 'phase space', whose coordinates are p and q. On M we define the two-form

$$\blacklozenge \quad \tilde{\omega} \equiv \tilde{d}q \wedge \tilde{d}p. \tag{5.17}$$

Consider a curve $\{q = f(t), p = g(t)\}$ on M which is a solution of (5.16). Its tangent vector, $\bar{U} = d/dt = f_{,t}\, \partial/\partial q + g_{,t}\, \partial/\partial p$, has the property

$$\blacklozenge \quad \pounds_{\bar{U}}\tilde{\omega} = 0, \tag{5.18}$$

as we shall now prove. Since $\tilde{d}\tilde{\omega} = 0$, we have from (4.67)

$$\pounds_{\bar{U}}\tilde{\omega} = \tilde{d}[\tilde{\omega}(\bar{U})]. \tag{5.19}$$

But since $\tilde{\omega} = \tilde{d}q \otimes \tilde{d}p - \tilde{d}p \otimes \tilde{d}q$, we have

$$\begin{aligned} \tilde{\omega}(\bar{U}) &= \langle \tilde{d}q, \bar{U}\rangle\, \tilde{d}p - \langle \tilde{d}p, \bar{U}\rangle\, \tilde{d}q \\ &= \frac{df}{dt}\tilde{d}p - \frac{dg}{dt}\tilde{d}q. \end{aligned} \tag{5.20}$$

On the other hand, since f and g satisfy (5.16), we have

$$\tilde{\omega}(\bar{U}) = \frac{\partial H}{\partial p}\tilde{d}p + \frac{\partial H}{\partial q}\tilde{d}q = \tilde{d}H. \tag{5.21}$$

Therefore $\tilde{d}[\tilde{\omega}(\bar{U})]$ vanishes, establishing (5.18). A vector field $\bar{U}$ that satisfies (5.18) is called a *Hamiltonian vector field*.

Exercise 5.2

(a) Prove that if $\bar{U}$ is a Hamiltonian vector field, there exists some $H(p, q)$ such that equations (5.16) are satisfied along the integral curves of $\bar{U}$.

(b) Prove that Hamiltonian vector fields form a Lie algebra.

By exercise 5.2(a), we interpret $\bar{U}$ as a tangent to the solution curves in phase space if $\bar{U}$ is Hamiltonian. Notice that the system is *conservative*, since (5.16) implies

$$\pounds_{\bar{U}}H \equiv \frac{dH}{dt} = 0. \tag{5.22}$$

5.5 Canonical transformation

Now the coordinates p and q are not unique. We define a *canonical transformation* as one which leaves $\tilde{\omega}$ in the same form. That is, new coordinates $P = P(q, p)$ and $Q = Q(q, p)$ are called canonical if

$$\tilde{\mathrm{d}}q \wedge \tilde{\mathrm{d}}p = \tilde{\mathrm{d}}Q \wedge \tilde{\mathrm{d}}P. \tag{5.23}$$

The necessary and sufficient condition for this is

$$\left(\frac{\partial Q}{\partial q}\frac{\partial P}{\partial p} - \frac{\partial Q}{\partial p}\frac{\partial P}{\partial q}\right) = 1. \tag{5.24}$$

One such transformation is $Q = p, P = -q$. A less trivial one is found if we follow a procedure similar to the one we used to deduce the Maxwell identities in thermodynamics: we write $p = p(q, Q), P = P(q, Q)$ and find from (5.23) that

$$\partial p/\partial Q = -\partial P/\partial q. \tag{5.25}$$

So if we take an arbitrary function $F(q, Q)$ and define

$$p = \partial F/\partial q, \qquad P = -\partial F/\partial Q,$$

then (5.25) is satisfied identically. Thus, $F(q, Q)$ is said to generate a canonical transformation. Since we could have chosen, instead of (q, Q), the pairs (q, P), (p, Q), or (p, P) to be independent in (5.23), there are clearly four types of such generating functions for canonical transformations. They are explored more fully in Goldstein (1950) (see bibliography).

5.6 Map between vectors and one-forms provided by $\tilde{\omega}$

One of the most important features of this geometrical point of view on Hamiltonian dynamics is that $\tilde{\omega}$ can be cast in a role similar to that which a metric plays on Riemannian manifolds: it provides an invertible 1–1 mapping between vectors and one-forms. If $\bar{V}$ is a vector field on M, we define a one-form field

$$\tilde{V} \equiv \tilde{\omega}(\bar{V}), \tag{5.26}$$

with components

$$(\tilde{V})_i = \omega_{ij} V^j. \tag{5.27}$$

Similarly, given a one-form field $\tilde{\alpha}$ we define a vector field $\bar{\alpha}$ as the (unique) vector such that

$$\tilde{\alpha} = \tilde{\omega}(\bar{\alpha}). \tag{5.28}$$

Exercise 5.3

Prove that $\langle \tilde{V}, \bar{V} \rangle = 0$, so that $\tilde{\omega}$ is not suitable as a metric.

Exercise 5.4

Prove that if $\tilde{\alpha} = f\,\tilde{\mathrm{d}}q + g\,\tilde{\mathrm{d}}p$, then

$$\bar{\alpha} = g\frac{\partial}{\partial q} - f\frac{\partial}{\partial p}. \tag{5.29}$$

Exercise 5.5

Prove that $\bar{X}$ is a Hamiltonian vector field on M if and only if $\tilde{X}$ is an exact one-form, i.e. if and only if there exists some function H such that $\tilde{X} = \tilde{d}H$, or $\bar{X} = \overline{dH}$.

5.7 Poisson bracket

Suppose there are two functions f and g on the manifold, and we define the vector fields $\bar{X}_f \equiv \overline{df}$ and $\bar{X}_g \equiv \overline{dg}$. Then consider the scalar

$$\{f, g\} \equiv \tilde{\omega}(\bar{X}_f, \bar{X}_g) = \langle \tilde{d}f, \bar{X}_g \rangle. \tag{5.30}$$

Since $\tilde{\omega} = \tilde{d}q \otimes \tilde{d}p - \tilde{d}p \otimes \tilde{d}q$, we have

$$\bar{X}_g = \frac{\partial g}{\partial q}\frac{\partial}{\partial p} - \frac{\partial g}{\partial p}\frac{\partial}{\partial q}, \tag{5.31}$$

which can be established by verifying that $\tilde{\omega}(\bar{X}_g) = \tilde{d}g$. Therefore we have

$$\{f, g\} = \langle \mathrm{d}f, \bar{X}_g \rangle = \frac{\partial g}{\partial q}\frac{\partial f}{\partial p} - \frac{\partial g}{\partial p}\frac{\partial f}{\partial q}.$$

This is what is usually called the *Poisson bracket* of the functions f and g. The definition (5.30) gives it a geometrical significance, and shows that the Poisson bracket is actually independent of the coordinates. It depends only on $\tilde{\omega}$.

Exercise 5.6

(a) Defining $\bar{X}_H \equiv \overline{dH}$, show that for any function K,

$$\{K, H\} = \bar{X}_H(K) = \mathrm{d}K/\mathrm{d}t, \tag{5.32}$$

where t is the parameter such that $\bar{X}_H = \mathrm{d}/\mathrm{d}t$. Thus, the Poisson bracket of a function with the Hamiltonian gives the time-derivative of that function along a solution curve. In particular, constants of the motion have vanishing Poisson bracket with H.

(b) Show that the Poisson brackets satisfy the Jacobi identity

$$\{f, \{g, h\}\} + \{g, \{h, f\}\} + \{h, \{f, g\}\} = 0 \tag{5.33}$$

for any C^2 functions f, g, h.

(c) Show from this that

$$[\bar{X}_f, \bar{X}_g] = -\bar{X}_{\{f,g\}}, \tag{5.34}$$

so that the Hamiltonian vector fields form a Lie algebra.

5.8 Many-particle systems: symplectic forms

In general one deals with systems which have more than one degree of freedom, so there are more than one q and p. A particle in three dimensions has 3 qs and 3 ps, so phase space is 6-dimensional. A system containing N such particles has a $6N$-dimensional phase space. If we consider now a general system with n degrees of freedom, then phase space is $2n$-dimensional, and all the above results still hold if we take the two-form $\tilde{\omega}$ to be

$$\tilde{\omega} = \sum_{A=1}^{n} \tilde{d}q^A \wedge \tilde{d}p_A. \tag{5.35}$$

Such an $\tilde{\omega}$ is called a *symplectic form*, and then phase space is a *symplectic manifold*.

Exercise 5.7

(a) Show that f is a constant of the motion if $\bar{X}_f = (\overline{\tilde{d}f})$ is an invariant of H, i.e.

$$\pounds_{\bar{X}_f} H = 0. \tag{5.36}$$

(Refer to exercise 5.6.)

(b) Define a volume-form $\tilde{\sigma}$ for phase space by

$$\tilde{\sigma} = \underbrace{\tilde{\omega} \wedge \ldots \wedge \tilde{\omega}}_{n \text{ times}}, \tag{5.37}$$

where $2n$ is the dimension of the space, Show that $\tilde{\sigma} \neq 0$ and that a Hamiltonian vector field $\bar{U}$ is divergence-free in this volume measure. Said another way, this volume in phase space is preserved by the time-evolution of the system. This is known as *Liouville's theorem*.

Exercise 5.8

We now prove the remarks made in §3.12 about the relation between Killing vectors and conserved quantities. For particle motion the coordinates of phase space are $\{q^A, p_A\} = \{x^i, p_i = mv_i\}$ and the Hamiltonian is $H = (1/2m)g^{ij}p_ip_j + \Phi(x^i)$. Prove that if $\bar{U}$ is a Killing vector and if Φ is constant along $\bar{U}$, then its conjugate momentum, $p_{\bar{U}} \equiv U^i p_i$, is a conserved quantity. Hint: using exercise 5.7, define $\bar{X}_f$ as the vector field in phase space whose space components equal $\bar{U}$ and whose momentum components vanish. Show that

$$\pounds_{\bar{X}_f} H = 0,$$

and find f from equation (5.31).

5.9. Linear dynamical systems: the symplectic inner product and conserved quantities

Even more strikingly simple ways of formulating conservation laws are possible for linear systems, by which we mean dynamical systems whose Hamiltonian has the form

$$H = \sum_{A,B=1}^{n} (T^{AB} p_A p_B + V_{AB} q^A q^B), \tag{5.38}$$

where T^{AB} and V_{AB} are independent of the p_As and q^As. This system is called linear because the equations of motion are linear in $\{q^A, p_A\}$:

$$\frac{\mathrm{d}p_A}{\mathrm{d}t} = -\frac{\partial H}{\partial q^A} = -\sum_B V_{AB} q^B, \tag{5.39}$$

$$\frac{\mathrm{d}q^A}{\mathrm{d}t} = \frac{\partial H}{\partial p_A} = \sum_B T^{AB} p_B. \tag{5.40}$$

Notice that we can take $T^{AB} = T^{BA}$ and $V_{AB} = V_{BA}$, since the antisymmetric part of, say, T^{AB} would make no contribution to H when contracted with the symmetric expression $p_A p_B$.

The linearity of the system ensures that if $\{q^A_{(1)}, p_{(1)A}\}$ and $\{q^A_{(2)}, p_{(2)A}\}$ are solutions then so is $\{\alpha q^A_{(1)} + \beta q^A_{(2)}, \alpha p_{(1)A} + \beta p_{(2)A}\}$ for arbitrary constants α and β. Thus, this phase space is not just a manifold; it has a natural vector-space structure as well. A vector space is, of course, a kind of manifold, since it has a map into R^n, but it is a manifold which can be *identified* with its tangent space at every point. That is, since a curve in a vector space is a sequence of vectors, the tangent to the curve is just the derivative of the vectors along the curve, which is another vector, i.e. another element of the vector space. A vector space is its own tangent space. More than this, *all* the tangent spaces T_P have a natural identification with each other: we are able to speak about vectors in different T_Ps as being equal or not, simply by whether or not their components are equal. (This means a vector space is a *flat* manifold: see chapter 6.)

Since a point in phase space is a vector, we can use the symplectic form $\tilde{\omega}$ to define an inner product between elements of phase space. If $\bar{Y}_{(1)}$ is the vector whose components are $\{q^A_{(1)}, p_{(1)A}, A = 1, \ldots, N\}$ and if $\bar{Y}_{(2)}$ similarly has components $\{q^A_{(2)}, p_{(2)A}\}$, then their *symplectic inner product* is defined as

$$\tilde{\omega}(\bar{Y}_{(1)}, \bar{Y}_{(2)}) = \sum_A (q^A_{(1)} p_{(2)A} - q^A_{(2)} p_{(1)A}). \tag{5.41}$$

If $\bar{Y}_{(1)}(t)$ and $\bar{Y}_{(2)}(t)$ are solution curves, then their symplectic inner product is independent of time t. To prove this, we simply substitute the equations of motion into the expression for $\mathrm{d}\tilde{\omega}(\bar{Y}_{(1)}, \bar{Y}_{(2)})/\mathrm{d}t$ (sum on repeated indices here):

$$\begin{aligned}\frac{\mathrm{d}}{\mathrm{d}t}\tilde{\omega}(\bar{Y}_{(1)}, \bar{Y}_{(2)}) &= \frac{\mathrm{d}}{\mathrm{d}t}(q^A_{(1)}) p_{(2)A} + q^A_{(1)} \frac{\mathrm{d}}{\mathrm{d}t} p_{(2)A} \\ &\quad - \frac{\mathrm{d}}{\mathrm{d}t}(q^A_{(2)}) p_{(1)A} - q^A_{(2)} \frac{\mathrm{d}}{\mathrm{d}t} p_{(1)A}\end{aligned}$$

$$= T^{AB}p_{(1)B}p_{(2)A} + V_{AB}q^A_{(1)}q^B_{(2)} - T^{AB}p_{(1)A}p_{(2)B} - V_{AB}q^A_{(2)}q^B_{(1)}.$$

From the symmetry of T^{AB} and V_{AB} we conclude:

$$\frac{\mathrm{d}}{\mathrm{d}t}\tilde{\omega}(\bar{Y}_{(1)}, \bar{Y}_{(2)}) = 0 \tag{5.42}$$

if $\bar{Y}_{(1)}(t)$ and $\bar{Y}_{(2)}(t)$ are solutions.

The symplectic inner produce enables us to define in an elegant way certain *conserved quantities* associated with solutions. At first sight this may not be obvious: although the symplectic inner product is conserved, the symplectic inner product of a solution with itself vanishes identically. The trick is to use an invariance of the system (i.e. of T^{AB} and V_{AB}) to generate from one solution $\bar{Y}$ another closely related one. For example, suppose T^{AB} and V_{AB} are independent of time. Then the equations of motion tell us that if $\bar{Y}(t)$ is a solution, so is $\mathrm{d}\bar{Y}/\mathrm{d}t$. We define the canonical energy E_c of the solution $\bar{Y}$ to be

♦ $$E_c(\bar{Y}) = \tilde{\omega}\left(\frac{\mathrm{d}\bar{Y}}{\mathrm{d}t}, \bar{Y}\right). \tag{5.43}$$

It is easy to verify that $E_c(\bar{Y})$ is just the value of the Hamiltonian on the solution $\bar{Y}$.

Other conserved quantities are just as easy to derive. It usually happens that T^{AB} and V_{AB} depend on the coordinates $\{x^i\}$ of the manifold in which the dynamical system is defined (Euclidean space for nonrelativistic dynamics). If, as in exercise 5.8, there is some vector field $\bar{U}$ for which

$$£_{\bar{U}}T^{AB} = 0 = £_{\bar{U}}V_{AB}, \tag{5.44}$$

then there is a conserved quantity associated with $\bar{U}$. (In computing $£_{\bar{U}}T^{AB}$ it is important to distinguish between indices A, B which refer to coordinates in phase space and the tensorial character of T^{AB} on the original manifold. The quantities T^{AB} may be scalars, or tensors on the original manifold, depending upon whether the quantities q^A are scalars or tensors of higher order. The indices A and B are *labels*; they do not imply that T^{AB} should be treated as a tensor of type $\binom{2}{0}$ when computing the Lie derivative with respect to $\bar{U}$, because $\bar{U}$ is a vector field in the original manifold, not in phase space.) As before, if $\bar{Y}$ is a solution, then so is $£_{\bar{U}}\bar{Y}$. (Again the same remark applies: this is a derivative in the original manifold, not in a phase space.) We therefore define the (conserved) *canonical $\bar{U}$-momentum*

♦ $$P_{\bar{U}}(\bar{Y}) = \tilde{\omega}(£_{\bar{U}}\bar{Y}, \bar{Y}). \tag{5.45}$$

The reader is invited to try a simple example, such as the one given in exercise 5.8, to verify that the usual conserved quantity does indeed appear.

Although our discussion has been confined to systems with a finite number

(N) of degrees of freedom, the formalism generalizes in a straightforward way to continuous systems, such as wave equations. Readers familiar with the Klein–Gordon equation may recognize the symplectic inner product: the integral of the conserved Klein–Gordon current density $\psi^*\dot{\psi} - \psi\dot{\psi}^*$ is just (to within constant factors) $\tilde{\omega}(\psi^*, \psi)$. A discussion of the canonical conserved quantities for waves in fluids, with application to questions of stability, can be found in Friedman & Schutz (1978) (see bibliography).

5.10 Fiber bundle structure of the Hamiltonian equations

Our original statement in §5.4 that we defined phase space to be the manifold whose coordinates are p and q, hid a lot of interesting and important structure. Suppose a dynamical system has the N coordinates $\{q^i\}$ corresponding to its N degrees of freedom. These define a manifold called *configuration space* M, and the evolution of the dynamical system in time is described by a curve $q^i(t)$ in M. The Lagrangian $\mathscr{L}$ is a function of q^i and dq^i/dt, and so is a function on TM, the tangent bundle of M. We now show that the momentum

$$p_i = \partial\mathscr{L}/\partial(q^i{}_{,t}), \tag{5.46}$$

is a one-form field on M, a cross-section of the cotangent bundle T^*M. We show this by its transformation properties. Let us define new coordinates for M

$$Q^{j'} = Q^{j'}(q^i). \tag{5.47}$$

Then the new momenta are

$$P_{j'} = \frac{\partial\mathscr{L}}{\partial Q^{j'}{}_{,t}} = \frac{\partial\mathscr{L}}{\partial q^k{}_{,t}}\frac{\partial q^k{}_{,t}}{\partial Q^{j'}{}_{,t}}. \tag{5.48}$$

Now, both $q^k{}_{,t}$ and $Q^{j'}{}_{,t}$ are elements of the fiber over any point P, and coordinates on this fiber undergo a natural change induced by (5.47). That is, if $\bar{V}$ is any vector at P its components change by

$$V^{j'} = \Lambda^{j'}{}_k V^k, \quad V^k = \Lambda^k{}_{j'} V^{j'}.$$

This applies as well to the velocity vector $q^k{}_{,t}$:

$$q^k{}_{,t} = \Lambda^k{}_{j'} Q^{j'}{}_{,t} \Longrightarrow \frac{\partial q^k{}_{,t}}{\partial Q^{j'}{}_{,t}} = \Lambda^k{}_{j'}.$$

Using this in (5.48) gives

$$P_{j'} = \Lambda^k{}_{j'} p_k, \tag{5.49}$$

so that the momentum is indeed a one-form.

It follows that phase space, whose coordinates are $\{q^i, p_i\}$, is nothing but the *cotangent bundle* T^*M, and the Hamiltonian is a function on this bundle. What is more, the symplectic form,

$$\tilde{\omega} = \tilde{d}q^i \wedge \tilde{d}p_i,$$

(summation convention employed) is independent of the coordinates in M. The transformation for it is

$$\begin{aligned} Q^{j'} &= Q^{j'}(q^i) \Longrightarrow \tilde{d}Q^{j'} = \Lambda^{j'}{}_i \tilde{d}q^i, \\ P_{j'} &= \Lambda^k{}_{j'} p_k \Longrightarrow \tilde{d}P_{j'} = \Lambda^k{}_{j',l} p_k \tilde{d}q^l + \Lambda^k{}_{j'} \tilde{d}p_k. \end{aligned} \tag{5.50}$$

(Remember that this $\tilde{d}$ operator acts in T^*M, not in M, and that the functions $\Lambda^k{}_{j'}$ are functions only of the coordinates of M). Then we find

$$\tilde{d}Q^{j'} \wedge \tilde{d}P_{j'} = \Lambda^{j'}{}_i \Lambda^k{}_{j',l} p_k \, \tilde{d}q^i \wedge \tilde{d}q^l + \Lambda^{j'}{}_i \Lambda^k{}_{j'} \tilde{d}q^i \wedge \tilde{d}p_k. \tag{5.51}$$

Now we also have

$$\Lambda^{j'}{}_i \Lambda^k{}_{j'} = \delta_i{}^k \Longrightarrow \Lambda^{j'}{}_i \Lambda^k{}_{j',l} = -\Lambda^{j'}{}_{i,l} \Lambda^k{}_{j'}.$$

So (5.51) becomes

$$\tilde{d}Q^{j'} \wedge \tilde{d}P_{j'} = -\Lambda^{j'}{}_{i,l} \Lambda^k{}_{j'} p_k \tilde{d}q^i \wedge \tilde{d}q^l + \tilde{d}q^i \wedge \tilde{d}p_i.$$

The first term on the right-hand side vanishes because

$$\Lambda^{j'}{}_{i,l} = \frac{\partial^2 Q^{j'}}{\partial q^i \partial q^l}$$

is symmetric in i and l and is contracted with the antisymmetric form $\tilde{d}q^i \wedge \tilde{d}q^l$. Therefore $\tilde{\omega}$ is independent of the coordinates of M and is a *natural* structure on the cotangent bundle T^*M. Moreover, T^*M is always orientable, since the volume-form $\tilde{\sigma}$ defined in exercise 5.7(b) is nowhere zero.

Clearly, although our examples treated the fiber structure as trivial (i.e. as a product of the q-space and p-space), it is possible to have nontrivial manifolds M and fiber bundles T^*M, in which all the coordinate-dependent formulae above are valid only in local coordinate patches. Even an example as simple as that of a bead constrained to move on the surface of a sphere has a nontrivial bundle structure for phase space, as we pointed out in §2.11.

C Electromagnetism

5.11 Rewriting Maxwell's equations using differential forms

Maxwell's equations, written in conventional form but with units where $c = \mu_0 = \epsilon_0 = 1$, are

$$\nabla \times \mathbf{B} - \frac{\partial}{\partial t}\mathbf{E} = 4\pi \mathbf{J}, \tag{5.52a}$$

$$\nabla \times \mathbf{E} + \frac{\partial}{\partial t}\mathbf{B} = 0, \tag{5.52b}$$

$$\nabla \cdot \mathbf{B} = 0, \tag{5.52c}$$

$$\nabla \cdot \mathbf{E} = 4\pi\rho. \tag{5.52d}$$

In writing these equations we have, of course, used the curl and divergence operations of ordinary flat three-space.

What we shall show below is that there exists a way of writing these equations using only the concepts of the metric and the exterior derivative. First we rewrite the equations in their relativistically invariant form† by first defining the *Faraday two-form* $\tilde{F}$, whose components are

$$(F_{\mu\nu}) \equiv \begin{pmatrix} 0 & -E_x & -E_y & -E_z \\ E_x & 0 & B_z & -B_y \\ E_y & -B_z & 0 & B_x \\ E_z & B_y & -B_x & 0 \end{pmatrix}. \qquad (5.53)$$

(Here, as in §2.31, Greek indices run over t, x, y, z.)

Exercise 5.9

Prove that under a spatial rotation $F_{\mu\nu}$ transforms in such a way that both **E** and **B** transform as three-vectors.

In terms of the Faraday tensor, Maxwell's equations take a particularly simple form. For instance, the four equations (5.52b, c) are just

$$F_{[\mu\nu,\gamma]} = 0 \Leftrightarrow \tilde{d}\tilde{F} = 0, \qquad (5.54)$$

where we have used the square-bracket notation to denote antisymmetrization.

Exercise 5.10

(a) Prove that (5.54) constitutes four linearly independent equations.
(b) Evaluate (5.54) for the components of $\tilde{F}$ given by (5.53) and prove their equality to (5.52b, c).

As for the rest of the equations, if we introduce the special-relativistic metric whose components in this coordinate system are

$$(g_{\mu\nu}) = \begin{pmatrix} -1 & 0 & 0 & 0 \\ 0 & 1 & 0 & 0 \\ 0 & 0 & 1 & 0 \\ 0 & 0 & 0 & 1 \end{pmatrix}, \qquad (5.55)$$

† For readers to whom this is unfamiliar, recall that Maxwell's equations are the correct theory for light and that special relativity was invented to explain certain properties of light, so the theory is *already* relativistically correct. All we do here is to find a convenient form for the equations.

then we can define an antisymmetric $\binom{2}{0}$ tensor F whose components are

$$F^{\mu\nu} = g^{\mu\alpha}g^{\nu\beta}F_{\alpha\beta},$$

$$(F^{\mu\nu}) = \begin{pmatrix} 0 & E_x & E_y & E_z \\ -E_x & 0 & B_z & -B_y \\ -E_y & -B_z & 0 & B_x \\ -E_z & B_y & -B_x & 0 \end{pmatrix}. \tag{5.56}$$

Exercise 5.11
Prove equation (5.56).

Then the remaining equations are

$$F^{\mu\nu}{}_{,\nu} = 4\pi J^{\mu}, \tag{5.57}$$

where we have defined the current four-vector to have components $\{J^t = \rho, J^i = (\mathbf{J})^i \text{ for } i = x, y, z\}$.

Exercise 5.12
Prove that the four equations (5.57) are just the same as (5.52a–d).

So far we have stuck to Lorentz coordinates because, while (5.54) is coordinate-independent, (5.57) is not a valid tensor equation in every coordinate system (recall exercise 4.15). On the other hand, we saw in exercise 4.23 how to define the divergence of an antisymmetric $\binom{2}{0}$ tensor (two-vector) if we have a volume-form. Because we have a metric, and because $\{\partial/\partial t, \partial/\partial x, \partial/\partial y, \partial/\partial z\}$ form an orthonormal basis in this metric, the preferred volume-form is

$$\tilde{\omega} = \tilde{\mathrm{d}}t \wedge \tilde{\mathrm{d}}x \wedge \tilde{\mathrm{d}}y \wedge \tilde{\mathrm{d}}z.$$

The following exercise develops the argument.

Exercise 5.13
(a) Define the two-form $^*\tilde{F}$ to be the contraction

$$^*\tilde{F} \equiv \tfrac{1}{2}\tilde{\omega}(\mathsf{F}), \tag{5.58}$$

i.e.

$$(^*\tilde{F})_{\mu\nu} = \tfrac{1}{2}\omega_{\alpha\beta\mu\nu}F^{\alpha\beta}.$$

This is, of course, the dual of F introduced in chapter 4. Find the components $(^*\tilde{F})_{\mu\nu}$ in terms of $\mathbf{E}$ and $\mathbf{B}$.

(b) Define the three-form $^*\tilde{J}$ by the contraction

$$^*\tilde{J} \equiv \tilde{\omega}(\bar{J}), \tag{5.59}$$

and show that (5.57) is equivalent to

$$\tilde{d}(^*\tilde{F}) = 4\pi {}^*\tilde{J}. \tag{5.60}$$

By exercise 4.23 this is also

$$\text{div}_\omega \mathsf{F} = 4\pi\bar{J}. \tag{5.61}$$

Note the great formal similarity between the two halves of our new form for Maxwell's equations:

♦ $$\tilde{d}\tilde{F} = 0, \tag{5.54}$$

♦ $$\tilde{d}^*\tilde{F} = 4\pi {}^*\tilde{J}. \tag{5.60}$$

Note also that they now are completely coordinate-free, so they have this form in *any* manifold with metric (because the metric was needed to obtain $^*\tilde{F}$ from $\tilde{F}$). The similarity between (5.54) and (5.60) is deep in Maxwell's equations. Note that the * operation on $\tilde{F}$ simply results in an exchange of **E** and **B** (cf. exercise 5.13(a)), and recall also that $\bar{J}$ was the *electrical* current density. If there were magnetic monopoles we would have two current densities, $\bar{J}_e$ and $\bar{J}_m$, and Maxwell's equations would take the symmetric form

$$\tilde{d}\tilde{F} = 4\pi {}^*\tilde{J}_m, \quad \tilde{d}^*\tilde{F} = 4\pi {}^*\tilde{J}_e. \tag{5.62}$$

Exercise 5.14

(a) Prove (5.62).

(b) Prove by exterior differentiation that equation (5.60) guarantees conservation of charge, i.e. that

$$\text{div}\,(\bar{J}) = 0. \tag{5.63}$$

Exercise 5.15

Establish the integral theorem for charge in the following way.

(a) Choose *any* oriented three-dimensional hypersurface $\mathscr{H}$ and restrict (5.60) to it. Prove that restriction commutes with exterior differentiation, i.e. that

$$\tilde{d}[(^*\tilde{F})|_{\mathscr{H}}] = (\tilde{d}^*\tilde{F})|_{\mathscr{H}}.$$

(b) Choose a region $\mathscr{D}$ of $\mathscr{H}$, with boundary $\partial\mathscr{H}$. Integrate the restriction of (5.60) over $\mathscr{D}$ and apply Stokes' theorem to find (appropriate restrictions implied)

$$\int_{\mathscr{D}} {}^*\tilde{J} = \frac{1}{4\pi}\oint_{\partial\mathscr{D}} {}^*\tilde{F}.$$

(c) In the case where $\mathscr{H}$ is a hypersurface $t = \text{const}$ in Minkowski spacetime and $\partial\mathscr{D}$ is a sphere, show that this gives the total charge in $\mathscr{D}$ as an integral of the normal component of the electric field over $\partial\mathscr{D}$.

5.12 Charge and topology

Since we can now formulate Maxwell's equations on any manifold with a metric, we can mention two attempts which have been made to resolve the puzzling question 'what is charge?' by answering 'charge is topology'. The first explanation, due to J. A. Wheeler (1962), is extremely simple. Consider figure 5.2, in which a hypersurface $t = \text{const}$ of some hypothetical spacetime is depicted. The lines drawn are integral curves of **E**. There is no charge density anywhere, and these integral curves are either closed (threading through the handle, out one hole, and down the other) or infinite (though they pass through the handle). Consider what an experimenter who measures **E** on the sphere S surrounding one hole will deduce: the integral $\int_S \tilde{F}|_S$ will certainly not vanish (**E** is outward-pointing all over S), and he will say the hole has positive charge. Likewise, a sphere around the other hole would give it negative charge, of exactly the same magnitude. (The calculation of exercise 5.15 fails because S does not divide the manifold into an inside and outside, cf. figure 4.10.) So this is a model for 'charge without charge', which has the bonus of explaining why negative charges equal positive charges. It has two drawbacks: first, no-one pretends to have a solution to, say, Einstein's equations which gives a geometry for spacetime that looks like this; and second, it is perhaps philosphically displeasing to think of

Fig. 5.2. A 'wormhole' or handle attached to a three-dimensional manifold with one dimension suppressed. Lines of force can thread through the handle, come out, and go backdown again to give each 'mouth' the appearance of charge in a charge-free space.

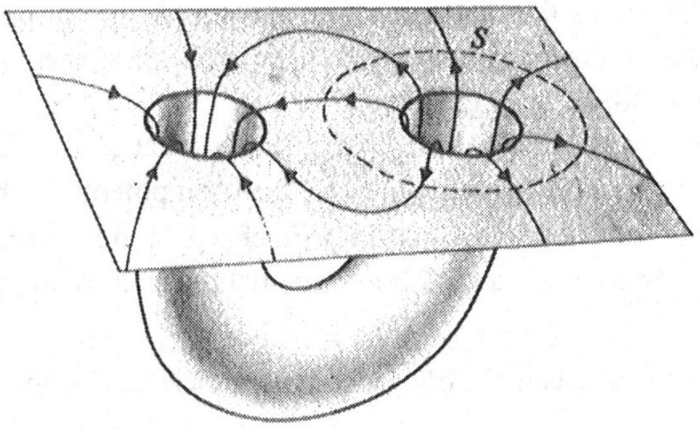

two charges, which may be separated by huge distances, linked together by their own special 'handle'.

The second explanation is more sophisticated, using a manifold made non-orientable by a special construction of the handle. This is due to Sorkin (1977) (reference in the bibliography of chapter 4). In this model, both holes have the *same* charge and so may be assumed to be close together, forming what to an outside observer looks like a single charge of twice the strength of each hole. Here the breakdown in exercise 5.15 occurs because the manifold is nonorientable. This model overcomes the second objection to Wheeler's picture, but not the first. And neither model explains why two unrelated charges should be equal. Nevertheless they illustrate a maxim which is becoming more convincing all the time: there is more to theoretical physics than just its local differential equations!

5.13 The vector potential

The existence of a 'vector potential' for Maxwell's equations follows naturally from (5.54). Since $\tilde{F}$ is a closed two-form, there is a one-form $\tilde{A}$ such that

$$\blacklozenge \qquad \tilde{F} = \tilde{d}\tilde{A} \tag{5.64}$$

in some neighborhood of any point. This one-form can be mapped into a vector by the metric, and this is called the vector potential. A more natural concept is, of course, the one-form potential. Note that $\tilde{A}$ is not uniquely defined: $\tilde{A}' = \tilde{A} + \tilde{d}f$, for an arbitrary function f, also gives $\tilde{F}$ in (5.64). This is a *gauge transformation*. Note also that if magnetic monopoles exist, then $\tilde{d}\tilde{F}$ does not vanish everywhere. By our discussion of exact forms in chapter 4, it will be possible to define $\tilde{A}$ only in simple regions which contain no magnetic monopoles. In particular, in a region of spacetime containing the world-line of a magnetic monopole, the one-form potential *cannot* be consistently defined everywhere.

Exercise 5.16

(a) Show that, if a one-form potential $\tilde{A}$ exists, then in nonrelativistic language it is related to the scalar potential ϕ and the vector potential A^i by $\phi = A_0$, A^i (vector potential) $= -A_i$ (one-form), where indices refer to the coordinates of (5.52).

(b) Show how ϕ and A^i defined in (a) change under a gauge transformation.

(c) To illustrate the problems caused to the one-form potential $\tilde{A}$ by magnetic monopoles, consider a situation with charges and *no* monopoles, but in which one defines a one-form potential $\tilde{\alpha}$ for $^*\tilde{F}$ by the equation $^*\tilde{F} = \tilde{d}\tilde{\alpha}$.

(By the duality between electric and magnetic fields under the

*-operation, $\tilde{\alpha}$ should have the same problems with electric charge as $\tilde{A}$ has with magnetic.) Write down Maxwell's equations in terms of $\tilde{\alpha}$ and show that $\tilde{\alpha}$ exists in regions that contain no charge and that can be shrunk to zero. Show this by finding an explicit solution for $\tilde{\alpha}$ in the case of a single isolated static charge q.

5.14 Plane waves: a simple example

Plane electromagnetic waves, as is well-known, travel at the speed of light. Consider a particular Faraday tensor $F^{\alpha\beta}$, all of whose components are functions only of $u \equiv t - x$ (recall that we are using units in which $c = 1$):

$$F^{\alpha\beta} = A^{\alpha\beta}(t-x) = A^{\alpha\beta}(u). \tag{5.65}$$

What are the conditions that this satisfy the empty-space equations $\mathrm{d}\tilde{F} = 0$, $\mathrm{d}{}^*\tilde{F} = 0$? From (5.65) we have

$$\begin{aligned} \mathrm{d}\tilde{F} &= \tilde{\mathrm{d}}(\tfrac{1}{2}F_{\mu\nu}\tilde{\mathrm{d}}x^{\mu} \wedge \tilde{\mathrm{d}}x^{\nu}) = \tfrac{1}{2}\tilde{\mathrm{d}}(F_{\mu\nu}) \wedge \tilde{\mathrm{d}}x^{\mu} \wedge \tilde{\mathrm{d}}x^{\nu} \\ &= \tfrac{1}{2}(\tilde{\mathrm{d}}A_{\mu\nu}/\mathrm{d}u)\tilde{\mathrm{d}}u \wedge \tilde{\mathrm{d}}x^{\mu} \wedge \tilde{\mathrm{d}}x^{\nu}. \end{aligned}$$

From (5.53) it is easy to deduce

$$\begin{aligned} \tilde{\mathrm{d}}\tilde{F} = \Big[&\frac{\mathrm{d}}{\mathrm{d}u}(B_z - E_y)\tilde{\mathrm{d}}t \wedge \tilde{\mathrm{d}}x \wedge \tilde{\mathrm{d}}y + \frac{\mathrm{d}}{\mathrm{d}u}(B_x)\tilde{\mathrm{d}}t \wedge \tilde{\mathrm{d}}y \wedge \tilde{\mathrm{d}}z \\ &+ \frac{\mathrm{d}}{\mathrm{d}u}(-B_x)\tilde{\mathrm{d}}x \wedge \tilde{\mathrm{d}}y \wedge \tilde{\mathrm{d}}z + \frac{\mathrm{d}}{\mathrm{d}u}(-B_y - E_z)\tilde{\mathrm{d}}t \wedge \tilde{\mathrm{d}}x \wedge \tilde{\mathrm{d}}z \Big], \end{aligned}$$

the vanishing of which implies (ignoring any static fields)

$$B_z = E_y, \qquad B_y = -E_z, \qquad B_x = 0. \tag{5.66}$$

Exercise 5.17

Show that the equation $\tilde{\mathrm{d}}{}^*\tilde{F} = 0$ implies

$$B_z = E_y, \qquad B_y = -E_z, \qquad E_x = 0. \tag{5.67}$$

By this exercise we see that a plane electromagnetic wave has transverse electric and magnetic fields (i.e. perpendicular to its direction of propagation), and that these are determined by two independent functions, $E_y(u)$ and $E_z(u)$, corresponding to the two independent polarizations of the wave.

D Dynamics of a perfect fluid

5.15 Role of Lie derivatives

By a 'perfect' fluid we mean one which has no viscosity and moves adiabatically, i.e. with no heat conduction. It is well-known that such a fluid obeys certain local conservation laws: during its motion any fluid element has a constant mass, entropy, and – in some sense – vorticity. These conservation laws are usually derived using ordinary vector calculus, and can seem rather complicated. From the geometric point of view, the existence of a flow suggests immediately the use of the Lie derivative, and we now show that the local conservation laws become much more transparent when framed with Lie derivatives.

5.16 The comoving time-derivative

We have seen in exercise 4.22 that the equation of continuity, whose conventional form is

$$\frac{\partial \rho}{\partial t} + \operatorname{div}(\rho \bar{V}) = 0$$

takes the form

$$\blacklozenge \qquad \left(\frac{\partial}{\partial t} + \pounds_{\bar{V}}\right)(\rho\tilde{\omega}) = 0, \tag{5.68}$$

where $\tilde{\omega} = \tilde{d}x \wedge \tilde{d}y \wedge \tilde{d}z$ is the volume three-form of Euclidean space. The operator $(\partial/\partial t + \pounds_{\bar{V}})$ is a natural time-derivative operator following a particular fluid element. To see this, think not of space but of the four-dimensional manifold called Galilean spacetime, whose coordinates are (x, y, z, t) (see §2.10). Any hypersurface $t = \text{const}$ is in fact Euclidean space. Then the motion of a fluid element describes a curve on spacetime, called the world-line of the element. In figure 5.3, two such world-lines (AA' and BB') are drawn. For an infinitesimal change in time dt, a point on this curve moves from the point with coordinates (x, y, z, t) to the one with coordinates $(x + V^x dt, y + V^y dt, z + V^z dt, t + dt)$. If we call $\bar{U}$ the tangent to the world line in the four-dimensional manifold, then it clearly has components $(V^x, V^y, V^z, 1)$. The time-derivative following a fluid element is simply $\pounds_{\bar{U}}$, the natural derivative along the world-line of the element.

Fig. 5.3. Two moments of Galilean time and the world lines AA' and BB' of two particles. The vector $\bar{U}$ is the tangent to AA' parameterized by time t.

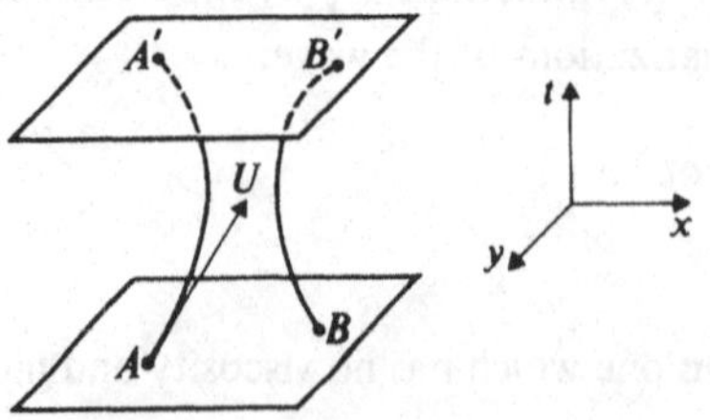

Exercise 5.18

Using equation (2.7) show that

$$£_{\bar{U}}\bar{W} = \left(\frac{\partial}{\partial t} + £_{\bar{V}}\right)\bar{W}, \tag{5.69}$$

where $\bar{W}$ is any vector field in the hypersurface $t = \text{const}$, i.e. any purely spatial vector field ($W^t \equiv 0$).

Equation (5.69) clearly holds if $\bar{W}$ is replaced by *any* $\binom{n}{0}$ tensor which is entirely in the three-space $t = \text{const}$. It might seem that the notion of a tensor being purely spatial is not invariant under coordinate changes in the four-dimensional manifold, since it simply says that all the t-components of the tensor vanish. This is acceptable here, however, because of the rigid distinction made in non-relativistic physics between space and time.

Exercise 5.19

The most general kind of coordinate transformation which remains 'natural' to the fiber-bundle structure of Galilean spacetime (§2.10) is

$$t' = g(t); \qquad x^{i'} = f^{i'}(x^i, t), i = 1, 2, 3. \tag{5.70}$$

Show that under this transformation a $\binom{n}{0}$ tensor **A** with no time-components ($\mathbf{A}(\ldots, \tilde{\omega}^t, \ldots) = 0$) remains one with no time-components, and a $\binom{0}{n}$ tensor **B** with no spatial components (i.e. only $B_{t\ldots t}$ is nonzero) remains one with no spatial components.

5.17 Equation of motion

The condition that the flow be adiabatic means that the total entropy of a fluid element must be conserved. It is convenient to work with S, the *specific* entropy (entropy per unit mass). This must clearly be *constant* during the flow:

♦ $$\left(\frac{\partial}{\partial t} + £_{\bar{V}}\right) S = 0. \tag{5.71}$$

The Euler equation of motion for a fluid whose pressure is p and which moves in a gravitational field whose potential is Φ can be written in Cartesian coordinates as

$$\frac{\partial}{\partial t} V^i + V^j \frac{\partial}{\partial x^j} V^i + \frac{1}{\rho}\frac{\partial}{\partial x^i} p + \frac{\partial}{\partial x^i}\Phi = 0. \tag{5.72}$$

There are two reasons that this equation is valid only in Cartesian coordinates: first, some indices i are up and some are down, and only in an orthonormal basis does this make no difference; second, the term $\partial V^i/\partial x^j$ transforms like a $\binom{1}{1}$ tensor *only* if the transformation matrix $\Lambda^{i'}{}_j$ is independent of position (exercise 4.5), which is true for a transformation from one Cartesian frame to another. The usual way to adapt it to arbitrary coordinates is to introduce the covariant derivative, which is defined in the chapter on Riemannian geometry. Here we show that there is a different, and very instructive, approach. First, note that the first two terms of (5.72) can be written as

$$\frac{\partial V_i}{\partial t} + V^j \frac{\partial V_i}{\partial x^j},$$

since there is no difference between V^i and V_i in Cartesian coordinates. (We use here, of course, the fact that the three-dimensional space has a metric tensor.) Next, replace the derivative $V^j\partial/\partial x^j$ with the Lie derivative (equation (3.14)) of the *one-form* $\tilde{V} = g(\bar{V}, \ )$:

$$\begin{aligned}(\pounds_{\bar{V}}\tilde{V})_i &= V^{j'}\frac{\partial}{\partial x^j} V_i + V_j \frac{\partial}{\partial x^i} V^j \\ &= V^j \frac{\partial}{\partial x^j} V_i + \frac{1}{2}\frac{\partial}{\partial x^i}(V_j V^j),\end{aligned}$$

where in obtaining the final expression we again used the fact that $V_j = V^j$. Therefore we find

$$V^j \frac{\partial}{\partial x^j} V_i = (\pounds_{\bar{V}}\tilde{V})_i - \frac{\partial}{\partial x^i}(\tfrac{1}{2}V^2). \qquad (5.73)$$

Both terms on the right-hand side are tensors in any coordinate system! Therefore (5.72) becomes the frame-independent expression

$$\blacklozenge \quad \left(\frac{\partial}{\partial t} + \pounds_{\bar{V}}\right)\tilde{V} + \frac{1}{\rho}\,\tilde{d}p + \tilde{d}(\Phi - \tfrac{1}{2}V^2) = 0. \qquad (5.74)$$

In this the role of the metric is crucial but hidden: it is required to form $\tilde{V}$ from $\bar{V}$, and hence to form $V^2 = \tilde{V}(\bar{V})$.

5.18 Conservation of vorticity

Now we are in a position to consider conservation of vorticity. In conventional terms, the vorticity is the curl of the velocity, $\nabla \times \bar{V}$. As we saw in chapter 4, this is properly the exterior derivative $\tilde{d}\tilde{V}$. Now, exterior differentiation and Lie differentiation commute (and of course $\tilde{d}$ and $\partial/\partial t$ commute since $\tilde{d}$ only involves spatial derivatives), so we find from (5.74)

$$\left(\frac{\partial}{\partial t} + \pounds_{\bar{V}}\right) \mathrm{d}V = \frac{1}{\rho^2}\,\mathrm{d}\rho \wedge \mathrm{d}p. \tag{5.75}$$

(We have dropped tildes over symbols for clarity.) There are two cases to be considered. The easier is when the fluid obeys an equation of state $p = p(\rho)$. Then $\mathrm{d}p \wedge \mathrm{d}\rho \equiv 0$ and we find that the vorticity two-form $\mathrm{d}V$ obeys the local (or convective) conservation law

$$\left(\frac{\partial}{\partial t} + \pounds_{\bar{V}}\right) \mathrm{d}V = 0. \tag{5.76}$$

This is the *Helmholtz circulation theorem*, written in its most natural form. A different result holds, however, if the more general equation of state $p = p(\rho, S)$ obtains. Then the right-hand side of (5.75) does not vanish, but its wedge product with $\mathrm{d}S$ does:

$$\mathrm{d}S \wedge \mathrm{d}\rho \wedge \mathrm{d}p = 0. \tag{5.77}$$

Exercise 5.20
Prove (5.77).

The exterior derivative of (5.71) gives

$$\left(\frac{\partial}{\partial t} + \pounds_{\bar{V}}\right) \mathrm{d}S = 0. \tag{5.78}$$

Therefore we can wedge $\mathrm{d}S$ with (5.75) to get

$$\mathrm{d}S \wedge \left(\frac{\partial}{\partial t} + \pounds_{\bar{V}}\right) \mathrm{d}V = 0,$$

or

♦ $$\left(\frac{\partial}{\partial t} + \pounds_{\bar{V}}\right) \mathrm{d}S \wedge \mathrm{d}V = 0. \tag{5.79}$$

This equation is the most general vorticity conservation law. It is called *Ertel's theorem*.

The meaning of the three-form $\mathrm{d}S \wedge \mathrm{d}V$ may not be immediately apparent, but it is possible to convert (5.79) into a conservation law for a scalar. The reason is that there is another conserved three-form, $\rho\omega$, and any two three-forms in a three-dimensional space are proportional. Therefore there is a scalar function α such that

$$\mathrm{d}S \wedge \mathrm{d}V = \alpha\rho\omega, \tag{5.80}$$

and (5.68) and (5.79) then give the scalar equation

$$\left(\frac{\partial}{\partial t} + £_{\bar{V}}\right)\alpha = 0.$$

It can be shown that, in conventional vector notation,

$$\alpha = \frac{1}{\rho}\nabla S \cdot \nabla \times V. \tag{5.81}$$

Exercise 5.21

Prove (5.81). (Hint: express both sides of (5.80) in terms of $dx \wedge dy \wedge dz$.)

In the notation introduced in chapter 4 we have

$$\alpha = \frac{1}{\rho}\epsilon^{ijk}S_{,i}V_{k,j}. \tag{5.82}$$

Therefore α is the *dual* of $dS \wedge dV$ with respect to $\rho\omega$. The conservation of α is then a natural consequence of the conservation of $dS \wedge dV$: the fact that $\rho\omega$ is conserved means that forming duals with respect to it is an operation which is also conserved, i.e. which commutes with the operator $\partial/\partial t + £_{\bar{V}}$.

Exercise 5.22

The shear of a velocity field $\bar{V}$ is defined in Cartesian coordinates by the equation

$$\sigma_{ij} = V_{i,j} + V_{j,i} - \tfrac{1}{3}\delta_{ij}\theta, \tag{5.83}$$

where θ is the *expansion*

$$\theta = \nabla \cdot \bar{V}. \tag{5.84}$$

Show that in an arbitrary coordinate system

$$\theta = \tfrac{1}{2}g^{ij}£_{\bar{V}}g_{ij}, \tag{5.85}$$

$$\sigma_{ij} = £_{\bar{V}}g_{ij} - \tfrac{1}{3}\theta g_{ij}. \tag{5.86}$$

E Cosmology

5.19 The cosmological principle

Most physicists are aware that Einstein's theory of general relativity has given modern physics a consistent and fruitful framework in which to study cosmology, the large-scale structure of our universe. Most are also aware that, at least at the simplest level, there are only three basic cosmological models: the 'closed', 'flat', and 'open' universes. What is probably less well known is that this simplicity of having only three models is not at all a prediction or consequence of Einstein's equations. Rather, it is simply a consequence of assuming that the universe is homogeneous and isotropic in its large-scale properties. (Homogeneity and isotropy will be defined precisely below.) General relativity, like all the fundamental theories of physics, is a dynamical theory: given initial conditions, it will predict their future evolution and past history. The uniformity of the universe is part of the initial conditions we put in to construct the simplest models. The important contribution of general relativity is that it permits us to choose the *geometry* of space – its metric tensor field – as a part of the initial conditions. This is not possible in Newtonian gravity, of course. Once we decide to choose the most uniform initial conditions, it is differential geometry that tells us that only three metric tensor fields are possible. Our aim in the next few sections is to find these metrics. We shall use the mathematics of symmetry and invariance developed in chapter 3, but we will not need to know anything about general relativity nor even about Riemannian geometry.

We begin with the physical problem: the universe. On a small scale the universe is certainly lumpy. On nearly any length scale from the nuclear (10^{-15} m) to the interstellar (10^{17} m), our world is characterized by clumping of matter into small regions with sharp demarcations between different kinds of matter or between matter and the vacuum. The stars themselves group into more or less isolated galaxies, galaxies congregate into clusters of several tens to thousands, and even clusters may associate in loose superclusters. But modern astronomy can see well beyond the supercluster length scale, and we find that in all directions the tendency is for greater and greater homogeneity in the properties of the universe when they are averaged over larger and larger length scales. Since it is these large-scale averaged properties (particularly the mean density and

Fig. 5.4. A slice of spacetime showing all the events labelled by coordinates t (time) and x, with $y = z = 0$. Because electromagnetic radiation travels at a finite speed, distant objects are seen at an earlier time in their own histories than nearby objects.

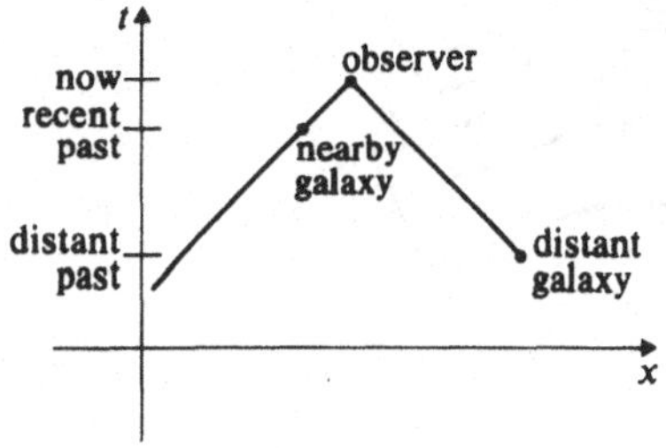

velocity) that are important for the dynamics of the universe, the cosmologist would like to incorporate this homogeneity into at least the simplest models. But what does homogeneity really mean? After all, in a dynamical universe, the more distant regions should look different from those nearby if only because they are seen at an earlier time in their history, as illustrated in figure (5.4). Indeed this is the case: the number of quasars, for instance, is much higher in distant regions than locally. The homogeneity one 'observes' is really an extrapolation to the present time of the condition of distant regions. Yet in relativity even 'the present time' is not an absolute concept. We cannot give a full discussion of these problems here, but we can say how they are resolved.

The basic idea is to split spacetime up into a family of three-dimensional spacelike submanifolds filling it up (a foliation). These are called hypersurfaces of constant time (see figure 5.5). This really amounts just to a choice of time-coordinate. The metric tensor $\mathbf{g}$ of spacetime has, like any $\binom{0}{m}$ tensor, a natural restriction to each hypersurface, and the hypersurface is space-like if $\mathbf{g}$ is positive-definite on all vectors tangent to it. The 'uniformity' of the cosmology depends on the Killing vectors or isometries of these hypersurfaces.

Let G be the Lie group of isometries of some manifold S with metric tensor field $\mathbf{g}$. The Lie algebra of G is that of the Killing vector fields of $\mathbf{g}$. Elements of G are mappings of S onto itself (diffeomorphisms). The action of G on S is said to be *transitive* on S if, for any two points P and Q of S, there is some element g of G for which $g(P) = Q$, i.e. which maps P to Q. The manifold S is said to be *homogeneous* if its isometry group acts transitively on it (see figure 5.6). What this means is just that the geometry is the same *everywhere* in S.

Suppose there are elements of G which leave some point P of S fixed. Then the product of any two also leaves P fixed, and since the identity e is one of them, they form a subgroup H_P of G called the *isotropy group* of P. These are, of course, the familiar rotations about an axis through P. The isotropy group of

Fig. 5.5. Slicing spacetime into spaces of constant time t.

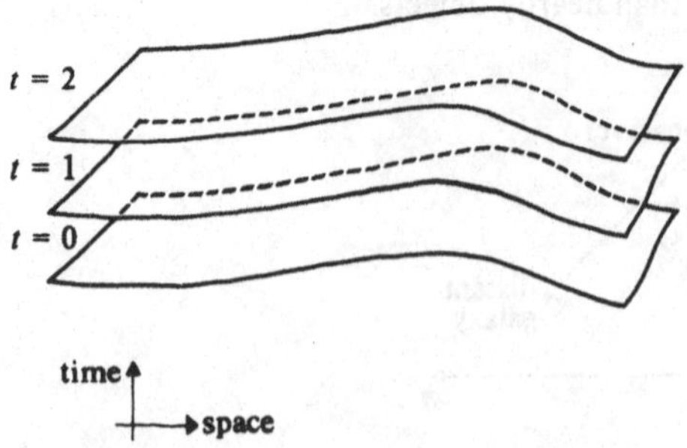

P keeps P fixed and therefore maps any curve through P to another curve through P (see figure 5.7). It consequently induces a map of tangent vectors at P to others at P: a map $T_P \to T_P$. This group of mappings is the linear isotropy group of P. (Recall the similar discussion of the adjoint representation of a Lie group, §3.17.) A manifold S of dimension m is said to be *isotropic about P* if its isotropy group H_P is just $SO(m)$, the group of rotations about arbitrary axes through P. If S is isotropic about every point P it is said to be *isotropic*.

A cosmological model M is said to be a *homogeneous cosmology* if it has a foliation of space-like hypersurfaces, each of which is homogeneous; and similarly for an isotropic cosmology. As discussed above, the evidence is strong that our universe is homogeneous, at least on large scales in our observable neighborhood. We also see no systematic variations in its structure in different directions in the sky. This suggests the universe is isotropic about us. But modern science does not like to assume that we live in a particularly favorable location in the universe. This is often elevated to the status of a principle, variously known as the *cosmological principle*, the Copernican principle, or the principle of mediocrity: the properties of the universe we see near us would be seen, on average, by any observer anywhere else in the universe. This principle enables cosmologists, in the absence of information to the contrary, to extend our local

Fig. 5.6. Some neighborhood U of P is mapped by g onto a neighborhood V of $Q = g(P)$ isometrically: there is no difference in the geometry near P from that near Q.

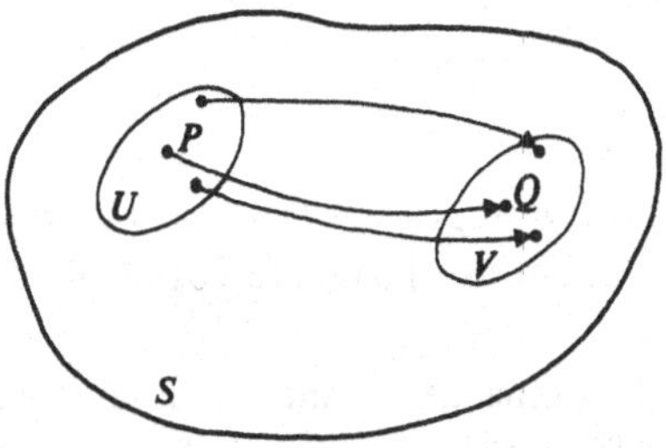

Fig. 5.7. The isotropy group of P maps $T_P \to T_P$ by mapping curves through P to other curves.

homogeneity and isotropy to the whole universe. This is not *necessary*, of course, and much current research is devoted to exploring inhomogeneous and/or anisotropic cosmologies. But the three basic models are the only three which have homogeneous, isotropic three-spaces. This is what we shall now prove.

Exercise 5.23
As we know from §3.9, the Killing vectors of the sphere S^2 are the vectors $\bar{l}_x, \bar{l}_y, \bar{l}_z$. These form a basis for the Lie algebra of the group of isometries of S^2, $SO(3)$. Prove that S^2 is a homogeneous and isotropic manifold.

5.20 Lie algebra of maximal symmetry

We shall begin by studying the Killing vector fields of a three-dimensional manifold S. If $\bar{\xi}$ is a Killing vector, its components in any coordinate system satisfy the equations

$$(\pounds_{\bar{\xi}}\mathsf{g})_{ij} = \xi^k g_{ij,k} + \xi^k{}_{,i}g_{kj} + \xi^k{}_{,j}g_{ik} = 0. \tag{5.87}$$

It will be more convenient to use the components of the one-form $\mathsf{g}(\bar{\xi},\ )$,

$$\xi_k = g_{kl}\xi^l. \tag{5.88}$$

These satisfy the equivalent equations

$$\xi_{i,j} + \xi_{j,i} - 2\xi_l\Gamma^l{}_{ij} = 0, \tag{5.89}$$

with the definition

$$\Gamma^l{}_{ij} = \tfrac{1}{2}g^{lm}(g_{mi,j} + g_{mj,i} - g_{ij,m}). \tag{5.90}$$

(The definition of $\Gamma^k{}_{ij}$, including its factor of $\frac{1}{2}$, is conventional and would make more sense after a reading of chapter 6. For us equation (5.90) simply defines a convenient shorthand notation.)

Equation (5.89) is symmetric under exchange of i and j, so it represents in n dimensions $\frac{1}{2}n(n+1)$ independent differential equations, six for $n = 3$. Since there are only three components of $\bar{\xi}$ to solve for, the system is overdetermined: a general metric tensor g has *no* Killing vectors. Our object is to find what form g must take in order that it allow the maximum number of Killing vectors. To see what this maximum number is, we differentiate (5.89) to get

$$\xi_{i,jk} + \xi_{j,ik} = 2(\xi_l\Gamma^l{}_{ij})_{,k}. \tag{5.91}$$

By adding (5.91) to itself with the index permutation $(i \to k, j \to i, k \to j)$ and subtracting the permutation $(i \to j, j \to k, k \to i)$ we arrive at the equation

$$\xi_{i,jk} = H_{ijk}{}^l\xi_l + K_{ijk}{}^{lm}\xi_{l,m}, \tag{5.92}$$

where $H_{ijk}{}^{l}$ is a complicated function of g_{ij} and its first and second derivatives, and $K_{ijk}{}^{lm}$ similarly depends on g_{ij} and its first derivatives. The key point about (5.92) is that if we know ξ_i and $\xi_{i,j}$ at any point P and if we know g_{ij} everywhere, then we can determine $\xi_{i,jk}$ at P from (5.92), and similarly all its higher derivatives at P by successively differentiating (5.92). On an analytic manifold (which we shall assume) this suffices to determine the vector field $\bar{\xi}$ everywhere. Moreover, we know that ξ_i at P determines the *symmetric part* of $\xi_{i,j}$ at P by equation (5.89). If follows that every Killing vector field on S is determined completely by giving the values of

$$\eta_i \equiv \xi_i(P) \text{ and } A_{ij} \equiv \xi_{[i,j]}(P) \qquad (5.93)$$

at *any* point P of S. It is important that a choice of $\{\eta_i, A_{ij}\}$ at P does not necessarily determine a Killing vector, because it may happen that (5.92) has no solutions: its right-hand side may not be symmetric under exchange of j and k. But the argument does show that there cannot be more Killing vectors than the number of independent choices of $\{\eta_i, A_{ij}\}$, which in m dimensions is

$$m + \tfrac{1}{2}m(m-1) = \tfrac{1}{2}m(m+1), \qquad (5.94)$$

by virtue of (5.93). A manifold is said to be *maximally symmetric* if it has the maximum number of Killing vector fields.

It is easy to show that a maximally symmetric connected manifold S is homogeneous. At any point P we can choose a Killing vector field having any tangent at P. The one-parameter subgroups associated with these Killing vectors can therefore map P to any point Q in some neighborhood U of P (see figure 5.8). By a succession of such maps we can clearly map P to any point in S whatever. It follows that the isometry group maps P to any point, and S is homogeneous.

Next we take a look at the isotropy group of P. Such transformations leave P fixed, so the associated Killing vector fields vanish at P. The Lie bracket of *any* two Killing fields $\bar{V}$ and $\bar{W}$ is

$$[\bar{V}, \bar{W}]^i = V^i{}_{,j}W^j - W^i{}_{,j}V^j,$$

Fig. 5.8. By choosing the appropriate one-parameter subgroup of the isometry group one can map P to any point Q or Q' in a neighborhood U.

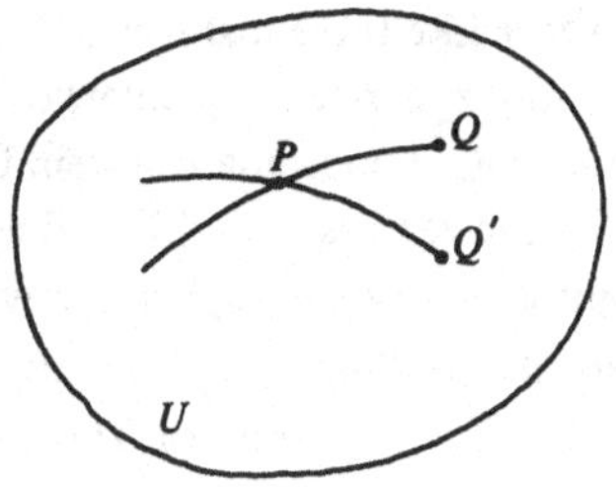

or

$$[\bar{V},\bar{W}]_i = V_{i,j}W^j - W_{i,j}V^j - g_{ik,j}(V^k W^j - W^k V^j). \tag{5.95}$$

If $\bar{V}$ and $\bar{W}$ both vanish at P, then so does $[\bar{V},\bar{W}]$. But $[\bar{V},\bar{W}]$ is a linear combination of Killing vector fields, so for it to vanish at P it must be a linear combination only of those fields which also vanish at P. So these fields form a Lie subalgebra, clearly the algebra of the isotropy group at P. The next exercise shows that the isotropy group is $SO(m)$ if S is space-like, i.e. that a maximally symmetric space-like manifold is isotropic.

Exercise 5.24

Choose at P the sort of coordinate system permitted by exercise 2.14, in which for a *space-like* manifold $g_{ij}(P) = \delta_{ij}$ and $g_{ij,k}(P) = 0$.

(a) Show that near P an isotropy Killing vector field is given by

$$V^i = A^i{}_j x^j + O(x^2), \tag{5.96}$$

where $A^i{}_j$ is an arbitrary antisymmetric matrix

$$A^i{}_j = -A^j{}_i. \tag{5.97}$$

(b) Let $\bar{W}$ be another isotropy Killing vector field,

$$W^i = B^i{}_j x^j + O(x^2),$$

and show that

$$[\bar{V},\bar{W}]^i = [A,B]^i{}_j x^j + O(x^2), \tag{5.98}$$

where $[A,B]^i{}_j$ denotes the elements of the matrix commutator of $A^i{}_j$ and $B^i{}_j$. This shows that the Lie algebra of the isotropy group is the same as the Lie algebra of $SO(m)$.

(c) Argue from this that the isotropy group of P is $SO(m)$.

(d) Show that if $\mathsf{g}|$ is *not* positive-definite (or negative-definite) then the isotropy group is not $SO(m)$. In particular show that the isotropy group of a point P in four-dimensional Minkowski space is the Lorentz group $L(4)$.

5.21 The metric of a spherically symmetric three-space

Now we restrict our attention to space-like three-manifolds. The isotropy group is $SO(3)$ and we say the manifold is spherically symmetric about any point. In this section we construct a convenient coordinate system for the rest of our calculation. We know that the Killing vectors of $SO(3)$ define spheres S^2 by their integral curves. Since every point is on one such sphere, they must foliate the manifold S. We will adopt spherical coordinates, with the usual θ and ϕ on each sphere and a third 'radial' coordinate labelling spheres. There is a particularly convenient choice for the radial coordinate. The metric of S induces a metric tensor on each sphere, which in turn defines a volume two-form and a total area (integral of the volume two-form). We *define* the radial coordinate r of a sphere by the equation

$$\text{area} = 4\pi r^2, \qquad r = (\text{area}/4\pi)^{1/2}. \tag{5.99}$$

This intrinsically defined coordinate need not be monotonically increasing everywhere, as figure 5.9 shows. But at least in some neighborhood of P it is guaranteed to be good by the local flatness theorem, exercise 2.14. (It is singular at $r = 0$, of course, but we know how to handle that.)

In addition to the radial coordinate we have to define θ and ϕ more precisely. We have placed θ and ϕ on each sphere but we have not said how the pole $\theta = 0$ of one sphere is related to that of another. That is, we are free to slide the coordinates of a sphere around as we move from one to another. We fix the pole in the following manner. At every point Q there is a vector $\bar{n}$ orthogonal to the sphere at that point ($\mathrm{g}(\bar{n}, \bar{V}) = 0$ for any $\bar{V}$ in $T_Q(S^2)$), normalized to unity ($\mathrm{g}(\bar{n}, \bar{n}) = 1$), and pointing away from P (which is well defined near P and extends to all of S by continuity). This vector field is called the unit normal vector field, and is C^∞ except at P. Choose the pole of any particular S^2 arbitrarily and then fix the poles of all the others by demanding they lie on the integral curve of $\bar{n}$ through the original pole. This is illustrated in figure 5.10. This clearly will imply that *any* integral curve of $\bar{n}$ is a curve of constant θ and ϕ, or in other words a coordinate line of the radial coordinate. Since $\partial/\partial\theta$ and $\partial/\partial\phi$ are tangent to the spheres this construction implies

$$g_{r\theta} = \mathrm{g}(\partial/\partial r, \partial/\partial\theta) = 0, \tag{5.100a}$$

Fig. 5.9. A radial coordinate labelling circles on a sphere, defined as the circumference ÷ 2π. This is the two-dimensional analogue of the situation described in the text. The radial coordinate increases away from P at first (say from A to B) but begins decreasing (from C to D) and becomes zero at P'.

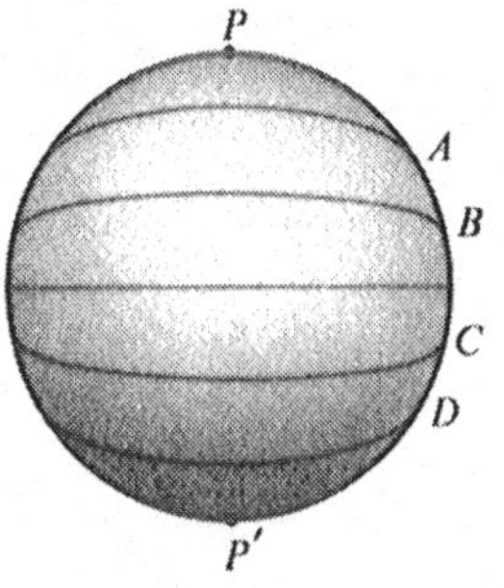

$$g_{r\phi} = \mathsf{g}(\partial/\partial r, \partial/\partial\phi) = 0. \tag{5.100b}$$

Moreover, on each sphere the metric is that of the unit sphere times r^2, the appropriate factor to make the area be $4\pi r^2$:

$$g_{\theta\theta} = r^2, \quad g_{\theta\phi} = 0, \quad g_{\phi\phi} = r^2 \sin^2\theta. \tag{5.100c}$$

We therefore have only one unknown metric component, g_{rr}.

Exercise 5.25

(a) Define the radial distance from P to a sphere with coordinate r to be the integral

$$\int_0^r (g_{rr})^{1/2}\, \mathrm{d}r \tag{5.101}$$

along a line $\theta = \text{const}$, $\phi = \text{const}$. Argue that g_{rr} must be independent of θ and ϕ.

(b) Show from exercise 2.14 that as one approaches P,

$$\lim_{r\to 0} g_{rr} = 1. \tag{5.102}$$

By exercise 5.25(a) we write $g_{rr} = f(r)$ and have the metric

$$(\mathsf{g}) = \begin{pmatrix} f(r) & 0 & 0 \\ 0 & r^2 & 0 \\ 0 & 0 & r^2\sin^2\theta \end{pmatrix}. \tag{5.103}$$

As we have used only the isotropy group of P to get this, we should not expect to be able to determine $f(r)$. For that we must use the rest of the isometries of S.

Fig. 5.10. Establishing the pole of each circle of constant r in figure 5.9 by requiring them all to lie on a single integral curve of the unit normal field $\vec{n}$.

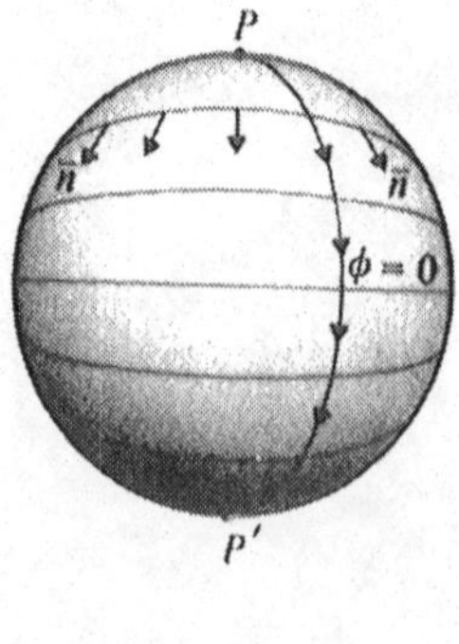

5.22 Construction of the six Killing vectors

There are a number of methods we could use to find the form of $f(r)$ that guarantees the homogeneity of S. The method we shall use is to construct all the Killing vector fields of S by using the vector spherical harmonics of §4.29.

Any vector field $\bar{V}$ on S can be written in the form

$$\bar{V} = \xi_{lm}(r)Y_{lm}\frac{\partial}{\partial r} + \eta_{lm}(r)\bar{Y}^{+}_{lm} + \zeta_{lm}(r)\bar{Y}^{-}_{lm}, \tag{5.104}$$

with an implied summation on l and m here and wherever they are repeated in the same term. We shall need the components of this equation. It is easy to deduce from equation (4.101) that

$$(\bar{Y}^{+}_{lm})^{\theta} = Y_{lm,\theta}; \quad (\bar{Y}^{+}_{lm})^{\phi} = \frac{1}{\sin^2\theta}Y_{lm,\phi}; \tag{5.105a}$$

$$(\bar{Y}^{-}_{lm})^{\theta} = \frac{1}{\sin\theta}Y_{lm,\phi}; \quad (\bar{Y}^{-}_{lm})^{\phi} = -\frac{1}{\sin\theta}Y_{lm,\theta}. \tag{5.105b}$$

It follows that

$$V^r = \xi_{lm}Y_{lm}, \tag{5.106a}$$

$$V^{\theta} = \eta_{lm}Y_{lm,\theta} + \zeta_{lm}Y_{lm,\phi}/\sin\theta, \tag{5.106b}$$

$$V^{\phi} = \eta_{lm}Y_{lm,\phi}/\sin^2\theta - \zeta_{lm}Y_{lm,\theta}/\sin\theta. \tag{5.106c}$$

These components have to satisfy Killing's equation

$$K_{ij} \equiv V^k g_{ij,k} + V^k{}_{,i}g_{kj} + V^k{}_{,j}g_{ik} = 0, \tag{5.107}$$

with g_{ij} from (5.103).

The three equations $\{K_{\theta\theta} = 0, K_{\theta\phi} = 0, K_{\phi\phi} = 0\}$ do not involve derivatives of ξ_{lm}, η_{lm}, or ζ_{lm}, so we shall tackle them first. First consider the combination (indices raised with (5.103))

$$0 = K^{\theta}{}_{\theta} + K^{\phi}{}_{\phi} = \frac{4}{r}\xi_{lm}Y_{lm} + 2\eta_{lm}L^2(Y_{lm}),$$

where L^2 is the operator defined by equation (3.33). Using (3.33) we get

$$[(2/r)\xi_{lm} - l(l+1)\eta_{lm}]Y_{lm} = 0.$$

By the linear independence of the spherical harmonics we have

$$\frac{2}{r}\xi_{lm} - l(l+1)\eta_{lm} = 0. \tag{5.108}$$

Next consider the combinations

$$0 = \tfrac{1}{2}(K^{\theta}{}_{\theta} - K^{\phi}{}_{\phi}) = F_{lm}\eta_{lm} + G_{lm}\zeta_{lm}, \tag{5.109a}$$

$$0 = -\frac{1}{r^2\sin\theta}K_{\theta\phi} = -G_{lm}\eta_{lm} + F_{lm}\zeta_{lm}, \tag{5.109b}$$

where F_{lm} and G_{lm} are abbreviations for the expressions

$$F_{lm} = Y_{lm,\theta\theta} - \cot\theta\, Y_{lm,\theta} - Y_{lm,\phi\phi}/\sin^2\theta,$$

$$G_{lm} = 2Y_{lm,\theta\phi}/\sin\theta - 2\cot\theta\, Y_{lm,\phi}/\sin\theta.$$

Equations (5.109) have the solution $\zeta_{lm} = \eta_{lm} = 0$ unless the determinant of their coefficients vanishes. But this is $(F_{lm})^2 + (G_{lm})^2$, so it vanishes only if both F_{lm} and G_{lm} vanish. It is easy to work out that this happens for $l = 0$ and $l = 1$ (any m) but not for $l \geqslant 2$. Moreover, it is obvious from (5.106) that $l = 0$ does not have a contribution from η or ζ (the fixed-point theorem for S^2 again!) so that we can conclude

$$\begin{aligned} &l = 1: \eta_{1m}, \zeta_{1m} \text{ arbitrary;} \\ &l \geqslant 2: \eta_{lm} = \zeta_{lm} = 0. \end{aligned} \tag{5.110}$$

Then (5.108) gives us

$$\begin{aligned} l &= 0: \ \xi_{00} = 0, \\ l &= 1: \ \xi_{1m} = r\eta_{1m}, \\ l &\geqslant 2: \ \xi_{lm} = 0. \end{aligned} \tag{5.111}$$

Now we turn to the other three equations in (5.107). The first is a scalar with respect to rotations:

$$0 = K_{rr} = (2f\xi_{lm,r} + f_{,r}\xi_{lm})Y_{lm},$$

which implies

$$f\xi_{lm,r} + \tfrac{1}{2}f_{,r}\xi_{lm} = 0. \tag{5.112}$$

The remaining two equations, $K_{r\theta} = K_{r\phi} = 0$, transform as a vector under rotations. The divergence of this vector (with respect to the volume of S^2) is

$$0 = (\sin\theta K_r{}^{\theta})_{,\theta} + (\sin\theta K_r{}^{\phi})_{,\phi} = \left(\eta_{lm,r} + \frac{1}{r^2}f\xi_{lm}\right)\sin\theta L^2(Y_{lm}),$$

which again implies (for $l > 0$)

$$\eta_{lm,r} + \frac{1}{r^2}f\xi_{lm} = 0. \tag{5.113}$$

The remaining equation can be taken to be the divergence of the dual of the vector in S^2,

$$0 = K_{r\theta,\phi} - K_{r\phi,\theta} = r^2\zeta_{lm,r}\sin\theta L^2(Y_{lm}),$$

which of course implies

$$\zeta_{lm,r} = 0. \tag{5.114}$$

We may conclude that $\{\zeta_{1m}, m = -1, 0, 1\}$ are three arbitrary constants, the only contribution from $\bar{Y}_{lm}$. The three equations (5.111) for the unknowns ξ_{1m}, η_{1m}, and f have the following solution in terms of the arbitrary constants K and V_m:

$$f = (1 - Kr^2)^{-1}, \tag{5.115}$$

$$\xi_{1m} = V_m(1 - Kr^2)^{1/2}, \tag{5.116}$$

$$\eta_{1m} = \frac{1}{r} V_m(1 - Kr^2)^{1/2}. \tag{5.117}$$

Exercise 5.26
Verify equations (5.105), (5.108), (5.109), (5.112), (5.113), (5.114), and (5.115–17).

Exercise 5.27
Show that the Killing vectors with $V_m = 0$ are those corresponding to the isotropy group of the origin $r = 0$.

Exercise 5.28
Show that the apparent singularity in η_{1m} as $r \to 0$ is a coordinate effect: the vector field is well-behaved at the origin.

Exercise 5.29
Set $K = 0$ in (5.115–17) and show that S is just E^3, Euclidean space. Find the constants V_m that define the Killing vectors $\{\partial/\partial x, \partial/\partial y, \partial/\partial z\}$, where the Cartesian coordinates are obtained from our polars in the usual way.

5.23 Open, closed, and flat universes

We now have a complete description of the geometry of the homogeneous and isotropic spaces of the cosmological model: they have the metric tensor

$$(g_{ij}) = \begin{pmatrix} (1 - Kr^2)^{-1} & 0 & 0 \\ 0 & r^2 & 0 \\ 0 & 0 & r^2 \sin^2\theta \end{pmatrix}. \tag{5.118}$$

It only remains to try to get a picture of this geometry. The following coordinate transformations are a help.

Exercise 5.30
Find a coordinate transformation from r to χ which produces the following metric components

for $K > 0$:

$$(g_{ij}) = \frac{1}{K}\begin{pmatrix} 1 & 0 & 0 \\ 0 & \sin^2\chi & 0 \\ 0 & 0 & \sin^2\chi \sin^2\theta \end{pmatrix} \tag{5.119a}$$

for $K<0$:

$$(g_{ij}) = \frac{1}{|K|}\begin{pmatrix} 1 & 0 & 0 \\ 0 & \sinh^2\chi & 0 \\ 0 & 0 & \sinh^2\chi\sin^2\theta \end{pmatrix}. \tag{5.119b}$$

This shows that the geometry really depends only on the sign of K. Its magnitude serves only as an overall scale factor.

In the case $K>0$, the sphere of radial coordinate χ has area $4\pi\sin^2\chi/K$, which increases away from $\chi=0$ to a maximum at $\chi=\pi/2$ and then decreases to zero at $\chi=\pi$. This is reminiscent of S^2 (figure 5.9). In fact, this is the metric of the sphere S^3 of radius $K^{-1/2}$. Because the space is finite, the universe is said to be *closed*.

Exercise 5.31

Find a coordinate transformation of E^4 from Cartesian coordinates $\{x^i\} = \{w, x, y, z\}$ to spherical coordinates $\{x^{i'}\} = \{r, \chi, \theta, \phi\}$ in which the metric $g_{ij} = \delta_{ij}$ has the components $g_{i'j'}$ given by (5.119a) when restricted to the sphere S^3, $w^2+x^2+y^2+z^2=K^{-1}$.

The case $K=0$ has been considered in exercise 5.29. It is the *flat* universe.

The case $K<0$ is the *open* universe, and it is the hardest to visualize. The surface area of a sphere of radial coordinate χ is $4\pi\sinh^2\chi/|K|$, and increases ever more rapidly with χ. This universe is unbounded.

Exercise 5.32

(a) By considering the relation between the areas of spheres $\chi=$ const and the distance of the sphere from the origin $\chi=0$, equation (5.101), prove that the metric (5.119b) is not the restriction of the Euclidean metric to *any* submanifold of any E^n.

(b) Find a submanifold of Minkowski space whose metric is that of (5.119b).

When Einstein's equations are supplied with initial data which are homogeneous and isotropic (and this includes not only the geometry but the matter variables as well), then the subsequent evolution of the universe maintains the symmetry. It follows that the only aspect of the geometry which can change with time is the scale factor K: the universe gets 'larger' or 'smaller' as time goes on. One must be careful, however, not to make coordinate-dependent statements. For the closed universe, whose total volume is finite, the change in K does cause a change in the total volume. But the flat and open universes are both infinite, so it is not meaningful to talk about their total volume. What general relativity tells us is that the coordinates of equation (5.119) are 'comoving': the local mean rest frame of the galaxies in any small region of the universe stays at constant $\{\chi, \theta, \phi\}$ as time evolves. It follows then that a change in K produces a change in the distance between galaxies, and this is what is meant by an expanding universe. In the 'standard model' of the universe, which assumes homogeneity and isotropy and a few other things, all three kinds of universe begin with zero 'volume' ($K = \infty$) and expand away from this 'big bang'. The closed universe expands to a maximum and recollapses, the flat universe expands at a rate which goes asymptotically to zero, and the open universe expands at a rate which goes asymptotically to a nonzero limit. All of these things are consequences of Einstein's equations. To understand these equations it is necessary to add one more level of structure to our manifolds: the affine connection. This is the subject of chapter 6.

5.24 Bibliography

A concise and well-written introduction to thermodynamics is E. Fermi, *Thermodynamics* (Dover, New York 1956). Caratheodory's theorem is discussed by S. Chandrasekhar, *An Introduction to the Study of Stellar Structure* (Dover, New York, 1958), and at a more advanced level in R. Hermann, *Differential Geometry and the Calculus of Variations* (Academic Press, New York, 1968).

Our discussion of Hamiltonian mechanics follows the spirit of R. Abraham & J. E. Marsden, *Foundations of Mechanics*, 2nd edn. (Benjamin/Cummings, Reading, Mass., 1978). See also W. Thirring, *A Course in Mathematical Physics I: Classical Dynamical Systems* (Springer Verlag, Vienna, 1978) and V. I. Arnold, *Mathematical Methods of Classical Mechanics* (Springer Verlag, Berlin, 1978). An introduction to the same ideas without the geometrical point of view may be found in H. Goldstein, *Classical Mechanics* (Addison-Wesley, Reading, Mass., 1950) or L. D. Landau & E. M. Lifshitz, *Mechanics* (Pergamon Press, London, 1959). The use of the canonical conserved quantities makes certain fluid instabilities easier to understand; see J. L. Friedman & B. F. Schutz, *Astrophys. J.* **221**, 937–57 (1978), and **222**, 281–96 (1978). A number of useful articles on Hamiltonian mechanics from a geometrical viewpoint may be found in *Topics in Nonlinear Dynamics – A Tribute to Sir Edward Bullard*, ed. S. Jorna (American Institute of Physics, 1978: A.I.P. Conference Proceeding no. 46). An interesting use of differential forms is the proof of the necessary and sufficient conditions that a set of dynamical equations possess a Hamiltonian (i.e. symplectic) structure, in R. M. Santilli, *Foundations of Theoretical*

Mechanics I – The Inverse Problem in Newtonian Mechanics (Springer, Berlin, 1978).

For an introduction to electromagnetic theory which includes a discussion of its relativistic version see J. D. Jackson, *Classical Electrodynamics* (Wiley, New York, 1976). A discussion which extends our own is C. W. Misner, K. S. Thorne & J. A. Wheeler, *Gravitation* (Freeman, San Francisco, 1973). Another is W. Thirring, *A Course in Mathematical Physics II: Classical Field Theory* (Springer Verlag, Vienna, 1979). An advanced discussion of relativistic wave equations is F. G. Friedlander, *The Wave Equation on a Curved Space–Time* (Cambridge University Press, 1976). Wheeler's 'charge without charge' is discussed in articles reprinted in J. A. Wheeler, *Geometrodynamics* (Academic Press, New York, 1962). Sorkin's nonorientable charge model is described in R. Sorkin, *J. Phys. A.* **12**, 403–21 (1979).

Introduction to fluid dynamics which include discussions of the Helmholtz theorem are *Fluid Mechanics* by L. D. Landau & E. M. Lifshitz (Pergamon Press, London, 1959) and *Hydrodynamics* by H. Lamb (Dover, New York, 1975). What we have called Ertel's theorem is really a special case of a more general result derived by H. Ertel, *Meteorologische Zeitschrift* **59**, 277 (1942). We touched on the properties of the manifold called 'Galilean spacetime', which is the arena for pre-relativistic physics. Discussions of its structure involve the concept of an affine connection, developed in the next chapter. See C. W. Misner, K. S. Thorne & J. A. Wheeler, *Gravitation* (Freeman, San Francisco, 1973) for a lucid discussion, or R. Hermann, *Topics in General Relativity* (Math-Sci Press, Brookline, Mass., 1973) for a more technical treatment. The use of Lie derivatives in continuum mechanics is very fruitful in elasticity theory. Indeed, it is very difficult to formulate the general-relativistic theory of elasticity without such techniques. See the treatment by B. Carter & H. Quintana, *Proc. Roy. Soc. London*, **A331**, 57 (1972), or B. Carter, *Proc. Roy. Soc. London, A*, to be published.

Cosmology is treated in most textbooks on general relativity. Our approach draws elements from both S. Weinberg, *Gravitation and Cosmology* (Wiley, New York, 1972), and Misner *et al.*, op. cit. For an easier introduction see M. Berry, *Principles of Cosmology and Gravitation* (Cambridge University Press, 1977). The astrophysical and observational sides of cosmology are dealt with in Weinberg and in P. J. E. Peebles, *Physical Cosmology* (Princeton University Press, 1971). Homogeneous but not necessarily isotropic cosmologies are developed using the techniques of group theory in M. P. Ryan & L. C. Shepley, *Homogeneous Relativistic Cosmologies* (Princeton University Press, 1975).

6 CONNECTIONS FOR RIEMANNIAN MANIFOLDS AND GAUGE THEORIES

6.1 *Introduction*

The subject of this chapter is outside the main theme of this book, which is the study of the differential structure of the manifold. The affine connection is an additional piece of structure which gives shape and curvature to a manifold; it does not arise naturally from the differential structure, nor is it even a tensor. For this reason, the chapter is marked as supplementary. Nevertheless, no treatment of differential geometry for physicists would be complete without this important and very topical subject. Connections are finding increasing popularity in physics, particularly in gauge theories in elementary particle physics. We shall mainly discuss *affine* connections (Riemannian manifolds), reserving an introductory section on gauge connections for the end.

In earlier chapters we have occasionally added extra structure to a manifold, in that we have singled out a particular tensor field as special, either to serve as a volume-element or as a metric. Volume-elements are not far removed from the differential structure of the manifold. The metric, on the other hand, creates even more structure than the affine connection, as we shall see below. But we have been able to avoid all that in our applications, only using the metric in its role as a mapping between $\binom{N}{M}$ tensors and $\binom{N-1}{M+1}$ tensors. The affine connection cannot be fitted into the structures we have already developed. From the point of view of the differential structure, it is a radical new addition to the manifold, and it has correspondingly rich possibilities for physical application.

6.2 *Parallelism on curved surfaces*

We have repeatedly emphasized that on a differentiable manifold there is no intrinsic notion of parallelism between vectors defined at different points. The affine connection is a *rule* whereby some notion of parallelism can be defined. To anticipate what kind of a rule may be possible, let us consider the notion of parallelism on an ordinary curved two-surface, the sphere. In figure 6.1, the vector $\bar{V}$ is the tangent to the great circle ABC at the north pole, point A. Suppose we carry, or *transport*, $\bar{V}$ along ABC to the south pole, C. In order to be defineable in two-dimensional terms it must be kept tangent to the sphere, so if we do not rotate it as we carry it, it will simply remain tangent to the curve *ABC*. It winds up as $\bar{V}'$ at *C*, pointing in what, to us three-dimensional beings, looks like the direction antiparallel to $\bar{V}$. Should we assume that, at least with respect to the sphere's geometry, $\bar{V}$ and $\bar{V}'$ are parallel? Before jumping to a conclusion, suppose we transport $\bar{V}$ from *A* to *C* on the path *ADC* shown in figure 6.2, where *ADC* is another great circle intersecting *ABC* at right angles at both poles. Since $\bar{V}$ starts out perpendicular to *ADC*, the natural way to move it without twisting is to keep it perpendicular to *ADC* and tangent to the sphere. This produces the vector $\bar{V}''$ at *C*, which, to us, is in fact parallel to $\bar{V}$. But $\bar{V}''$ and $\bar{V}'$, both vectors at *C*, are antiparallel! Which is parallel to $\bar{V}$? Clearly, if we simply consider the intrinsic properties of the sphere, neither vector deserves to be called parallel to $\bar{V}$. There is *no* global notion of parallelism. All one can do – and this is what we have done – is to define a notion of parallel *transport*, of moving the vector along a curve without changing its direction. *The affine connection is a rule for parallel transport.*

Fig. 6.1. Parallel transport of a vector $\bar{V}$ along a great circle of the sphere.

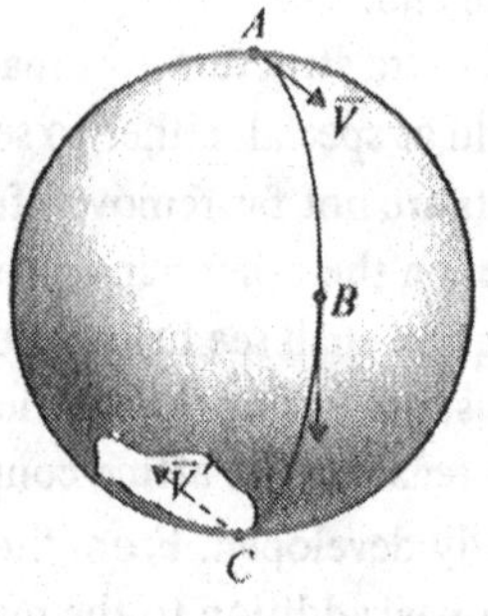

Fig. 6.2. An alternative path for parallel transport, with a different result.

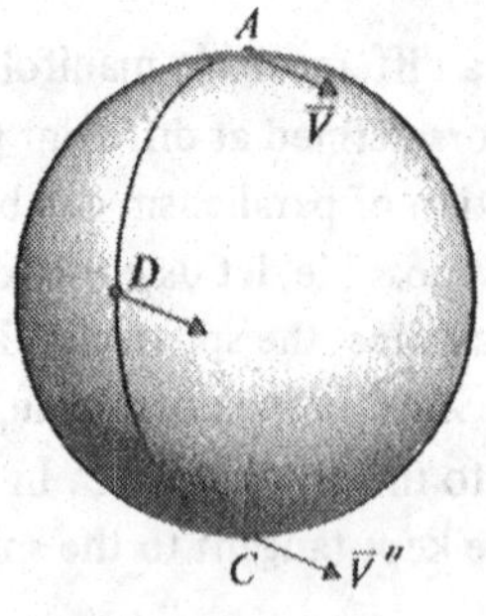

6.3 *The covariant derivative*

We shall for the moment view the affine connection in an abstract sense; it will become more concrete when we introduce components in the next section. For now, suppose we have a curve $\mathscr{C}$ and a connection, a rule for parallel transport. Let the tangent to $\mathscr{C}$ be $\bar{U} = \mathrm{d}/\mathrm{d}\lambda$. At the point P, pick an arbitrary vector $\bar{V}$ from T_P. Then the connection allows us to define a vector field $\bar{V}$ along the curve $\mathscr{C}$, which is obtained by parallel-transporting $\bar{V}$ (see figure 6.3). Since we can now say that $\bar{V}$ does not change along $\mathscr{C}$, we can define a derivative with respect to which $\bar{V}$ has zero rate of change. This is called the *covariant derivative along* $\bar{U}$, $\nabla_{\bar{U}}$, and we write

♦ $$\nabla_{\bar{U}}\bar{V} = 0 \Leftrightarrow \bar{V} \text{ is parallel-transported along } \mathscr{C}. \tag{6.1}$$

If $\bar{W}$ is a vector field defined everywhere on $\mathscr{C}$, we can define its covariant derivative along $\mathscr{C}$ in much the same way as we did for Lie derivatives (see figure 6.4). To define $\nabla_{\bar{U}}\bar{W}$ at P, it will be convenient to express all vectors as functions of λ. If P has parameter value λ_0, then we define the field $\bar{W}^*_{\lambda_0+\epsilon}(\lambda)$ to be that parallel-transported field ($\nabla_{\bar{U}}\bar{W}^* = 0$) which equals $\bar{W}$ at $\lambda_0 + \epsilon$. The vector $\bar{W}^*_{\lambda_0+\epsilon}(\lambda_0)$ is the vector $\bar{W}(\lambda_0 + \epsilon)$ parallel-transported back to λ_0. Then the derivative may be evaluated entirely in the vector space T_P:

$$(\nabla_{\bar{U}}\bar{W})_P = \lim_{\epsilon \to 0} \frac{\bar{W}_{\lambda_0+\epsilon}(\lambda_0) - \bar{W}(\lambda_0)}{\epsilon}. \tag{6.2}$$

Although this procedure resembles the one we used for defining the Lie derivative, it is important to understand the significant difference: 'dragging back' a

Fig. 6.3. The affine connection permits us to *define* $V(Q)$ for any point Q on $\mathscr{C}$ by parallel transport from P.

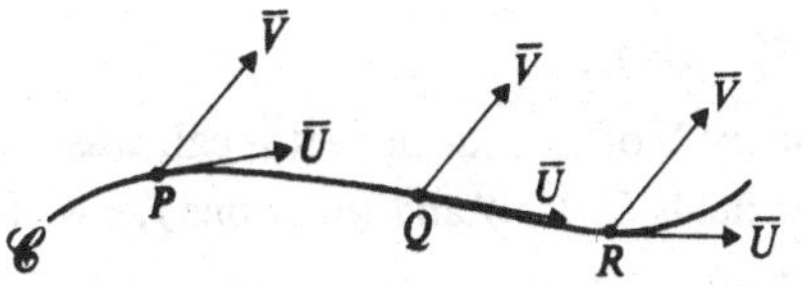

Fig. 6.4. The vector field $\bar{W}$ on $\mathscr{C}$ is not parallel transported. Comparison with one which is permits a definition of the covariant derivative of $\bar{W}$.

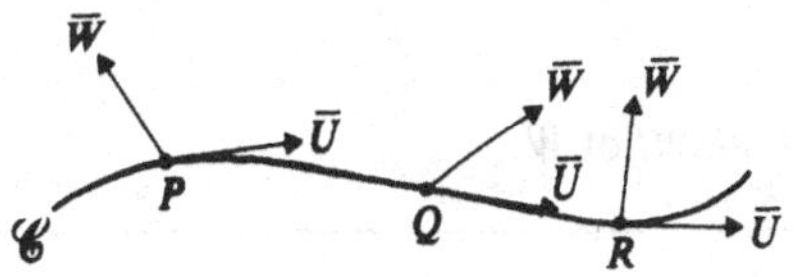

vector for the Lie derivative required the entire congruence, so that $\bar{U}$ and $\bar{W}$ had to be defined in a neighborhood of the curve $\mathscr{C}$; parallel-transport, by contrast, requires only the curve $\mathscr{C}$, the fields $\bar{U}$ and $\bar{W}$ on the curve, and of course the connection on the curve.

It is clear from (6.2) that $\nabla_{\bar{U}}$ is a differential operator:

$$\begin{aligned}\nabla_{\bar{U}}(f\bar{W}) &= f\nabla_{\bar{U}}\bar{W} + \bar{W}\nabla_{\bar{U}}f \\ &= f\nabla_{\bar{U}}\bar{W} + \bar{W}\frac{\mathrm{d}f}{\mathrm{d}\lambda},\end{aligned} \tag{6.3a}$$

where the last step is the obvious extension to scalars. The covariant derivative can also be extended to tensors of arbitrary type by the Leibniz rules

♦ $$\nabla_{\bar{U}}(\bar{A}\otimes\bar{B}) = (\nabla_{\bar{U}}\bar{A})\otimes\bar{B} + \bar{A}\otimes(\nabla_{\bar{U}}\bar{B}), \tag{6.3b}$$

♦ $$\nabla_{\bar{U}}\langle\tilde{\omega},\bar{A}\rangle = \langle\nabla_{\bar{U}}\tilde{\omega},\bar{A}\rangle + \langle\tilde{\omega},\nabla_{\bar{U}}\bar{A}\rangle. \tag{6.3c}$$

Equations (6.3) guarantee *compatibility* of the connection with the differential structure.

Suppose that we were to change the parameter along our curve from λ to μ. Then the new tangent would be $g\bar{U}$, where $g = \mathrm{d}\lambda/\mathrm{d}\mu$. From (6.2) it is clear that the covariant derivative would also be multiplied by g, since ϵ would be replaced by $\delta\mu = \epsilon\,\mathrm{d}\mu/\mathrm{d}\lambda$ while $\bar{W}^*_{\mu_0+\delta\mu}(\mu_0)$ is the same as $\bar{W}^*_{\lambda_0+\epsilon}(\lambda_0)$. (This is, strictly speaking, part of the definition of what we mean by a connection: the notion of parallel-transport along a curve must be independent of the parameter on the curve.) Therefore we conclude that for any function g

$$\nabla_{g\bar{U}}\bar{W} = g\nabla_{\bar{U}}\bar{W}. \tag{6.4a}$$

We must also put another restriction on the affine connection, which is that at a point the covariant derivatives in different directions should have the additive property

$$(\nabla_{\bar{U}}\bar{W})_P + (\nabla_{\bar{V}}\bar{W})_P = (\nabla_{\bar{U}+\bar{V}}\bar{W})_P. \tag{6.4b}$$

This makes ∇ behave like the ordinary ∇ of Euclidean vector calculus. Together (6.4a, b) imply that for any vector fields $\bar{U}$, $\bar{V}$, $\bar{W}$ and functions f, g we have

♦ $$\nabla_{f\bar{U}+g\bar{V}}\bar{W} = f\nabla_{\bar{U}}\bar{W} + g\nabla_{\bar{V}}\bar{W}. \tag{6.4c}$$

Exercise 6.1

Show that (6.4c) and the fact that $\nabla_{\bar{U}}\bar{W}$ is a vector imply that $\nabla\bar{W}$ is a $\binom{1}{1}$ tensor field whose value on arguments $\bar{U}$ and $\tilde{\omega}$ is

$$\nabla\bar{W}(\tilde{\omega};\bar{U}) = \langle\tilde{\omega},\nabla_{\bar{U}}\bar{W}\rangle. \tag{6.5}$$

This tensor is called the ***gradient*** of W.

The fact that $\nabla\bar{W}$ is a tensor field means that we have been able to remove the curve entirely from the definition of the covariant derivative. The tensor $\nabla\bar{W}$ is defined only by $\bar{W}$ and the connection. One might be tempted to go further and say that ∇ is itself a $\binom{0}{1}$ tensor field which is just the connection; but this would be wrong. While ∇ may symbolize the connection, it is *not* a tensor field, since $\nabla(f\bar{W}) \neq f\nabla\bar{W}$ (cf. (6.3a)). For this reason, the connection cannot be regarded as a tensor field.

6.4 *Components: covariant derivatives of the basis*

Since any tensor can be expressed as a linear combination of basis tensors, and these basis tensors are all derivable from the vector basis $\{\bar{e}_i\}$, the connection can be completely described by giving the gradients of the basis vectors. So we define

♦ $$\nabla_{\bar{e}_i}\bar{e}_j = \Gamma^k{}_{ji}\bar{e}_k. \tag{6.6}$$

The functions $\Gamma^k{}_{ji}$ are called *Christoffel symbols*. For fixed (i, j) $\Gamma^k{}_{ji}$ is the kth component of the vector field $\nabla_{\bar{e}_i}\bar{e}_j$. Note carefully the order of the indices on Γ: the one associated with the derivative goes last. We shall often use the short-hand

$$\nabla_{\bar{e}_i} \equiv \nabla_i. \tag{6.7}$$

In an n-dimensional manifold, the n^3 functions $\Gamma^k{}_{ji}$ completely determine the affine connection, and this is often the most convenient way of describing the connection. Notice that $\Gamma^k{}_{ji}$ is not a component of a tensor; under a basis transformation the indices k and i transform like tensor indices (by (6.5)) but the index j does not (by (6.3a)).

Exercise 6.2

Show that

$$\Gamma^{k'}{}_{j'i'} = \Lambda^{k'}{}_k\Lambda^i{}_{i'}\Lambda^i{}_{j'}\Gamma^k{}_{ji} + \Lambda^{k'}{}_k\Lambda^i{}_{i'}(\nabla_i\Lambda^k{}_{j'}),$$

where by $\nabla_i\Lambda^k{}_{j'}$ we mean $\mathrm{d}\Lambda^k{}_{j'}/\mathrm{d}\lambda_i$, in which $\bar{e}_i = \mathrm{d}/\mathrm{d}\lambda_i$ and $\Lambda^k{}_{j'}$ is treated as a function on the integral curves of $\bar{e}_i$.

Exercise 6.3

Show, by exercise 6.1, that

$$\{\Gamma^k{}_{ji}\bar{e}_k \otimes \tilde{\omega}^i\}$$

is a collection of n $\binom{1}{1}$ tensors. (Here $\{\tilde{\omega}^i\}$ is the one-form basis dual to $\{\bar{e}_j\}$.)

Exercise 6.4

On the unit sphere the usual spherical coordinates θ and ϕ define the basis $\{\bar{e}_\theta = \partial/\partial\theta, \bar{e}_\phi = \partial/\partial\phi\}$. Extend the reasoning used in §6.2 to deduce that

$$\Gamma^\theta{}_{\phi\phi} = -\sin\theta\cos\theta, \Gamma^\phi{}_{\theta\phi} = \Gamma^\phi{}_{\phi\theta} = \cot\theta,$$

and all other Γs vanish. (N.b. this is a difficult problem. You should make maximum use of the symmetry of the sphere, and make intelligent guesses about how vectors behave under parallel transport.)

Exercise 6.5

From (6.6) and (6.3c) deduce that

$$\blacklozenge \quad \nabla_i\tilde{\omega}^j = -\Gamma^j{}_{ki}\tilde{\omega}^k. \tag{6.8}$$

Now that we have the derivatives of basis vectors, we can find the derivatives of arbitrary tensors. For example, if $U = \mathrm{d}/\mathrm{d}\lambda$ then

$$\begin{aligned}\nabla_{\bar{U}}\bar{V} &= U^i\nabla_{\bar{e}_i}(V^j\bar{e}_j)\\ &= U^i(\nabla_{\bar{e}_i}V^j)\bar{e}_j + U^iV^j\nabla_{\bar{e}_i}\bar{e}_j.\end{aligned}$$

In the first term, V^j is simply a function, so $U^i\nabla_i(V^j) = \mathrm{d}V^j/\mathrm{d}\lambda$. Therefore we have

$$\begin{aligned}\nabla_{\bar{U}}\bar{V} &= \frac{\mathrm{d}V^j}{\mathrm{d}\lambda}\bar{e}_j + U^iV^j\Gamma^k{}_{ji}\bar{e}_k\\ &= \left(\frac{\mathrm{d}V^j}{\mathrm{d}\lambda} + \Gamma^j{}_{ki}V^kU^i\right)\bar{e}_j.\end{aligned}$$

To get the final expression we had to redefine some summation indices in the final term. Since $\nabla\bar{V}$ is itself a tensor, it has components

$$(\nabla\bar{V})^j{}_i = \nabla_i(V^j) + \Gamma^j{}_{ki}V^k.$$

A word about the term $\nabla_i(V^j)$. If $\bar{e}_i$ is the vector $\mathrm{d}/\mathrm{d}\mu$, then $\nabla_i(V^j) = \mathrm{d}V^j/\mathrm{d}\mu$, with V^j simply a function along the curve whose parameter is μ. If $\bar{e}_i$ is a coordinate basis vector then $\bar{e}_i = \partial/\partial x^i$ and we have

$$\nabla_iV^j = \frac{\partial}{\partial x^i}V^j = V^j{}_{,i},$$

using the comma notation introduced for differential forms. It is customary even where $\bar{e}_i$ is not a coordinate basis vector to use the comma notation:

$$\nabla_{\bar{e}_i}f = \bar{e}_i[f] \equiv f_{,i} \tag{6.9}$$

on any function f. When $\bar{e}_i$ is a coordinate basis vector, this is the usual partial derivative; when $\bar{e}_i$ is not, then this is simply the derivative of f along $\bar{e}_i$. We can thus write

$$\blacklozenge \quad (\nabla\bar{V})^j{}_i = V^j{}_{,i} + \Gamma^j{}_{ki}V^k \equiv V^j{}_{;i}. \tag{6.10}$$

We have here introduced the semicolon notation for the covariant derivative. Whereas neither $V^j{}_{,i}$ nor $\Gamma^j{}_{ki}V^k$ transforms like a tensor, their sum clearly does.

Exercise 6.6

Show that if $\tilde{\omega}$ is a one-form

$$(\nabla\tilde{\omega})_{ij} \equiv \omega_{i;j} = \omega_{i,j} - \Gamma^k{}_{ij}\omega_k. \tag{6.11}$$

Exercise 6.7

Show that if $\mathbf{T}$ is a $\binom{N}{M}$ tensor,

$$\begin{aligned} T^{i\ldots j}{}_{k\ldots l;m} = {} & T^{i\ldots j}{}_{k\ldots l,m} + \Gamma^i{}_{nm}T^{n\ldots j}{}_{k\ldots l} + \ldots \\ & + \Gamma^j{}_{nm}T^{i\ldots n}{}_{k\ldots l} - \Gamma^n{}_{km}T^{i\ldots j}{}_{n\ldots l} - \ldots \\ & - \Gamma^n{}_{lm}T^{i\ldots j}{}_{k\ldots n}. \end{aligned} \tag{6.12}$$

6.5 Torsion

The two quantities $[\bar{U}, \bar{V}]$ and $\nabla_{\bar{U}}\bar{V} - \nabla_{\bar{V}}\bar{U}$ are both vector fields and are both antisymmetric in $\bar{U}$ and $\bar{V}$. A connection is said to be *symmetric* if they are equal:

♦ $$\nabla_{\bar{U}}\bar{V} - \nabla_{\bar{V}}\bar{U} = [\bar{U}, \bar{V}] \Leftrightarrow \text{symmetric connection.} \tag{6.13}$$

The name 'symmetric' is used because of the property proved in the following exercise.

Exercise 6.8

Show that in a *coordinate* basis, (6.13) implies that a connection is symmetric if and only if

♦ $$\Gamma^k{}_{ij} = \Gamma^k{}_{ji}. \tag{6.14}$$

For a nonsymmetric connection we define the *torsion* $T^k{}_{ji}$:

♦ $$\nabla_{\bar{e}_i}\bar{e}_j - \nabla_{\bar{e}_j}\bar{e}_i - [\bar{e}_i, \bar{e}_j] \equiv T^k{}_{ji}\bar{e}_k. \tag{6.15}$$

Exercise 6.9

Show that $\{T^k{}_{ji}\}$ are the components of a $\binom{1}{2}$ *tensor*, which we call the torsion tensor $\mathbf{T}$:

$$\nabla_{\bar{U}}\bar{V} - \nabla_{\bar{V}}\bar{U} - [\bar{U}, \bar{V}] = \mathbf{T}(\ ;\bar{U}, \bar{V}).$$

The empty slot in $\mathbf{T}$ is for a one-form argument.

Exercise 6.10

Suppose a manifold has two connections defined on it, with Christoffel symbols $\Gamma^k{}_{ij}$ and $\Gamma'^k{}_{ij}$. Show that

$$D^k{}_{ij} \equiv \Gamma^k{}_{ij} - \Gamma'^k{}_{ij}$$

are the components of a $\binom{1}{2}$ tensor. Show that the tensor **D** is symmetric in its vector arguments if and only if both connections have the same torsion tensor.

Exercises 6.9 and 6.10 show that we can always define the symmetric part $\nabla_{(S)}$ of any connection ∇ by defining the Christoffel symbols

$$\Gamma^k{}_{(S)ij} = \Gamma^k{}_{ij} - \tfrac{1}{2}T^k{}_{ij}.$$

While torsion is in principle a useful part of the connection, it has not had as much popularity as the symmetric part in constructing mathematical models for physical laws. From now on we will deal with symmetric connections unless otherwise specified. One reason for this will be apparent in exercise 6.18 below. Notice that the definition (6.13) immediately guarantees the following.

Exercise 6.11

A manifold has a symmetric connection. Show that in any expression for the components of the Lie derivative of a tensor, all commas can be replaced by semicolons. An example:

$$\begin{aligned}(\pounds_{\bar{U}}\tilde{\omega})_i &= \omega_{i,j}U^j + \omega_j U^j{}_{,i}\\ &= \omega_{i;j}U^j + \omega_j U^j{}_{;i}.\end{aligned}$$

(Naturally, *all* commas must be changed, not just some.)

6.6 Geodesics

A geodesic curve is a curve that parallel-transports its own tangent vector. The geodesic equation is

♦ $$\nabla_{\bar{U}}\bar{U} = 0. \qquad (6.16a)$$

If λ is the parameter of the curve and $\{x^i\}$ is any coordinate system, this becomes

$$\frac{dU^i}{d\lambda} + \Gamma^i{}_{jk}U^jU^k = 0, \qquad (6.16b)$$

or

♦ $$\frac{d^2x^i}{d\lambda^2} + \Gamma^i{}_{jk}\frac{dx^j}{d\lambda}\frac{dx^k}{d\lambda} = 0. \qquad (6.16c)$$

The last equation is a quasi-linear system of differential equations for $x^i(\lambda)$, the equation of the curve.

Exercise 6.12

Recall that our definition of a curve includes its parameter. If λ is a parameter for which (6.16c) is true, show that a change of parameter to

$$\mu = a\lambda + b, \tag{6.17}$$

where a and b are constants, also gives a solution to (6.16c). The parameter of a geodesic curve is called an *affine* parameter.

Notice that only the symmetric part of a connection contributes to the geodesic equation. This provides a way of displaying the geometrical effect of torsion. Take a geodesic through a point P with tangent vector $\bar{U}$. In T_P choose a linear subspace R_P of dimension $n-1$ (the manifold's dimension being n) which is linearly independent of $\bar{U}$. Pick a vector $\bar{\xi}$ in R_P and construct a geodesic through P tangent to $\bar{\xi}$. Using the symmetric part of the connection, parallel-transport $\bar{U}$ along $\bar{\xi}$ a small affine parameter distance ϵ. Construct a new geodesic through this new point tangent to $\bar{U}$ there (see figure 6.5). This geodesic will be roughly parallel to the first one. In this manner, any point in the neighborhood of P can be given a geodesic 'parallel' to $\bar{U}$. Along this congruence of geodesics we can transport the original 'linking' vector $\bar{\xi}$ in two ways, either by parallel-transport or by Lie dragging. Let $\bar{\xi}$ be parallel-transported. Then by (6.15) we have

$$(\pounds_{\bar{U}}\bar{\xi})^i = -(\nabla_{\bar{\xi}}\bar{U})^i - T^i{}_{jk}\xi^j U^k.$$

The initial vector $\bar{\xi}$, however, had the property that $\nabla_{(S)\bar{\xi}}\bar{U} = 0$, so that $(\nabla_{\bar{\xi}}\bar{U})^i = \frac{1}{2}T^i{}_{jk}U^j\xi^k$ initially. Therefore we have the initial value

$$(\pounds_{\bar{U}}\bar{\xi})^i = -\tfrac{1}{2}T^i{}_{jk}\xi^j U^k.$$

Fig. 6.5. Two parallel geodesics $\bar{U}$ and $\bar{U}'$ and the vector $\bar{\xi}$ which connects them in the plane R_P and is parallel-transported along $\bar{U}$. If there is torsion it rotates away from $\bar{U}'$.

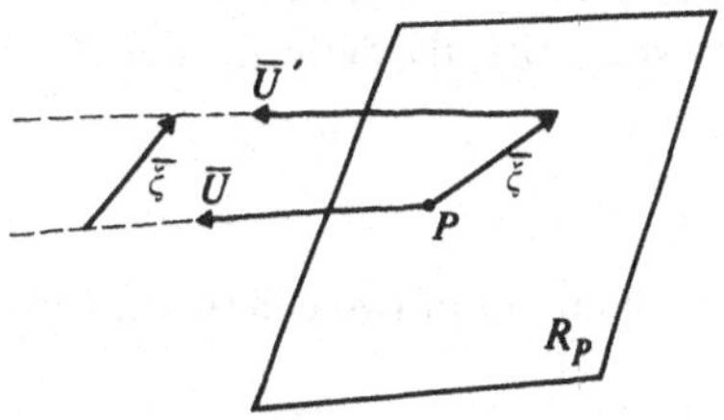

What this means is that a vector ξ parallel-transported by a *symmetric* connection would stay 'attached' to the parallel congruence of geodesics we have constructed. But if the connection is not symmetric, the vector does not remain fixed in this congruence. Speaking loosely, it is 'rotated' relative to nearby geodesics by the action of torsion. Conversely, if we regard the parallel-transported vector ξ as defining a 'fixed' direction as it moves along, then the congruence of 'parallel' geodesics twists around the one carrying ξ. (One cannot, however, define precisely the notions of 'rotation' and 'twist' without a metric.)

6.7 *Normal coordinates*

It will be helpful below to use a coordinate system based on geodesics. To construct this, we note that the geodesic curves through a point P give a 1–1 mapping of a neighborhood of P onto a neighborhood of the origin of T_P. This map arises because each element of T_P defines a unique geodesic curve through P, so we can associate the vector in T_P with the point an affine parameter distance $\Delta\lambda = 1$ along the curve from P. (Recall that if two elements of T_P are parallel, their geodesic curves have the same path but different parameters, and so the map picks out different points along the path.) Using this map and choosing an arbitrary basis for T_P, one defines the normal coordinates of a point Q to be the components of the vector in T_P it is associated with. This map will generally be 1–1 only in some neighborhood of P, since geodesics may cross on a curved manifold. For some connections, such as that of flat space, the map is 1–1 over the entire manifold. (The map *from* T_P *to* the manifold is well-defined even if geodesics cross. It is called the *exponential map*. If it is defined for all elements of T_P at all points P then the manifold is said to be geodesically complete.) For our purposes the principal interest in the normal coordinates is that $\Gamma^i{}_{jk} = 0$ at P (but not elsewhere in the neighborhood of P). To see this, note that if a vector $\bar{U}$ with components $U^i(P)$ defines a geodesic curve, then the coordinates of the point with affine parameter λ along that curve are simply $x^i = \lambda U^i(P)$, with the convention that $\lambda = 0$ at P. Therefore $\mathrm{d}^2x^i/\mathrm{d}\lambda^2$ vanishes, and (6.16c) tells us that $\Gamma^i{}_{jk}U^k(P)U^j(P)$ must vanish along the whole curve. At P, however, U^i had an arbitrary direction, which means that $\Gamma^i{}_{jk}(P) = 0$.

The fact that it is always possible to choose a coordinate system to make $\Gamma^i{}_{jk}$ vanish at a point will be of great help in proving several theorems below. Since it is not necessary for $\Gamma^i{}_{jk}$ to vanish anywhere else, the derivatives of $\Gamma^i{}_{jk}$ at P do not vanish.

6.8 *Riemann tensor*

One might expect that the commutator of two convariant derivatives,

$$[\nabla_{\bar{U}}, \nabla_{\bar{V}}] \equiv \nabla_{\bar{U}}\nabla_{\bar{V}} - \nabla_{\bar{V}}\nabla_{\bar{U}},$$

should be a differential operator. In fact, however, it has the following remarkable property: the operator $\mathbf{R}$, defined by

$$\blacklozenge \quad [\nabla_{\bar{U}}, \nabla_{\bar{V}}] - \nabla_{[\bar{U}, \bar{V}]} \equiv \mathbf{R}(\bar{U}, \bar{V}), \tag{6.18}$$

is a *multiplicative* operator. Even more remarkably, $\mathbf{R}$ does not depend on derivatives of $\bar{U}$ and $\bar{V}$ either. These properties are explained and proved in the following exercise.

Exercise 6.13

Prove, for an arbitrary function f, that

(a) $\mathbf{R}(\bar{U}, \bar{V})f\bar{W} = f\mathbf{R}(\bar{U}, \bar{V})\bar{W}$,

(b) $\mathbf{R}(f\bar{U}, \bar{V})\bar{W} = f\mathbf{R}(\bar{U}, \bar{V})\bar{W}$.

Because of these properties, (6.18) actually defines a *tensor*, which is called the Riemann tensor. Given vectors $\bar{U}$, $\bar{V}$, (6.18) shows that $\mathbf{R}(\bar{U}, \bar{V})$ is a $\binom{1}{1}$ tensor, since the left-hand side operates on a vector to give a new vector. With $\bar{U}$ and $\bar{V}$ also regarded as variable arguments, the Riemann tensor becomes a $\binom{1}{3}$ tensor. (N.b. the conventions used for defining the Riemann tensor, (6.18) and (6.19), are by no means universal. Other definitions may differ in sign and the ordering of arguments. When consulting other books, make sure you find what convention is being used. We follow Misner, Thorne & Wheeler (1973).)

Exercise 6.14

The components of the Riemann tensor, $R^i{}_{jkl}$, are defined by

$$[\nabla_i, \nabla_j]\bar{e}_k - \nabla_{[\bar{e}_i, \bar{e}_j]}\bar{e}_k = R^l{}_{kij}\bar{e}_l. \tag{6.19}$$

(a) Show that in a *coordinate* basis

$$\blacklozenge \quad R^l{}_{kij} = \Gamma^l{}_{kj,i} - \Gamma^l{}_{ki,j} + \Gamma^m{}_{kj}\Gamma^l{}_{mi} - \Gamma^m{}_{ki}\Gamma^l{}_{mj}. \tag{6.20}$$

(b) In a noncoordinate basis define the commutation coefficients $C^i{}_{jk}$ by

$$[\bar{e}_j, \bar{e}_k] = C^i{}_{jk}\bar{e}_i. \tag{6.21}$$

Show that

$$R^l{}_{kij} = \Gamma^l{}_{kj,i} - \Gamma^l{}_{ki,j} + \Gamma^m{}_{kj}\Gamma^l{}_{mi} - \Gamma^m{}_{ki}\Gamma^l{}_{mj} - C^m{}_{ij}\Gamma^l{}_{km}, \tag{6.22}$$

where $f_{,i} \equiv \bar{e}_i[f]$.

(c) Show that

$$R^l{}_{k(ij)} \equiv \tfrac{1}{2}(R^l{}_{kij} + R^l{}_{kji}) = 0, \tag{6.23a}$$

and

$$R^l{}_{[kij]} = 0. \tag{6.23b}$$

(Hint: for (6.23b), use normal coordinates. The result, of course, is independent of the basis.)

(d) Using (c) show that in an n-dimensional manifold, the number of linearly independent components of $R^l{}_{kij}$ is

$$n^4 - n^2\,\frac{n(n+1)}{2} - n\,\frac{n(n-1)(n-2)}{3!} = \tfrac{1}{12}n^2(n^2-1). \tag{6.24}$$

Exercise 6.15

Show that

$$R^l{}_{k[ij;m]} = 0. \tag{6.25}$$

These are called the *Bianchi identities*. (Hint: again work in normal coordinates.) Show that this result is equivalent in a coordinate basis to the Jacobi identity for covariant derivatives

$$[\nabla_i, [\nabla_j, \nabla_k]] + [\nabla_j, [\nabla_k, \nabla_i]] + [\nabla_k, [\nabla_i, \nabla_j]] = 0.$$

Cf. equations (2.14) and (3.9).

6.9 *Geometric interpretation of the Riemann tensor*

Like the interpretation of the other commutator we have studied, $[\bar{U}, \bar{V}]$, this involves a closed or almost-closed loop. Our approach will be based upon the exponentiation of the covariant derivative, and so will closely parallel that for the Lie bracket. If a vector field $\bar{A}$ is defined along a curve whose tangent is $\bar{U}$, then parallel-transport permits us to bring $\bar{A}$ from any point Q on the curve to any other point P. The vector so produced, $\bar{A}(Q\to P)$ in T_P (not in general equal to $\bar{A}(P)$) is called the image at P of $\bar{A}(Q)$, and it depends of course on the curve. In fact, if $\bar{A}$ and $\bar{U}$ are analytic we can write the Taylor series

$$\begin{aligned}\bar{A}(Q\to P) &= \bar{A}(P) + \lambda\nabla_{\bar{U}}\bar{A}(P) + \tfrac{1}{2}\lambda^2\nabla_{\bar{U}}\nabla_{\bar{U}}\bar{A}(P) + \ldots \\ &= \exp[\lambda\nabla_{\bar{U}}]\bar{A}|_P,\end{aligned} \tag{6.26}$$

where λ is the curve's parameter ($\bar{U} = \mathrm{d}/\mathrm{d}\lambda$) and the 'exp' notation is again just a shorthand for the line above it.

Now consider two congruences with tangents $\bar{U} = \mathrm{d}/\mathrm{d}\lambda$ and $\bar{V} = \mathrm{d}/\mathrm{d}\mu$, for which $[\bar{U}, \bar{V}] = 0$. Their intersections therefore form closed loops, as shown in figure 6.6. If we parallel-transport a vector from some point R to Q along a curve $\bar{V}$ as shown, we thereby define a vector at Q

$$\bar{A}(R\to Q) = \exp[\mu\nabla_{\bar{V}}]\bar{A}|_Q,$$

where μ is the parameter distance from Q to R. If we then parallel-transport the resulting vector from Q to P, we get at P a vector we call

$$\bar{A}(R\to Q\to P) = \exp[\lambda\nabla_{\bar{U}}]\exp[\mu\nabla_{\bar{V}}]\bar{A}|_P,$$

where λ is the parameter distance from P to Q. We could have done the transporting another way, namely by first going to S (a distance λ along a $\bar{U}$-curve) and then to P (a distance μ along a $\bar{V}$-curve). The values of λ and μ are the same as above because $\bar{U}$ and $\bar{V}$ commute. The second method would produce

$$\bar{A}(R \to S \to P) = \exp\,[\mu\nabla_{\bar{V}}]\,\exp\,[\lambda\nabla_{\bar{U}}]\bar{A}|_P.$$

Their difference, which we shall call $\delta\bar{A}$, can be found for small λ and μ by using the Taylor expansion:

$$\begin{aligned}\delta\bar{A} &= [\mathrm{e}^{\lambda\nabla_{\bar{U}}}, \mathrm{e}^{\mu\nabla_{\bar{V}}}]\bar{A} \\ &= [1 + \lambda\nabla_{\bar{U}} + \tfrac{1}{2}\lambda^2\nabla_{\bar{U}}\nabla_{\bar{U}},\quad 1 + \mu\nabla_{\bar{V}} + \tfrac{1}{2}\mu^2\nabla_{\bar{V}}\nabla_{\bar{V}}]\bar{A} + O(3),\end{aligned}$$

where $O(3)$ means terms in $\mu^n\lambda^m$, where $n + m \geqslant 3$. Evaluating this gives

$$\delta\bar{A} = \lambda\mu[\nabla_{\bar{U}}, \nabla_{\bar{V}}]\bar{A} + O(3), \tag{6.27}$$

which is of course just the Riemann tensor, and does *not* involve derivatives of $\bar{A}$. Viewed another way, this is the change in $\bar{A}$ that would be produced if we were to parallel-transport it around the loop *PQRSP*. This change is just the Riemann tensor times the 'area' of the loop, $\lambda\mu$:

$$\delta A^i = \lambda\mu R^i{}_{jkl}A^jU^kV^l.$$

Another important geometrical aspect of the Riemann tensor involves *geodesic deviation*, the fact that geodesics begun parallel do not stay parallel. To measure this precisely, we consider a congruence of geodesics with tangent $\bar{U}$ ($\nabla_{\bar{U}}\bar{U} = 0$), and a connecting vector ξ which is Lie dragged by the congruence ($\pounds_{\bar{U}}\xi = 0$) (see figure 6.7). The manner in which ξ changes along $\bar{U}$ will be our measure of geodesic deviation. Its first derivative, $\nabla_{\bar{U}}\xi$, depends upon initial conditions, upon whether the geodesics are set up initially parallel or not. The geometry enters into the second derivative $\nabla_{\bar{U}}\nabla_{\bar{U}}\xi$, which tells how the initial rate of separation of the geodesics changes. So we have

Fig. 6.6. Parallel transport around a closed loop generally does not return the same vector as it began with.

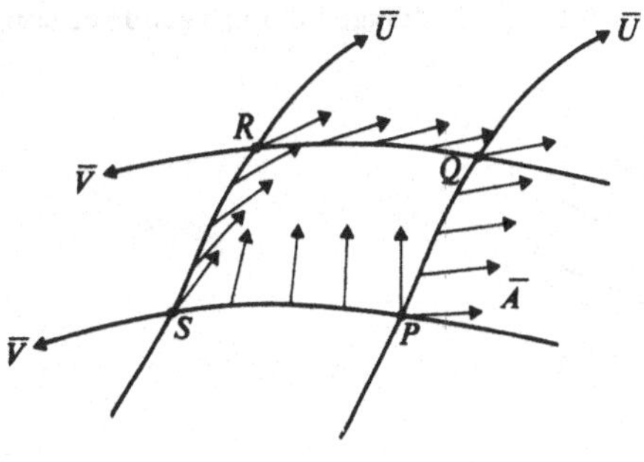

$$\nabla_{\bar{U}}\nabla_{\bar{U}}\xi = \nabla_{\bar{U}}(\pounds_{\bar{U}}\xi + \nabla_{\xi}\bar{U})$$
$$= \nabla_{\bar{U}}\nabla_{\xi}\bar{U}$$
$$= [\nabla_{\bar{U}}, \nabla_{\xi}]\bar{U} + \nabla_{\xi}\nabla_{\bar{U}}\bar{U}.$$

The first step used exercise 6.11. The last term in the last line vanishes because $\bar{U}$ is a geodesic, so we have

$$\nabla_{\bar{U}}\nabla_{\bar{U}}\xi = \mathsf{R}(\bar{U}, \xi)\bar{U}, \tag{6.28a}$$

or in component form

$$(\xi^i{}_{;j}U^j)_{;k}U^k = R^i{}_{jkl}U^jU^k\xi^l.$$

Notice that the left-hand side can be simplified because $U^j{}_{;k}U^k = 0$, and so we get

$$\xi^i{}_{;j;k}U^jU^k = R^i{}_{jkl}U^jU^k\xi^l. \tag{6.28b}$$

Equation (6.28) is called the equation of geodesic deviation.

6.10 *Flat spaces*

Euclid's axiom that parallel lines when extended never meet is the defining axiom for a *flat* space. From (6.28) it is clear that this means that a space is *flat* if and only if the Riemann tensor vanishes. Thus, the Riemann tensor is the measure of the *curvature* of a manifold with a connection. A flat space, by (6.27), has a *global* notion of parallelism: a vector at point R can be said to be parallel to one at P, because it can be parallel-transported to P in a manner independent of the path. Thus in a flat space all tangent spaces T_P may be identified with each other. Moreover, the exponential map is extendible indefinitely (provided the manifold's global topology is not artificially complicated by 'cutting and pasting') and the entire manifold may be identified with its tangent space. Notice that none of this requires a metric tensor. Minkowski space is just as flat as Euclidean space.

Fig. 6.7. A connecting vector ξ Lie dragged along a geodesic congruence.

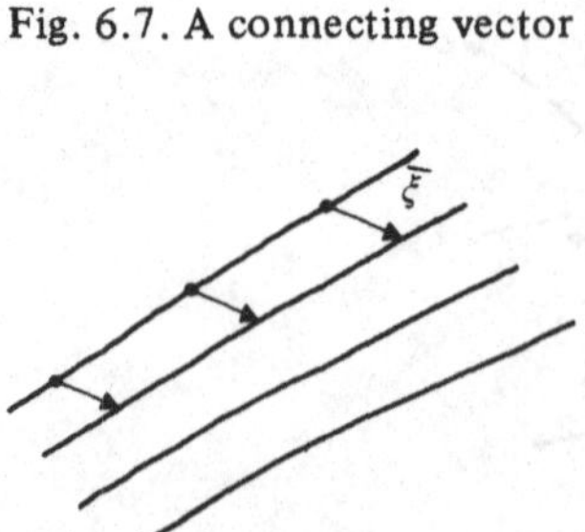

Exercise 6.16

Consider a two-dimensional flat space with Cartesian coordinates x, y and polar coordinates r, θ.

(a) Use the fact that $\bar{e}_x$ and $\bar{e}_y$ are globally parallel vector fields ($\bar{e}_x(P)$ is parallel to $\bar{e}_x(Q)$ for arbitrary P, Q) to show that

$$\Gamma^r{}_{\theta\theta} = -r, \quad \Gamma^\theta{}_{r\theta} = \Gamma^\theta{}_{\theta r} = 1/r,$$

and all other Γs are zero in polar coordinates.

(b) For an arbitrary vector field $\bar{V}$, evaluate $\nabla_i V^j$ and $\nabla_i V^i$ for polar coordinates in terms of the components V^r and V^θ.

(c) For the basis $\hat{\mathbf{r}} \equiv \partial/\partial r$, $\hat{\boldsymbol{\theta}} \equiv (1/r)\partial/\partial\theta$ find all the Christoffel symbols.

(d) Same as (b) for the basis in (c).

This exercise makes the important point that, although on a flat manifold coordinates exist in which $\Gamma^i{}_{jk} = 0$ everywhere, it is possible to choose coordinates in which they do *not* vanish.

6.11 *Compatibility of the connection with the volume-measure or the metric*

If a manifold has not only a connection but also a volume form or a metric, one usually makes certain compatibility demands. For example, both the connection and the volume-form can define the divergence of a vector field $\bar{V}$. The covariant divergence is $\nabla \cdot \bar{V} \equiv \nabla_i V^i$. The volume-form divergence is defined by

$$£_{\bar{V}}\tilde{\omega} = (\mathrm{div}_{\tilde{\omega}}\bar{V})\tilde{\omega}.$$

We say that ∇ and $\tilde{\omega}$ are *compatible* if $\mathrm{div}_{\tilde{\omega}}\bar{V} = \nabla \cdot \bar{V}$ for all $\bar{V}$.

Exercise 6.17

(a) Show that ∇ and $\tilde{\omega}$ are compatible if and only if $\nabla\tilde{\omega} = 0$. (Hint: use exercise 6.11 to evaluate $£_{\bar{V}}\tilde{\omega}$.)

(b) In coordinates $(x^1, \ldots, x^n)$ suppose $\omega_{12\ldots n} = f$. Show that ∇ and $\tilde{\omega}$ are compatible if and only if for all k

$$(\ln f)_{,k} = \Gamma^j{}_{jk}.$$

In a similar way, there is a natural compatibility demand if the manifold has a metric tensor g. Two vectors $\bar{A}$ and $\bar{B}$ have the inner product $\mathsf{g}(\bar{A}, \bar{B})$ at a point P. We say that ∇ and g are compatible if this inner product is preserved by parallel-transport of $\bar{A}$ and $\bar{B}$ along any curve, for any vectors $\bar{A}$ and $\bar{B}$.

Exercise 6.18

(a) Show that ∇ and $\mathsf{g}|$ are compatible if and only if

$$\nabla \mathsf{g}| = 0. \quad (6.29)$$

(b) In coordinates $(x^1, \ldots, x^n)$ show that ∇ and $\mathsf{g}|$ are compatible if and only if

$$\Gamma^i{}_{jk} = \tfrac{1}{2} g^{il}(g_{lj,k} + g_{lk,j} - g_{jk,l}). \quad (6.30)$$

Here g^{ij} are the elements of the matrix inverse to the matrix of components g_{lm} (cf. equation 2.55)). (Hint: use the symmetry $\Gamma^i{}_{jk} = \Gamma^i{}_{kj}$.)

Exercise 6.19

Recall that exercise 4.13 enables one to define a preferred volume-form if one has a metric. (This is another compatibility, that of the metric and volume-form). Show that if the metric and connection are compatible, then the preferred volume-form and the connection are compatible. (Hint: you will have to show that $g_{,k} = g^{ji} g_{ij,k}$. Use equation (4.39) for this purpose.)

Equation (6.30) shows the remarkable fact that a metric actually determines the compatible symmetric connection uniquely. Such a connection is called a *metric connection.*

Exercise 6.20

Show that for an arbitrary vector $\bar{V}$

$$(\pounds_{\bar{V}} \mathsf{g}|)_{ij} = \nabla_i V_j + \nabla_j V_i.$$

Therefore a Killing vector (cf. §3.11) obeys *Killing's equation*

$$\nabla_i V_j + \nabla_j V_i = 0.$$

Cf. equation (5.89).

6.12 *Metric connections*

Because (6.30) is such a strong constraint on the connection, metric connections have additional properties that general symmetric connections do not. To derive some of them it is easiest to work in a normal coordinate system. Notice that (6.29) and (6.30) imply

$$\Gamma^i{}_{jk} = 0 \text{ at } P \Leftrightarrow g_{lm,n} = 0 \text{ at } P. \quad (6.31)$$

Exercise 6.21

Show that (6.20), (6.30), and (6.31) imply that in normal coordinates at a point P

$$R_{ijkl} \equiv g_{im}R^m{}_{jkl} = \tfrac{1}{2}(g_{il,jk} - g_{ik,jl} + g_{jk,il} - g_{jl,ik}). \tag{6.32}$$

Exercise 6.22

(a) Show that (6.32) implies the identity

$$R_{ijkl} = R_{klij}. \tag{6.33}$$

(b) Show that (6.33) and (6.23) imply that in an n-dimensional manifold the number of linearly independent components of R_{ijkl} is

$$\tfrac{1}{8}n(n-1)(n^2-n+2) - \tfrac{1}{24}n(n-1)(n-2)(n-3) = \tfrac{1}{12}n^2(n^2-1).$$

Exercise 6.23

(a) Define the tensor R_{kl}, called the *Ricci tensor*, by

$$R_{kl} = R^i{}_{kil}, \tag{6.34}$$

and the *Ricci scalar* R by

$$R = g^{kl}R_{kl}. \tag{6.35}$$

Show that R_{ij} is *symmetric*.

(b) Show that the contracted Bianchi identities

$$R^i{}_{j[il;m]} = 0 \quad \text{and} \quad g^{jl}R^i{}_{j[il;m]} = 0$$

imply

$$(R^{ij} - \tfrac{1}{2}Rg^{ij})_{;j} = 0. \tag{6.36}$$

(Raising indices on R^{ij} is accomplished by the metric: $R^{ij} = g^{il}g^{jm}R_{lm}$.)

(c) Define the Weyl tensor:

$$C^{ij}{}_{kl} = R^{ij}{}_{kl} - 2\delta^{[i}{}_{[k}R^{j]}{}_{l]} + \tfrac{1}{3}\delta^{[i}{}_{[k}\delta^{j]}{}_{l]}R. \tag{6.37}$$

Show that every contraction between indices of C_{ijkl} gives zero: it is a 'pure' fourth rank tensor.

Equation (6.36) plays a fundamental role in Einstein's theory of gravitation (general relativity). Spacetime is represented as a four-dimensional manifold with metric, a generalization of flat Minkowski spacetime. The empty-space (source-free) gravitational field (i.e. metric) is found by solving the differential equations

$$G^{ij} \equiv R^{ij} - \tfrac{1}{2}Rg^{ij} = 0, \tag{6.38}$$

where G^{ij} is called the Einstein tensor. The identities (6.36) reduce the number of independent equations in (6.38) from 10 ($= \tfrac{1}{2}n(n+1)$ because G^{ij} is symmetric) to 6. This guarantees that the solution, g_{ij}, which also has 10 independent components, is determined only up to the four functional degrees of freedom represented by the coordinate transformations of g_{ij}.

Exercise 6.24

Show that a geodesic joining points P and Q is a curve of extremal length among all curves joining P and Q. Do this by showing that

$$\int_P^Q \left| \mathrm{g}\left(\frac{\mathrm{d}\bar{x}}{\mathrm{d}\lambda}, \frac{\mathrm{d}\bar{x}}{\mathrm{d}\lambda} \right) \right|^{1/2} \mathrm{d}\lambda$$

is unchanged to first order by changes in $x^i(\lambda)$ away from a geodesic curve. (Bear in mind that any curve has a unique parameter; a geodesic curve's parameter must be affine.) Discuss the need for the absolute value signs above if the metric is indefinite, and in particular discuss separately the case of a null geodesic (length zero).

6.13 *The affine connection and the equivalence principle*

We all learned our basic geometry and physics by studying flat manifolds: Euclidean three-space, Galilean spacetime (though it probably was not given that name) and later (if at all), Minkowski spacetime. General relativity, on the other hand, uses a curved spacetime. It seems natural to think of a flat space as the simplest kind of space. But from the point of view of manifold theory, even a flat space is by no means simple: it has far more structure than the ordinary differentiable manifold, for it has an affine connection. The existence of this connection does not intrude into elementary geometry and physics, because one usually adopts rectangular coordinates, in which the Christoffel symbols vanish. But if the physical laws are framed in flat space using curvilinear coordinates, then the Christoffel symbols must be used, and the connection becomes visible.

It may seem that this is a complication to be avoided, but consider its potential for generalization. Most physical laws written in this way involve the Christoffel symbols but not the Riemann tensor, so their equations are meaningful – identical – whether the manifold is flat or curved. It is therefore natural to postulate that in the curved spacetime of general relativity the laws of physics have exactly the same mathematical form as they have in a curved coordinate system in the flat spacetime of Minkowski space. This is called the principle of minimal coupling (of physical fields to the curvature of spacetime), or the strong principle of equivalence. It is a postulate, widely adopted, which is consistent with experiment. A full discussion of it can be found in Misner *et al.* (1973). The point that needs to be exphasized here is the rather remarkable circumstance that, by expressing the flat-space laws of physics in a curved coordinate system, one obtains the curved-space form of the laws. This circumstance can be traced to the fact that flat space, though having zero curvature, has a perfectly definite connection and is therefore only a special kind of 'curved' space.

6.14 *Connections and gauge theories: the example of electromagnetism*

'Gauge theories' is the collective name for a large variety of theories of elementary particle interactions, all of which share one feature: invariance of their physical predictions under a group of transformations of the basic variables of the field theory. Electromagnetism is the best-known example: if the basic variable is taken to be the one-form ('vector') potential $\tilde{A}$, then the physical predictions of the theory are invariant under the gauge transformation $\tilde{A} \to \tilde{A} + \tilde{d}f$. The word 'gauge' is applied by analogy to the transformations of all these theories. A general discussion of gauge theories is beyond our scope (see the lectures by Trautman (1973) in the bibliography). We will confine our remarks to electromagnetism, illustrated with the equation for a charged particle of mass m and zero spin. We will see that a connection different from but in the same spirit as the affine connection arises in a natural way and in particular leads us to 'invent' the electromagnetic field!

Consider first the neutral scalar particle of mass m whose wave-function ψ obeys the Klein–Gordon equation and the (conserved) normalization condition

$$(\nabla_\mu \nabla^\mu - m^2)\psi = 0, \quad \int d^3x(\psi^* \dot{\psi} - \psi \dot{\psi}^*) = 1, \tag{6.39}$$

where Greek indices run over (t, x, y, z), and where we assume for simplicity the metric of Minkowski spacetime. Clearly, if ψ is a solution then so is $\psi e^{i\phi}$, where ϕ is any real constant. This is a gauge transformation: $\psi \to \psi e^{i\phi}$. We shall now make an analogy which will carry through our whole discussion. The gauge transformations are of a very restricted sort, since ϕ cannot depend on position. This is analogous to the coordinate-freedom in the description of some (any) physical system in rectangular coordinates in special relativity. The permissible coordinate transformations are the rotations, Lorentz boosts, and translations, and these are all rigid: one cannot make one transformation at one point and a different one somewhere else. Relaxing this restriction in special relativity in order to permit arbitrary coordinates forces one, as we have seen in the last section, to introduce the affine connection in order to preserve a coordinate-independent derivative, the covariant derivative. Once the equations of motion of the physical system are written down with a connection in them, it is natural to use them when the connection is not flat. These turn out to be the appropriate equations for that system in *general* relativity. This procedure of generalizing the coordinate freedom thus leads to a theory of the way the system interacts with a gravitational field. In a similar manner, we will now generalize the gauge freedom of the field ψ and find, automatically, a theory for how the field interacts with electromagnetism.

The generalization is obvious: we would like a general gauge transformation

$$\psi \rightarrow \psi e^{i\phi(\bar{x})}, \tag{6.40}$$

where now ϕ is an arbitrary real function of position $\bar{x}$ in Minkowski spacetime. But because the field equations involve derivatives, this produces a change

$$\tilde{d}\psi \rightarrow (\tilde{d}\psi + i\psi\,\tilde{d}\phi)e^{i\phi(\bar{x})}. \tag{6.41}$$

To see how to eliminate this extra term, let us look at the situation more geometrically. The factor $e^{i\phi}$ is a complex number on the unit circle; the gauge transformation is a representation of the action of the group $U(1)$ (unitary group in one complex dimension) on ψ. So the transformation $\psi \rightarrow \psi e^{i\phi(\bar{x})}$ can be thought of as picking out an element of $U(1)$ at each $\bar{x}$ and allowing it to act on ψ. The natural geometrical structure here is the *fiber bundle*, whose base manifold is Minkowski spacetime and whose fibers are the group $U(1)$ (which can be visualized as the unit circle in the complex plane). A gauge transformation (6.40) is then a cross-section of the fiber bundle. We shall call this bundle the *$U(1)$-bundle*.

Now, the thing we want to look at is not ψ itself but $\nabla_\mu \psi$, which at any point P is an element of $T^*{}_P$, the vector space of one-forms at P. Consider a curve $\mathscr{C}$ with parameter λ in the base manifold. As we move along the curve we encounter a sequence of one-forms $\tilde{d}\psi$, one at each point. If ψ satisfies the Klein–Gordon equation, (6.39), we will say that $\tilde{d}\psi$ changes along $\mathscr{C}$ in the 'correct' manner. There is a restricted set of gauge transformations (ϕ = const) for which the new $\tilde{d}\psi$ is also 'correct'. Suppose we make an arbitrary gauge transformation. Then to the curve $\mathscr{C}$ in the base manifold there corresponds a curve $\mathscr{C}^*$ in the $U(1)$-bundle which passes through each fiber above a point of $\mathscr{C}$ at the point on the fiber (element of $U(1)$) which corresponds to the gauge transformation at that point of $\mathscr{C}$. If this transformation is not constant (if $\mathscr{C}^*$ is not 'parallel' to $\mathscr{C}$) then the gradient of the transformed ψ will not be 'correct': it will not equal, to within a phase, that of the original ψ. So we will define a *connection one-form* $\tilde{A}$ on the base manifold, which will depend upon the curve $\mathscr{C}^*$ in such a way as to correct the derivative of ψ. The definition is:

(i) If ψ solves (6.39) then $\tilde{A} = 0$.

(ii) Under a transformation $\psi \rightarrow \psi e^{i\phi(\bar{x})}$ the connection one-form transforms as

$$\tilde{A} \rightarrow \tilde{A} + \tilde{d}\phi. \tag{6.42}$$

(iii) The *gauge-covariant derivative* of ψ is

$$\tilde{D}\psi = \tilde{d}\psi - i\psi\tilde{A}. \tag{6.43}$$

Properties (ii) and (iii) mean that $\tilde{D}\psi$ changes, under a gauge transformation, to $e^{i\phi(\bar{x})}\tilde{D}\psi$; property (i) guarantees that $\tilde{D}\psi$ is 'correct' on $\mathscr{C}$.

Let us now understand why $\tilde{A}$ is called a connection. The *affine* connection is represented by the Christoffel symbols, which are added to the ordinary partial derivative in order to give a 'correct' derivative: one which gives parallel transport (cf. (6.43) with (6.10)). In order to preserve the 'correctness' of the derivative, the Christoffel symbols must transform under a coordinate change in a manner which depends on the coordinate change (exercise 6.2) in a way very similar to the way $\tilde{A}$ changes under a gauge transformation (6.42). The difference between the connections is what they set out to preserve: an affine connection preserves parallelism; our one-form connection preserves the gradient under a gauge transformation.

We can now write the gauge-covariant form of the Klein–Gordon equation:

$$D_\mu D^\mu \psi - m^2\psi = (\nabla_\mu - iA_\mu)(\nabla^\mu - iA^\mu)\psi - m^2\psi = 0. \qquad (6.44)$$

This equation reduces to the usual Klein–Gordon equation if the phase of ψ is 'correct'; and any ψ obtained by an arbitrary gauge transformation of a 'correct' ψ solves (6.44).

The curvature tensor of an affine connection can be defined in a coordinate system by an equation like (6.18):

$$[\nabla_\mu, \nabla_\nu]V^\alpha = R^\alpha{}_{\beta\mu\nu}V^\beta.$$

The analogue here is

$$[D_\mu, D_\nu]\psi = F_{\mu\nu}\psi. \qquad (6.45)$$

It is a straightforward calculation to show that the *gauge-curvature two-form* $\tilde{F}$ whose components are $F_{\mu\nu}$ is simply

$$\tilde{F} = -i\tilde{d}\tilde{A}. \qquad (6.46)$$

Clearly $\tilde{F}$ is *gauge-invariant* (cf. (ii) above). (The Riemann tensor was coordinate-invariant.) The Klein–Gordon equation is *gauge-flat* ($\tilde{F} = 0$) because there exists a gauge in which $\tilde{A} = 0$. But because of the obvious analogy with electromagnetism ($\tilde{A}$ = one-form potential, $i\tilde{F}$ = Faraday tensor: see chapter 5), it is clearly tempting to regard (6.44) as a generalization of the Klein–Gordon equation to the case where the particle has charge and interacts with an external electromagnetic field $\tilde{F}$. This is in fact correct, and (6.44) can be derived more directly from the fact that the canonical momentum of a classical particle in an external electromagnetic field is $\bar{p}_c = \bar{p} + (q/c)\bar{A}$, where $\bar{p}$ is the 'true' four-momentum of the particle. By the correspondence principle the equation $\bar{p}\cdot\bar{p} + m^2 = 0$ becomes (in units where $\hbar = c = 1$)

$$(-i\nabla_\mu - qA_\mu)(-i\nabla^\mu - qA^\mu) + m^2\psi = 0.$$

This shows us that (6.44) *is* the wave equation for such a particle with charge

$q = 1$. We can summarize what we have learned in the following way: a scalar particle of mass m and charge q in the presence of an external electromagnetic field with one-form potential $\tilde{A}$ obeys the equation

$$(\nabla_\mu - iqA_\mu)(\nabla^\mu - iqA^\mu)\psi - m^2\psi = 0. \tag{6.47}$$

A gauge-transformation consists of the following:

$$\tilde{A} \to \tilde{A} + \tilde{d}\phi, \tag{6.48a}$$

$$\psi \to \psi e^{i\phi/q}. \tag{6.48b}$$

We can regard $\tilde{A}$ as a connection on the $U(1)$-bundle and $\tilde{F}$ as its curvature.

Exercise 6.25

(a) Verify that $\tilde{D}\psi \to e^{i\phi(\bar{x})}\tilde{D}\psi$ under a gauge transformation.

(b) Verify (6.46).

6.15 *Bibliography*

A very complete reference for Riemannian geometry is S. Kobayashi & K. Nomizu, *Foundations of Differential Geometry*, two volumes (Interscience, New York, 1963 and 1969).

Good modern introductions to pseudo-Riemannian geometry (a metric connection with an indefinite metric) are in *Gravitation* by C. W. Misner, K. S. Thorne & J. A. Wheeler (Freeman, San Francisco, 1973); *The Large-Scale Structure of Space–Time*, by S. W. Hawking & G. F. R. Ellis (Cambridge University Press, 1973); and *Gravitation and Cosmology*, by S. Weinberg (Wiley, New York, 1972).

A more complete exposition on the role of connections in gauge theories is A. Trautman, Infinitesimal connections in physics, in the *Proceedings of the International Symposium on New Mathematical Methods in Physics*, ed. K. Bleuler & A. Reetz (Bonn, 1973). See also R. Hermann, *Vector Bundles in Mathematical Physics*, two volumes (Benjamin, Reading, Mass., 1970). For a 'physicist's' description of gauge theories see J. C. Taylor, *Gauge Theories of the Weak Interactions* (Cambridge University Press, 1976). A mathematical discussion of connections in their wider meaning can be found in Y. Choquet-Bruhat, C. Dewitt-Morette & M. Dillard-Bleick, *Analysis, Manifolds, and Physics* (North-Holland, Amsterdam, 1977). Connections on fiber bundles are reviewed in B. Carter, Underlying mathematical structures of classical gravitation theory, in *Recent Developments in Gravitation*, ed. M. Levy & S. Deser (Plenum, New York, 1979).

There are many other topics one can study once one has introduced the affine connection. For example, Lie groups have a natural affine connection which makes the one-parameter subgroups into geodesics in the group manifold. This construction is explored in a step-by-step fashion in a number of exercises in Misner *et al.*, op. cit.

Torsion may play a role in gravitation, a suggestion first made by Cartan. The Einstein–Cartan theory of gravitation has received considerable attention lately: see A. Trautman, *Bull. de l'Academie Polonaises des Sciences* (math., astr., phys.) **20**, 185–90 (1972).

The Riemann tensor is not an easy tensor to calculate from a knowledge of the metric tensor's components. The task is made somewhat easier by Cartan's method of moving frames, which makes use of the calculus of differential forms. See Misner *et al.*, op. cit.

Although we have not considered it, the differential geometry of *complex* manifolds is interesting and may play a large role in future physical applications. For an introduction aimed at physicists, see E. J. Flahery, *Hermitian and Kählerian Geometry in Relativity* (Springer, Berlin, 1976). A standard mathematical reference is S. S. Chern, *Complex Manifolds Without Potential Theory* (D. Van Nostrand, New York, 1967).

APPENDIX: SOLUTIONS AND HINTS FOR SELECTED EXERCISES

2.1 $[\hat{\mathbf{r}}, \hat{\boldsymbol{\theta}}] = -\hat{\boldsymbol{\theta}}/r \neq 0$.

2.2(b) Since $a\mathrm{d}/\mathrm{d}\lambda + b\mathrm{d}/\mathrm{d}\mu = b\mathrm{d}/\mathrm{d}\mu + a\mathrm{d}/\mathrm{d}\lambda$, equation (2.13) implies $\exp[a\mathrm{d}/\mathrm{d}\lambda] \exp[b\mathrm{d}/\mathrm{d}\mu] = \exp[b\mathrm{d}/\mathrm{d}\mu] \exp[a\mathrm{d}/\mathrm{d}\lambda]$, which certainly implies $[\mathrm{d}/\mathrm{d}\lambda, \mathrm{d}/\mathrm{d}\mu] = 0$. Conversely, if the order of $\mathrm{d}/\mathrm{d}\lambda$ and $\mathrm{d}/\mathrm{d}\mu$ on the right-hand side of (2.13) does not matter, then they may be manipulated exactly as real numbers, for which (2.13) is true.

2.3 Expand out each term, e.g. $[[\bar{X}, \bar{Y}], \bar{Z}] = \bar{X}\bar{Y}\bar{Z} - \bar{Y}\bar{X}\bar{Z} - \bar{Z}\bar{X}\bar{Y} + \bar{Z}\bar{Y}\bar{X}$. Each such term is to be interpreted as a differential operator on functions. The C^2 requirement guarantees each term exists. When all three terms in (2.14) are so expanded, the result follows.

2.4 Each matrix is a $\binom{1}{1}$ tensor requiring one vector and one one-form to give a real number. Since there are two matrices involved, transformation can produce a number when supplied with two vectors and two one-forms. Linearity is easy to check.

2.5(a) In n dimensions a $\binom{2}{0}$ tensor has n^2 independent components, while two vectors have between them only $2n$ components. In general this is inadequate.

2.6 A linear combination of two $\binom{2}{0}$ tensors is defined in terms of their values on arbitrary one-forms $\tilde{p}$ and $\tilde{q}$: $(\alpha\mathbf{h} + \beta\mathbf{r})(\tilde{p}, \tilde{q}) = \alpha\mathbf{h}(\tilde{p}, \tilde{q}) + \beta\mathbf{r}(\tilde{p}, \tilde{q})$. This is still a linear function of $\tilde{p}$ and $\tilde{q}$, and so is a $\binom{2}{0}$ tensor. The zero tensor has value zero on any $\tilde{p}$ and $\tilde{q}$, and the other axioms are also obvious. The space has n^2 dimensions because each tensor $\mathbf{h}$ is completely defined by its n^2 components $\mathbf{h}(\tilde{\omega}^i, \tilde{\omega}^j) = h^{ij}$. The n^2 tensors $\{\bar{e}_i \otimes \bar{e}_j\}$ are a basis because they are linearly independent: the linear combination $\beta^{ij}\bar{e}_i \otimes \bar{e}_j$ vanishes if and only if all β^{ij} vanish, as can be seen by allowing the tensor to operate on all pairs of basis one-forms. In fact it is easy to verify that $\mathbf{h} = h^{ij}\bar{e}_i \otimes \bar{e}_j$.

2.7 Six. Six.

2.8 Linearity: $\mathbf{C}(a\bar{V} + b\bar{W}) = \mathbf{B}(\mathbf{A}(a\bar{V} + b\bar{W})) = \mathbf{B}(a\mathbf{A}(\bar{V}) + b\mathbf{A}(\bar{W}))$, $a\mathbf{B}(\mathbf{A}(\bar{V})) + b\mathbf{B}(\mathbf{A}(\bar{W})) = a\mathbf{C}(\bar{V}) + b\mathbf{C}(\bar{W})$.
Components: if $\mathbf{A}(\bar{e}_j) = A^k{}_j\bar{e}_k$ then $\mathbf{C}(\bar{e}_j) = \mathbf{B}(A^k{}_j\bar{e}_k) = \mathbf{B}(\bar{e}_k)A^k{}_j = B^i{}_k A^k{}_j \bar{e}_i$, from which the result follows.

Each $\binom{1}{1}$ tensor is a linear transformation of T_P. Our result shows that linear transformations form a group under the operation of composition (by which **C** was produced from **A** and **B**).

2.9 $\mathbf{T}(\tilde{\omega}^{i'}, \tilde{\omega}^{j'}) = \mathbf{T}(\Lambda^{i'}{}_k\tilde{\omega}^k, \Lambda^{j'}{}_l\tilde{\omega}^l) = \Lambda^{i'}{}_k\Lambda^{j'}{}_l\mathbf{T}(\tilde{\omega}^k, \tilde{\omega}^l)$. This generalizes to $T^{i'\ldots j'}{}_{k'\ldots l'} = \Lambda^{i'}{}_r \ldots \Lambda^{j'}{}_s\Lambda^t{}_{k'} \ldots \Lambda^u{}_{l'}T^{r\ldots s}{}_{t\ldots u}$, where $(i' \ldots j')$ are N indices and $(k' \ldots l')$ are N' indices.

2.10 True of any vector space: the zero element is unique.

2.11 The value of the tensor (in the original basis) on a one-form $\tilde{p}$ and a vector $\bar{V}$ is $A^i{}_jV^jp_i$. In the new basis it is $A^{i'}{}_{j'}V^{j'}p_{i'} = (\Lambda^{i'}{}_r\Lambda^s{}_{j'}A^r{}_s)$ $(\Lambda^{j'}{}_kV^k)$ $(\Lambda^l{}_{i'}p_l)$, where we have used the transformed components of everything. Doing the sums on i' and j' first and using (2.34) gives that the new value is the same as the old, i.e. that the rule gives a real number associated with the vector and one-form, independent of the basis.

2.12 (a) $\Lambda = \begin{pmatrix} 1/\sqrt{6} & -1/\sqrt{2} \\ 1/\sqrt{6} & 1/\sqrt{2} \end{pmatrix}$ produces the canonical form $\begin{pmatrix} 1 & 0 \\ 0 & 1 \end{pmatrix}$.

(b) $\Lambda = \begin{pmatrix} 1/\sqrt{2} & 1/\sqrt{2} \\ -1/\sqrt{2} & 1/\sqrt{2} \end{pmatrix}$ produces the canonical form $\begin{pmatrix} -1 & 0 \\ 0 & 1 \end{pmatrix}$.

(c) $\Lambda = \begin{pmatrix} 0 & 1/2 \\ 1 & 0 \end{pmatrix}$ produces the canonical form $\begin{pmatrix} -1 & 0 \\ 0 & 1 \end{pmatrix}$.

2.13 (a) The transformation law can be deduced from equation (2.55). As a function of one-forms, $\mathsf{g}^{-1}(\tilde{p}, \tilde{q}) = \mathsf{g}(\bar{p}, \bar{q})$ leads to (2.55) and clearly shows the linearity property.

(b) In such a basis, the metric has components $\pm\,\delta_{ij}$, so g^{-1} has components $\pm\,\delta^{ij}$ as well, being just the inverse matrix.

2.14 Make a Taylor expansion of g_{ij} and $\Lambda^i{}_{j'}$ about P. Try to satisfy (i) and (ii) by choosing the coefficients of the $\Lambda^i{}_{j'}$ expansion appropriately for arbitrary coefficients of the g_{ij} expansion. Show, by counting the number of coefficients, that (i)–(iii) hold. Remember that not all $\partial\Lambda^i{}_{j'}/\partial x^{k'}$ at P are independent, since $\Lambda^i{}_{j'} = \partial x^i/\partial x^{j'}$ implies $\partial\Lambda^i{}_{j'}/\partial x^{k'} = \partial\Lambda^i{}_{k'}/\partial x^{j'}$.

2.15 (a) $g_{rr} = 1, g_{r\theta} = 0, g_{\theta\theta} = r^2$.

(b) orthonormal. $\hat{\mathbf{r}} = \partial/\partial r$, $\hat{\boldsymbol{\theta}} = r^{-1}\,\partial/\partial\theta$.

2.16 (a) $\tilde{\mathrm{d}}f$ has components $(\partial f/\partial r, \partial f/\partial\theta)$; $\bar{\mathrm{d}}f$ has components $(\partial f/\partial r, r^{-2}\,\partial f/\partial\theta)$.

(b) On the orthonormal basis, both $\tilde{\mathrm{d}}f$ and $\bar{\mathrm{d}}f$ have components $(\partial f/\partial r, r^{-1}\,\partial f/\partial\theta)$.

3.1 (a) On functions, equation (3.3) shows that each side of (3.8) reduces to the operator $[\bar{V}, \bar{W}]$. On a vector field $\bar{U}$, equation (3.6) gives the left-hand side as $[\bar{V}, [\bar{W}, \bar{U}]] - [\bar{W}, [\bar{V}, \bar{U}]]$ and the right-hand side as $[[\bar{V}, \bar{W}], \bar{U}]$. The Jacobi identity (exercise 2.3) and the antisymmetry of the Lie bracket establish the result.

(b) On functions, we have the Jacobi identity (2.14) again. On vectors, equation (3.8) converts (3.9) into the Lie derivative with respect to $[[\bar{X}, \bar{Y}], \bar{Z}] + [[\bar{Y}, \bar{Z}], \bar{X}] + [[\bar{Z}, \bar{X}], \bar{Y}]$, which vanishes.

3.2(b) $$\pounds_{\bar{V}} \bar{U} = \pounds_{\bar{V}}(U^i \bar{e}_i) = (\pounds_{\bar{V}} U^i) \bar{e}_i + U^i \pounds_{\bar{V}} \bar{e}_i = [V^j \bar{e}_j(U^i)] \bar{e}_i - U^i \pounds_{\bar{e}_i}(V^j \bar{e}_j) = [V^j \bar{e}_j(U^i) - U^j \bar{e}_j V^i] \bar{e}_i - U^k V^j \pounds_{\bar{e}_k} \bar{e}_j .$$

The last step required relabelling of indices. Use of (3.7) in the last term produces the desired result.

3.3 Follows from (2.7) with $V^i = \delta^i{}_1$.

3.4 From (3.13) we have

$$(\pounds_{\bar{V}} \tilde{\omega})_i W^i = V^j \frac{\partial}{\partial x^j}(\omega_i W^i) - \omega_i \left(V^j \frac{\partial}{\partial x^j} W^i - W^j \frac{\partial}{\partial x^j} V^i \right)$$
$$= \left(V^j \frac{\partial}{\partial x^j} \omega_i + \omega_j \frac{\partial}{\partial x^i} V^j \right) W^i,$$

where indices have been relabelled in the second term. The arbitrariness of W^i gives the result.

3.5 (a) $$\left[\sum_a \alpha_{(a)} \bar{A}_{(a)}, \sum_b \beta_{(b)} \bar{A}_{(b)} \right] = \sum_{a,b} [\alpha_{(a)} \bar{A}_{(a)}, \beta_{(b)} \bar{A}_{(b)}]$$
$$= \sum_{a,b} \{\alpha_{(a)} \beta_{(b)} [\bar{A}_{(a)}, \bar{A}_{(b)}] + \alpha_{(a)} [\bar{A}_{(a)}(\beta_{(b)})] \bar{A}_{(b)} - \beta_{(b)} [\bar{A}_{(b)}(\alpha_{(a)})] \bar{A}_{(a)}\}.$$

3.6 $$(\pounds_{\bar{V}} \mathsf{T})^{i\ldots j}{}_{k\ldots l} = V^r \frac{\partial}{\partial x^r} T^{i\ldots j}{}_{k\ldots l} - T^{r\ldots j}{}_{k\ldots l} \frac{\partial}{\partial x^r} V^i - \ldots - T^{i\ldots r}{}_{k\ldots l} \frac{\partial}{\partial x^r} V^j + T^{i\ldots j}{}_{r\ldots l} \frac{\partial}{\partial x^k} V^r + \ldots + T^{i\ldots j}{}_{k\ldots r} \frac{\partial}{\partial x^l} V^r .$$

Set $V^i = \delta^i{}_1$ and get $(\pounds_{\bar{V}} \mathsf{T})^{i\ldots j}{}_{k\ldots l} = \partial T^{i\ldots j}{}_{k\ldots l}/\partial x^1$.

3.7 $$[L^2, \pounds_{\bar{l}_z}] = \pounds_{\bar{l}_x} [\pounds_{\bar{l}_x}, \pounds_{\bar{l}_z}] + [\pounds_{\bar{l}_x}, \pounds_{\bar{l}_z}] \pounds_{\bar{l}_x} + \pounds_{\bar{l}_y} [\pounds_{\bar{l}_y}, \pounds_{\bar{l}_z}] + [\pounds_{\bar{l}_y}, \pounds_{\bar{l}_z}] \pounds_{\bar{l}_y} = \pounds_{\bar{l}_x} \pounds_{\bar{l}_y} + \pounds_{\bar{l}_y} \pounds_{\bar{l}_x} - \pounds_{\bar{l}_y} \pounds_{\bar{l}_x} - \pounds_{\bar{l}_x} \pounds_{\bar{l}_y} = 0.$$

The second step used equation (3.8). To prove (3.33), derive the following relations: $\bar{l}_x = -\sin\phi\, \partial/\partial\theta - \cos\phi \cot\theta\, \partial/\partial\phi$; $\bar{l}_y = -\cos\phi\, \partial/\partial\theta + \sin\phi \cot\theta\, \partial/\partial\phi$; $\bar{l}_z = \partial/\partial\phi$.

3.8 Follows trivially from $\pounds_{a\bar{V}} = a\pounds_{\bar{V}}$ if a is *constant*. (This is not true if a is a general function.)

3.9 When Lie dragged along ϕ from $\phi = 0$, $\bar{e}_z$ is unchanged but $\bar{e}_x$ becomes

$\bar{e}_r = \cos\phi\,\bar{e}_x + \sin\phi\,\bar{e}_y$. The third basis vector, $\bar{e}_\phi$, has Cartesian representation $-\sin\phi\,\bar{e}_x + \cos\phi\,\bar{e}_y$. The three vector harmonics are $\exp(2i\phi)\bar{e}_z$, $\exp(2i\phi)(\cos\phi\,\bar{e}_x + \sin\phi\,\bar{e}_y)$, $\exp(2i\phi)(-\sin\phi\,\bar{e}_x + \cos\phi\,\bar{e}_y)$. It might be more useful to use the more compact linear combinations $\exp(2i\phi)\bar{e}_z$, $\exp(i\phi)(\bar{e}_x + i\bar{e}_y)$, $\exp(3i\phi)(\bar{e}_x - i\bar{e}_y)$. It is obvious from these that $\bar{e}_x + i\bar{e}_y$ has eigenvalue $+1$ and $\bar{e}_x - i\bar{e}_y$ eigenvalue -1, but these are easy to verify directly. The one-form $\tilde{d}z$ is unchanged by Lie dragging, and $\tilde{d}x$ becomes $\tilde{d}r = \cos\phi\,\tilde{d}x + \sin\phi\,\tilde{d}y$. The third basis one-form, $\tilde{d}\phi$, has Cartesian representation $-\sin\phi\,\tilde{d}x + \cos\phi\,\tilde{d}y$. The three one-form harmonics are, then, $\exp(2i\phi)\tilde{d}z$, $\exp(i\phi)(\tilde{d}x + i\tilde{d}y)$, $\exp(3i\phi)(\tilde{d}x - i\tilde{d}y)$. Since $£_{\bar{e}_\phi} f = 2if$, $\tilde{d}(£_{\bar{e}_\phi} f) = 2i\tilde{d}f$. But it is easy to show (using cylindrical coordinates) that $\tilde{d}(£_{\bar{e}_\phi} f) = £_{\bar{e}_\phi}(\tilde{d}f)$, which completes the proof. This is a special case of a general theorem proved in §4.21 below.

3.10 A right-invariant vector field is invariant under the map R_g, generated by right translations analogously to L_g. Figure 3.10 applies here, too, so they form a Lie algebra. The integral curves through e of a right-invariant vector field are one-parameter subgroups for the same reason as for left-invariant integral curves. But the subgroups are in 1–1 correspondence with the two sets of integral curves, so the curves are the same. The integral curves of a left-invariant field not passing through e are obtained by left-translation of those which do, i.e. of the one-parameter subgroups. The curve of $\bar{V}$ through, say, h is $hg_{\bar{V}_e}(t)$. The right-invariant curve through h is $g_{\bar{V}_e}(t)h$. This is not the same unless h and $g_{\bar{V}_e}(t)$ commute.

3.11 (a) Because left-translation by h^{-1} is a 1–1 map of neighborhoods of h on to neighborhoods of e, the vector-field map L_h is also 1–1 and invertible. It follows that if $\{\bar{V}_i(e)\}$ is linearly independent then so is $\{L_h\bar{V}_i(e) \equiv \bar{V}_i(h)\}$ for any h.

(b) The point is that the fields $\{\bar{V}_i\}$ are globally a basis, so any vector field is defined by giving $\{\alpha_i(g)\}$ for all g. This maps TG on to $G \times R^n$.

3.12 (b) The key step is that $(B^{-1}AB)^n = B^{-1}ABB^{-1}AB \ldots B^{-1}AB = B^{-1}A^nB$.

(c) Block-diagonal matrices are easy to exponentiate, since

$$\begin{pmatrix} P_1 & 0 & 0 & \cdots \\ 0 & P_2 & 0 & \cdots \\ 0 & 0 & P_3 & \cdots \\ \vdots & \vdots & \vdots & \ddots \end{pmatrix}^n = \begin{pmatrix} (P_1)^n & 0 & 0 & \cdots \\ 0 & (P_2)^n & 0 & \cdots \\ 0 & 0 & (P_3)^n & \cdots \\ \vdots & \vdots & \vdots & \ddots \end{pmatrix}.$$

In case (i), $\exp(t\lambda_j)$ is the usual exponential function. In (ii),

$$\exp\left[t\begin{pmatrix} r_j & s_j \\ -s_j & r_j \end{pmatrix}\right]$$
$$= \exp\left[t\frac{1}{\sqrt{2}}\begin{pmatrix} 1 & -\mathrm{i} \\ \mathrm{i} & -1 \end{pmatrix}\begin{pmatrix} r_j+\mathrm{i}s_j & 0 \\ 0 & r_j-\mathrm{i}s_j \end{pmatrix}\frac{1}{\sqrt{2}}\begin{pmatrix} 1 & -\mathrm{i} \\ \mathrm{i} & -1 \end{pmatrix}\right]$$
$$= \frac{1}{\sqrt{2}}\begin{pmatrix} 1 & -\mathrm{i} \\ \mathrm{i} & -1 \end{pmatrix}\exp\left[t\begin{pmatrix} r_j+\mathrm{i}s_j & 0 \\ 0 & r_j-\mathrm{i}s_j \end{pmatrix}\right]\frac{1}{\sqrt{2}}\begin{pmatrix} 1 & -\mathrm{i} \\ \mathrm{i} & -1 \end{pmatrix}$$
$$= \frac{1}{\sqrt{2}}\begin{pmatrix} 1 & -\mathrm{i} \\ \mathrm{i} & -1 \end{pmatrix}\exp(tr_j)\begin{pmatrix} \cos tr_j + \mathrm{i}\sin ts_j & 0 \\ 0 & \cos tr_j - \mathrm{i}\sin ts_j \end{pmatrix}$$
$$\times\frac{1}{\sqrt{2}}\begin{pmatrix} 1 & -\mathrm{i} \\ \mathrm{i} & -1 \end{pmatrix},$$

from which the answer follows. For (iii), some experimentation will verify that

$$\begin{pmatrix} x & 1 & 0 & 0 & \cdots \\ 0 & x & 1 & 0 & \cdots \\ 0 & 0 & x & 1 & \cdots \\ \vdots & \vdots & \vdots & \vdots & \ddots \end{pmatrix}^n$$
$$= \begin{pmatrix} x^n & \frac{\mathrm{d}}{\mathrm{d}x}x^n & \frac{1}{2!}\frac{\mathrm{d}^2}{\mathrm{d}x^2}x^n & \frac{1}{3!}\frac{\mathrm{d}^3}{\mathrm{d}x^3}x^n & \cdots \\ 0 & x^n & \frac{\mathrm{d}}{\mathrm{d}x}x^n & \frac{1}{2!}\frac{\mathrm{d}^2}{\mathrm{d}x^2}x^n & \cdots \\ 0 & 0 & x^n & \frac{\mathrm{d}}{\mathrm{d}x}x^n & \cdots \\ \vdots & \vdots & \vdots & \vdots & \ddots \end{pmatrix}$$

When multiplied by t^n and put into the exponentiation sum, this gives (3.59c).

3.13 The sequence of matrices

$$\begin{pmatrix} \cos t & \sin(t/2) \\ -\sin t & \cos t \end{pmatrix}$$

is a continuous path containing e (for $t=0$) and

$$\begin{pmatrix} -1 & 1 \\ 0 & -1 \end{pmatrix}$$

(for $t=\pi$). The matrix is not in a one-parameter subgroup because it is not the exponential of any matrix. This follows from exercise 3.12. It is easy to verify that none of the forms (3.59) can be transformed as in (3.56) to give the desired matrix, because of the negative elements on the main diagonal.

3.14 (a) An eigenvalue λ of A satisfies the equation $\det(A-\lambda I)=0$ $=\det(A-\lambda I)^{\mathrm{T}}=\det(A^{\mathrm{T}}-\lambda I)=\det(A^{-1}-\lambda I)$ and so is an eigenvalue of A^{-1}. The converse also holds.

(b) $\det(A-\lambda I)=0 \Rightarrow 0=\det(A^{-1})\det(A-\lambda I)=\det(I-\lambda A^{-1})$ $=\det(-\lambda I)\det(A^{-1}-\lambda^{-1}I)$. None of the eigenvalues is zero since $\det(A)\neq 0$, so we conclude $\det(A^{-1}-\lambda^{-1}I)=0$. Thus, if A is in $O(n)$ and λ is an eigenvalue of A, so is $1/\lambda$. But the equation $\det(A-\lambda I)=0$ is real, so its solutions come in complex-conjugate pairs. In order for these pairs to be inverse they must have the form $(e^{i\theta}, e^{-i\theta})$.

(c) These forms are just (3.58a, b) for the given eigenvalues, with (3.62b) being a special case of (3.62c). The only case to exclude is (3.58c) when $\mu_j=\pm 1$. This form is impossible because $B^{-1}AB$ is in $O(n)$ while (3.58c) is easily seen not to be.

(d) The Lie algebra can be found by looking at the tangent space of any element, in particular of e, and so we can restrict attention to the generators of the one-dimensional subgroups of $SO(n)$. The problem may be solved by examining canonical forms, but the following method is quicker. Consider the element $\exp(tA)$, where A is in the Lie algebra of $O(n)$. Then $[\exp(tA)]^{-1}=\exp(-tA)$ and $[\exp(tA)]^{\mathrm{T}}$ $=\exp(tA^{\mathrm{T}})$. These are equal for any t, so $A^{\mathrm{T}}=-A$. The converse is proved in the same way. The dimension of $O(n)$ is the maximum number of linearly independent antisymmetric $n\times n$ matrices, $\frac{1}{2}n(n-1)$.

3.15 A matrix A is in $SO(n)$ if and only if its canonical form (3.62) has an even number of blocks (-1). An element of $O(n)$ not in $SO(n)$ has an odd number of blocks (-1), and may obviously be obtained from one in $SO(n)$ by the given transformation.

3.16 As in the previous problem, the canonical form of A in $SO(n)$ has an even number of blocks (-1), which may be ordered to be a special case of (3.62c) with $\theta=\pi$. Any canonical form (3.62a) or (3.62c) is a special case of the exponentials (3.59a) or (3.59b). For $SO(3)$, the canonical form must be one block (3.62a) and another (3.62c). The eigenvector for (3.62a) is the axis of rotation.

3.17 Use equation (3.60).

3.18 The matrix $\mathrm{diag}[\exp(ia_1t), \exp(ia_2t), \ldots]$ is the exponential of $\mathrm{diag}(ia_1t, ia_2t, \ldots)$. The first matrix has determinant $\exp(it\,\Sigma_j a_j)$, and if this equals 1 the second matrix is traceless. Let us establish a correspondence between complex numbers and real 2 × 2 matrices, defined by $a + ib \Leftrightarrow \begin{pmatrix} a & b \\ -b & a \end{pmatrix}$. Then multiplication preserves this: $(a + ib)(c + id) \Leftrightarrow \begin{pmatrix} a & b \\ -b & a \end{pmatrix}\begin{pmatrix} c & d \\ -d & c \end{pmatrix}$. There is thus a group isomorphism between the complex numbers and matrices of this special form. (It is in fact an algebra isomorphism, since it is preserved by addition as well.) This generalizes to a group isomorphism between $GL(n, C)$ and the subgroup of $GL(2n, R)$ consisting of matrices built of 2 × 2 blocks of the form $\begin{pmatrix} a & b \\ -b & a \end{pmatrix}$. Hermitian conjugation in $GL(n, C)$ is simply the transpose operation in $GL(2n, R)$. Thus we may regard $U(n)$ as a subgroup of $O(2n)$. Since $O(2n)$ is generated by antisymmetric $2n \times 2n$ matrices, $U(n)$ is generated by anti-Hermitian matrices. These must be trace-free by our first observation and by exercise 3.20(a).

3.20 (a) $(B^{-1}AB)^j{}_k = (B^{-1})^j{}_l A^l{}_m B^m{}_k$. But $(B^{-1})^j{}_l B^m{}_j = \delta_l{}^m$, so $\mathrm{tr}(B^{-1}AB) = \mathrm{tr}(A)$.

(b) Since $\det(B^{-1}) = 1/\det(B)$, we have $\det(\exp(A)) = \det(B^{-1}\exp(A)B) = \det(\exp(B^{-1}AB))$; moreover, $\exp(\mathrm{tr}(A)) = \exp(B^{-1}(\mathrm{tr}\,A)B) = \exp(\mathrm{tr}(B^{-1}AB))$. Thus, we need only prove (3.67) for the various canonical forms. The form (3.58a) is trivial. For (3.58b), inspection of (3.59b) proves the result. The same is true of (3.58c), for the matrix written in (3.59c) has unit determinant.

3.21 (a) Use the identity $(\bar{a} \times \bar{b}) \times \bar{c} = (\bar{a} \cdot \bar{c})\bar{b} - (\bar{b} \cdot \bar{c})\bar{a}$.

3.22 (a) Note that $\det\begin{pmatrix} a & b \\ -\bar{b} & \bar{a} \end{pmatrix} = |a|^2 + |b|^2$ only vanishes for the zero matrix.

(b) The dimension is 4 because there are 4 real numbers freely chosen to define an element of H.

(d) Equation (3.73) is the equation for S^3, so the 1–1 mapping is established by associating the point $(\alpha_1, \alpha_2, \alpha_3, \alpha_4)$ of S^3 in R^4 with the matrix A.

3.23 Since $[g(s)]^{-1} = \exp(-s\bar{Y})$ we can write (3.79) as

$$\exp(s\bar{Y})\exp(t\bar{X})\exp(-s\bar{Y}) = \exp[tAd_{g(s)}(\bar{X})].$$

Differentiating both sides with respect to t at $t = 0$ gives

$$\exp(s\bar{Y})\bar{X}\exp(-s\bar{Y}) = Ad_{g(s)}(\bar{X}).$$

Expanding the left-hand side in powers of s gives

$$\bar{X} + s[\bar{Y}, \bar{X}] + \tfrac{1}{2}s^2[\bar{Y}, [\bar{Y}, \bar{X}]] + \tfrac{1}{3}s^3[\bar{Y}, [\bar{Y}, [\bar{Y}, \bar{X}]]] + \ldots,$$

proving the result.

3.24 (b) $\bar{l}_x(Y_{1-1}) = iY_{10}/\sqrt{2}$; $\bar{l}_y(Y_{1-1}) = Y_{10}/\sqrt{2}$; $\bar{l}_z(Y_{1-1}) = iY_{1-1}$; $\bar{l}_x(Y_{10}) = i(Y_{1-1} - Y_{11})/\sqrt{2}$; $\bar{l}_y(Y_{10}) = -(Y_{1-1} + Y_{11})/\sqrt{2}$; $\bar{l}_z(Y_{10}) = 0$; $\bar{l}_x(Y_{11}) = -iY_{10}/\sqrt{2}$; $\bar{l}_y(Y_{11}) = Y_{10}/\sqrt{2}$; $\bar{l}_z(Y_{11}) = iY_{11}$.

Any complex linear combination of the three functions will thus be transformed into another one by differentiation along $\bar{l}_x$, $\bar{l}_y$, or $\bar{l}_z$.

3.25 The first of (3.30) implies $\bar{l}_z(f) = 0$. The remaining two then imply $\bar{l}_x(f) = \bar{l}_y(f) = 0$. Thus, f must be constant on the sphere.

3.26

$$\Lambda^{j'}{}_k = \left(\frac{3}{8\pi}\right)^{1/2} \begin{pmatrix} 1 & -i & 0 \\ 0 & 0 & \sqrt{2} \\ 1 & i & 0 \end{pmatrix};$$

$$\Lambda^k{}_{j'} = \frac{1}{2}\left(\frac{8\pi}{3}\right)^{1/2} \begin{pmatrix} 1 & 0 & 1 \\ i & 0 & -i \\ 0 & \sqrt{2} & 0 \end{pmatrix};$$

$$X^{j'}{}_{k'} = \frac{i}{\sqrt{2}} \begin{pmatrix} 0 & 1 & 0 \\ 1 & 0 & -1 \\ 0 & -1 & 0 \end{pmatrix}; \quad X^j{}_k = L_1.$$

4.1 $\mathbf{B}(\bar{U} + \bar{W}, \bar{U} + \bar{W}) = 0 = \mathbf{B}(\bar{U}, \bar{U}) + \mathbf{B}(\bar{U}, \bar{W}) + \mathbf{B}(\bar{W}, \bar{U}) + \mathbf{B}(\bar{W}, \bar{W})$ $= \mathbf{B}(\bar{U}, \bar{W}) + \mathbf{B}(\bar{W}, \bar{U})$.

4.2 (a) $\tilde{p}(\ldots, \bar{U}, \ldots, \bar{W}, \ldots) = p_{\ldots i \ldots j \ldots} U^i W^j = -p_{\ldots j \ldots i \ldots} U^i W^j$ $= -\tilde{p}(\ldots, \bar{W}, \ldots, \bar{U}, \ldots)$.

(b) Follows from $A_{ijk} = -A_{jik} = A_{kij}$, etc.

(c) $A_{ij}B^{ij} = \frac{1}{2}A_{ij}B^{ij} + \frac{1}{2}A_{ji}B^{ji} = \frac{1}{2}(A_{ij}B^{ij} - A_{ij}B^{ji}) = A_{ij}B^{[ij]}$. First step merely relabels dummy indices.

(d) $B^{[ij]} = \frac{1}{2}[\mathbf{B}(\tilde{\omega}^i, \tilde{\omega}^j) - \mathbf{B}(\tilde{\omega}^j, \tilde{\omega}^i)] = 0$.

4.3 The components of $\tilde{\omega}$ are its values on sets of p basis vectors. If $p > n$ there must be at least two vectors which are the same. Exchanging these makes no change at all, but at the same time must change the sign of the component. The only number which equals its negative is zero.

4.4 One need only demonstrate that the sum of two p-forms is a p-form, i.e. totally antisymmetric, and that a p-form times a number is likewise a p-form. The dimension of this space is the number of independent components a p-form can have, C^n_p from equation (4.7).

4.5 $\tilde{p} \wedge \tilde{q}(\bar{U}, \bar{V}) = \tilde{p}(\bar{U})\tilde{q}(\bar{V}) - \tilde{q}(\bar{U})\tilde{p}(\bar{V}) = -\tilde{p} \wedge \tilde{q}(\bar{V}, \bar{U})$. Obviously $\tilde{p} \wedge \tilde{p}(\bar{U}, \bar{V}) = 0$ for any $\bar{U}$, $\bar{V}$.

4.6 Check equation (4.9): $\tilde{\alpha}(\bar{U}, \bar{V}) = \frac{1}{2}\alpha_{ij}[\tilde{\omega}^i(\bar{U})\tilde{\omega}^j(\bar{V}) - \tilde{\omega}^j(\bar{U})\tilde{\omega}^i(\bar{V})]$ $= \frac{1}{2}\alpha_{ij}(U^iV^j - V^iU^j) = \frac{1}{2}\alpha_{ij}U^iV^j + \frac{1}{2}\alpha_{ji}V^iU^j = \alpha_{ij}U^iV^j$. The number of independent two-forms $\tilde{\omega}^j \wedge \tilde{\omega}^k$ is $\frac{1}{2}n(n-1)$, which is the dimension of the space of two-forms.

4.7 $(1 + 1)^n = \Sigma^n_{p=0} C^n_p$ by the binomial theorem. Notice that this includes $p = 0$, the one-dimensional space of zero-forms.

4.8 By equation (4.9) and its generalization we know that $\tilde{q} = (1/2!) \times q_{jk}\tilde{\omega}^j \wedge \tilde{\omega}^k$ and $\tilde{p} \wedge \tilde{q} = (1/3!)(\tilde{p} \wedge \tilde{q})_{ijk}\tilde{\omega}^i \wedge \tilde{\omega}^j \wedge \tilde{\omega}^k$. But $\tilde{p} \wedge \tilde{q} = (1/2!)p_i q_{jk}\tilde{\omega}^i \wedge (\tilde{\omega}^j \wedge \tilde{\omega}^k) = (1/2!)p_i q_{jk}\tilde{\omega}^i \wedge \tilde{\omega}^j \wedge \tilde{\omega}^k = (1/2!) \times p_{[i}q_{jk]}\tilde{\omega}^i \wedge \tilde{\omega}^j \wedge \tilde{\omega}^k$. Since $\{\tilde{\omega}^i \wedge \tilde{\omega}^j \wedge \tilde{\omega}^k\}$ form a basis, we conclude $(\tilde{p} \wedge \tilde{q})_{ijk} = (3!/2!)p_{[i}q_{jk]} = \frac{1}{2}(p_i q_{jk} + p_k q_{ij} + p_j q_{ki} - p_i q_{kj} - p_k q_{ji} - p_j q_{ik}) = p_i q_{jk} + p_k q_{ij} + p_j q_{ki}$. The generalization to equation (4.11) is straightforward. The student is reminded that $C_p^{p+q} = C_q^{p+q}$, so that (4.11) treats $\tilde{p}$ and $\tilde{q}$ even-handedly.

4.9 Equation (4.16) is bilinear in $\tilde{\beta}$ and $\tilde{\alpha}$, so it suffices to prove it for the case $\tilde{\beta} = \tilde{\omega}^1 \wedge \ldots \wedge \tilde{\omega}^p$ and $\tilde{\alpha} = \tilde{\omega}^{p+1} \wedge \ldots \wedge \tilde{\omega}^{p+q}$. Then $\tilde{\beta} \wedge \tilde{\alpha}(\bar{\xi}) = (\tilde{\omega}^1 \wedge \ldots \wedge \tilde{\omega}^p \wedge \tilde{\omega}^{p+1} \wedge \ldots \wedge \tilde{\omega}^{p+q})(\bar{\xi})$. This wedge product has $(p+q)!$ terms, all possible permutations of the indices. The ones which have $\tilde{\omega}^1$ first contract with $\bar{\xi}$ to give $\tilde{\omega}^1(\bar{\xi})(\tilde{\omega}^2 \wedge \ldots \wedge \tilde{\omega}^p \wedge \tilde{\omega}^{p+1} \wedge \ldots \wedge \tilde{\omega}^{p+q})$. Each term with $\tilde{\omega}^2$ first is an odd permutation of one with $\tilde{\omega}^1$ first, obtained by exchanging $\tilde{\omega}^1$ with $\tilde{\omega}^2$. Therefore these terms contract with $\bar{\xi}$ to give $-\tilde{\omega}^2(\bar{\xi})(\tilde{\omega}^1 \wedge \tilde{\omega}^3 \wedge \ldots \wedge \tilde{\omega}^p \wedge \tilde{\omega}^{p+1} \wedge \ldots \wedge \tilde{\omega}^{p+q})$. Similarly there is a contraction $\tilde{\omega}^3(\bar{\xi})(\tilde{\omega}^1 \wedge \tilde{\omega}^2 \wedge \tilde{\omega}^4 \wedge \ldots)$ and so on. Now the first p such contractions are just $\tilde{\beta}(\bar{\xi}) \wedge \tilde{\alpha}$, since they involve only the one-forms in $\tilde{\beta}$.
The remaining q contractions are the contractions of $\bar{\xi}$ with the one-forms in $\tilde{\alpha}$, with $\tilde{\beta}$ wedged in front, except that their overall sign is governed by the degree of $\tilde{\beta}$. That is, the first such term is $(-1)^p\tilde{\omega}^{p+1}(\bar{\xi})(\tilde{\omega}^1 \wedge \ldots \wedge \tilde{\omega}^p \wedge \tilde{\omega}^{p+2} \wedge \ldots \wedge \tilde{\omega}^{p+q})$, and all other terms are also $(-1)^p$ times the terms that appear in $\tilde{\beta} \wedge \tilde{\alpha}(\bar{\xi})$.

4.10 In Cartesian coordinates, $\bar{W} = \bar{U} \times \bar{V}$ has components $(U^2V^3 - U^3V^2, U^3V^1 - U^1V^3, U^1V^2 - U^2V^1)$. Equation (4.20), with $\omega_{123} = 1$, becomes ${}^*(\bar{U} \times \bar{V})_{12} = (U \times V)^3$, etc., which proves equation (4.22).

4.11 (a) The right-hand side of the equation after (4.35) is simply a sum of $(p+1)!$ terms. The terms in which the lower index i is first number $p!$, and are the first in the next line. There are also $p!$ terms in which the lower m is first, and these are given (with the correct sign) by the second term. The final line before (4.36) performs the sum on i ($\delta^i{}_i = n$).
(b) Applying (4.36) $(n-p)$ times to reduce the n-delta to a p-delta gives a factor $1 \cdot 2 \ldots (n-p)$, as in (4.37).

4.12 (a) The keypoint to observe is that $\epsilon_{ij\ldots k}A^{1i}A^{2j}\ldots A^{nk} = A^{11} \times (\epsilon_{1j\ldots k}A^{2j}\ldots A^{nk}) + A^{12}(\epsilon_{2j\ldots k}A^{2j}\ldots A^{nk}) + \ldots = A^{11}(\epsilon_{\alpha\ldots\beta} \times A^{2\alpha}\ldots A^{n\beta}) - A^{12}(\epsilon_{\alpha\ldots\beta}A^{2\alpha}\ldots A^{n\beta}) + - \ldots$, where in the final line Greek indices assume only $(n-1)$ values in each sum, missing out 1 in the first sum, 2 in the second, etc. The pth term in parenthesis is, by (4.39), the determinant of the $(n-1)\times(n-1)$ matrix obtained by excluding the first row and the pth column of the original matrix. (b) Equation (4.39) is obviously antisymmetric if 2 is exchanged with 1, and this leads to the desired result.

4.13 Let the matrix $\Lambda^i{}_{j'}$ transform basis one-forms, $\tilde{\omega}^i = \Lambda^i{}_{j'}\tilde{d}x^{j'}$. Then $\tilde{\omega} = \Lambda^1{}_{j'}\ldots\Lambda^n{}_{k'}\tilde{d}x^{j'}\wedge\ldots\wedge\tilde{d}x^{k'} = \Lambda^1{}_{j'}\ldots\Lambda^n{}_{k'}\epsilon^{j'\ldots k'}\tilde{d}x^{1'}\wedge\ldots\wedge\tilde{d}x^{n'} = \det(\Lambda)\,\tilde{d}x^{1'}\wedge\ldots\wedge\tilde{d}x^{n'}$. But the metric's components transform by $g_{i'j'} = \Lambda^k{}_{i'}\Lambda^l{}_{j'}g_{kl}$, which as a matrix equation has the determinant $\det(g_{i'j'}) = [\det(\Lambda)]^2\det(g_{ij})$. (Note that the determinant of the transpose of a matrix equals that of the original matrix, by exercise 4.12(b).) Moreover, the original basis was orthonormal, so $\det(g_{ij}) = \pm 1$. It follows that $\det(\Lambda) = |\det(g_{i'j'})|^{1/2}$, proving the result.

4.14 (a) Special case of property (2) with $\tilde{\alpha}$ the zero-form f, using property (3) to eliminate $\tilde{d}\tilde{d}g$.
(b) Obvious.

4.15 $V^{i'}{}_{,k'} = \Lambda^j{}_{k'}(\Lambda^{i'}{}_l V^l)_{,j} = \Lambda^j{}_{k'}\Lambda^{i'}{}_l V^l{}_{,j} + \Lambda^j{}_{k'}\Lambda^{i'}{}_{l,j}V^l$. The second term is not part of the tensor transformation law. Now, in $[\bar{U},\bar{V}]^{i'}$ the two 'wrong' terms are $U^jV^l\Lambda^{i'}{}_{l,j} - V^jU^l\Lambda^{i'}{}_{l,j}$. This vanishes because $\Lambda^{i'}{}_{l,j} = \Lambda^{i'}{}_{j,l} = \partial^2x^{i'}/\partial x^j\partial x^l$.

4.16 $\text{Curl}(\text{grad} f) = {}^*\tilde{d}(\tilde{d}f) = {}^*(\tilde{d}\tilde{d}f) = 0$.
$\text{Div}(\text{curl}\,\bar{a}) = \tilde{d}^*({}^*\tilde{d}\tilde{a}) = \tilde{d}({}^{**}\tilde{d}\tilde{a}) = \tilde{d}\tilde{d}\tilde{a} = 0$.

4.17 The first expression in (4.64) is the component version of $\tilde{d}\tilde{\alpha} = 0$. The next step is identical to that in exercise 4.11(a). The second term on the right-hand side of (4.64) is obtained from the first by simply exchanging j and l; this brings in a sign change and shows they are equal.

4.18 $\text{Div}(\bar{a}) = \tilde{d}^*\tilde{a} = 0 \Rightarrow {}^*\tilde{a} = \tilde{d}\tilde{b}$ for some $\tilde{b} \Rightarrow \tilde{a} = {}^*\tilde{d}\tilde{b} = \text{curl}(\bar{b})$.
$\text{Curl}(\bar{a}) = {}^*\tilde{d}\tilde{a} = 0 \Rightarrow \tilde{d}\tilde{a} = 0 \Rightarrow \tilde{a} = \tilde{d}f = \text{grad}(f)$.

4.19 Choose a vector basis $(\bar{e}_1, \bar{e}_2, \ldots, \bar{e}_n)$ for which $(\bar{e}_2, \ldots, \bar{e}_n)$ are tangent to ∂U: $\tilde{n}(\bar{e}_p) = 0$ for $2 \leqslant p \leqslant n$. This means that $\tilde{n}(\bar{e}_1) \neq 0$, since $\tilde{n}$ must have at least one nonzero component. Since $\tilde{\omega}$ is an n-form, it has only one independent component, which is $\tilde{\omega}(\bar{e}_1, \bar{e}_2, \ldots, \bar{e}_n) = (\tilde{n}\wedge\tilde{\alpha})(\bar{e}_1, \bar{e}_2, \ldots, \bar{e}_n)$. The only nonzero term in this occurs when $\tilde{n}$ contracts with $\bar{e}_1$, so we have $\tilde{n}(\bar{e}_1)\tilde{\alpha}(\bar{e}_2, \ldots, \bar{e}_n)$. But $\tilde{\alpha}(\bar{e}_2, \ldots, \bar{e}_n)$ is the one independent component of $\tilde{\alpha}|_{\partial U}$ in this basis, which is therefore found to be $\tilde{\omega}(\bar{e}_1, \ldots, \bar{e}_n)/\tilde{n}(\bar{e}_1)$. But $\tilde{\alpha}$ itself is not unique: $\tilde{\alpha} + f\tilde{n}$ for any function f works just as well. Now, the only requirement of $\tilde{n}$ is that it be normal to ∂U. On our basis this means $\tilde{n}$ has components $(n^1, 0, \ldots, 0)$. Clearly any two normals $\tilde{n}$ and $\tilde{n}'$ are related by $\tilde{n} = f\tilde{n}'$. By our previous result, this changes $\tilde{\alpha}|_{\partial U}$ to $f^{-1}\tilde{\alpha}|_{\partial U}$, so that $\tilde{n}(\bar{\xi})\,\tilde{\alpha}|_{\partial U}$ is unchanged.

4.20 Following the steps from (4.76) we find $\tilde{d}[\tilde{\omega}(\xi)] = (f\xi^i)_{,i} \times \tilde{d}x^1 \wedge \ldots \wedge \tilde{d}x^n = f^{-1}(f\xi^i)_{,i}\tilde{\omega}$.

4.21 Use exercise 4.13, showing that the metric in spherical polars is $\mathrm{diag}(1, r^2, r^2 \sin^2\theta)$. Then div ξ follows from setting $f = r^2 \sin\theta$ in (4.80).

4.22 From (4.67), $£_{\bar{V}}(\rho\tilde{\omega}) = \tilde{d}[\rho\tilde{\omega}(\bar{V})] = \tilde{d}[\tilde{\omega}(\rho\bar{V})] = \mathrm{div}(\rho\bar{V})\tilde{\omega}$.

4.23 (a) Since $\tilde{\omega}(\xi) = {}^*\xi$, the dual of (4.77) gives (4.81) immediately.

(b) ${}^*\mathbf{F}$ is a $(n-p)$-form, $\mathrm{d}^*\mathbf{F}$ is a $(n-p+1)$-form, so ${}^*\mathrm{d}^*\mathbf{F}$ is a $(p-1)$-vector. Equation (4.83) is proved by a simple generalization of (4.76).

(c) $(\mathrm{div}_{\tilde{\omega}}\,\mathbf{F})^{i\ldots j} = f^{-1}(fF^{ki\ldots j})_{,k}$.

4.24 (a) If $\tilde{\omega} = \tilde{d}\tilde{\alpha}$ then $\int \tilde{\omega} = \oint \tilde{\alpha}$; but the second integral vanishes since there is no boundary.

(b) $\tilde{d}\tilde{\omega} = \tilde{d}x^1 \wedge \tilde{d}x^2 \wedge \tilde{d}x^3$ is the usual volume-form, so if B is the unit ball (the interior of S^2 in R^3) then $\int_B \mathrm{d}\tilde{\omega} =$ volume of ball $= 4\pi/3$. By Stokes' theorem, $\int_B \tilde{d}\tilde{\omega} = \int_{S^2} \tilde{\omega}|_{S^2}$. Now, any two-form on a two-dimensional manifold is closed, since all three-forms vanish identically. Thus, $\tilde{\omega}$ is closed, but violates (a) above, so it is not exact.

(c) We are given $\tilde{\beta}$ defined everywhere on S^2 with $\tilde{d}\tilde{\beta} = 0$. Integrate $\tilde{d}\tilde{\beta}$ over any region of S^2 bounded by a single closed curve $\mathscr{C}$ to find $\oint_{\mathscr{C}} \tilde{\beta} = 0$ for any $\mathscr{C}$. This can only be true if $\tilde{\beta} = \tilde{d}f$ for some f: otherwise some curve $\mathscr{C}$ could be found on which $\tilde{\beta}$ would have a non-vanishing integral. In fact f can be constructed by choosing an arbitrary value f_0 at a point P and integrating $\tilde{\beta}$ on any curve from P to any point Q, defining $f(Q) = f_0 + \int \tilde{\beta}$. The condition $\oint \tilde{\beta} = 0$ guarantees that $f(Q)$ is independent of the path from P to Q.

4.25 (a) (i) and (ii) are trivial. For (iii) suppose $\tilde{\alpha} - \tilde{\beta} = \tilde{d}\tilde{\mu}_1$, $\tilde{\beta} - \tilde{\gamma} = \tilde{d}\tilde{\mu}_2$. Then $\tilde{\alpha} - \tilde{\gamma} = \tilde{d}(\tilde{\mu}_1 + \tilde{\mu}_2)$.

(b) First prove that if $\tilde{\beta}_1 \approx \tilde{\alpha}_1$ and $\tilde{\beta}_2 \approx \tilde{\alpha}_2$ then $a\tilde{\beta}_1 + b\tilde{\beta}_2 \approx a\tilde{\alpha}_1 + b\tilde{\alpha}_2$ for any real numbers a, b. This is trivial since there exist $\tilde{\mu}_1$ and $\tilde{\mu}_2$ such that $\tilde{\beta}_1 = \tilde{\alpha}_1 + \tilde{d}\tilde{\mu}_1$ and $\tilde{\beta}_2 = \tilde{\alpha}_2 + \tilde{d}\tilde{\mu}_2 \Rightarrow a\tilde{\beta}_1 + b\tilde{\beta}_2 = a\tilde{\alpha}_1 + b\tilde{\alpha}_2 + \tilde{d}(a\tilde{\mu}_1 + b\tilde{\mu}_2)$. Thus we can consistently define the linear combination $aA_1 + bA_2$ of equivalence classes A_1 and A_2 to be the equivalence class of the same linear combination of any of their elements. We can thus regard the equivalence classes themselves as vectors in a vector space: the identity is the equivalence class of the zero vector of Z^p, the inverse of any class A is $-A$, etc.

(c) Take the vector $(0, b)$ in R^2. What is its equivalence class? It is all vectors of the form $(0, b) + (a, 0)$ for arbitrary a and fixed b. The locus of points is a straight line parallel to the x-axis a distance b from it. In this fashion we can identify the space of equivalence classes with the congruence of such lines.

4.26 The vector field cannot vanish anywhere because the mapping leaves no point fixed: a fixed point would correspond to a block (+ 1) in the canonical form of T, but T is already in canonical form with no such blocks. Notice that it is crucial that the sphere be odd-dimensional.

4.27 (a) Trivial.

(b) $H^{n-1}(S^{n-1})$ is a one-dimensional vector space (R^1), so any equivalence class is a multiple of any other. Since $\tilde{\omega}$ is not exact, it is in a nonzero equivalence class. By exercise 4.25(b) it follows that *every* equivalence class has a multiple of $\tilde{\omega}$ in it, so that for any $\tilde{\alpha}$ there exists a number of such that $\tilde{\alpha} - a\tilde{\omega} \approx 0$, i.e. is exact. Integrating over S^{n-1} gives the value of a.

(c) If $\tilde{\alpha} - a\tilde{\omega} = \tilde{d}\tilde{\beta}$, then $\tilde{\beta}$ is a $(n-2)$-form. Let its dual with respect to $\tilde{\omega}$ be $\bar{V}$, $\bar{V} = {}^*\tilde{\beta}$, or $\tilde{\beta} = (-1)^n\, {}^*\bar{V}$. Then $\tilde{d}\tilde{\beta} = (-1)^n\,(\mathrm{div}_{\tilde{\omega}}\,\bar{V})\,\tilde{\omega}$. Let f equal the dual of $\tilde{\alpha}$, so that we get $(f-a)\,\tilde{\omega} = (-1)^n\,(\mathrm{div}_{\tilde{\omega}}\,\bar{V})\tilde{\omega}$. This is equivalent to what was to have been proved.

(d) In this case $\tilde{\omega} = x\tilde{d}y - y\tilde{d}x = \tilde{d}\theta$ on the unit circle, where θ is the polar angle. Any other one-form $\tilde{\alpha}$ can be written as $g(\theta)\tilde{d}\theta$, so we wish to find $f(\theta)$ such that $\tilde{d}f = [g(\theta) - a]\,\tilde{d}\theta$ everywhere. Since $\tilde{d}f = (\mathrm{d}f/\mathrm{d}\theta)\tilde{d}\theta$, f solves $\mathrm{d}f/\mathrm{d}\theta = g(\theta) - a$ or $f = \int g\mathrm{d}\theta - a\theta$. To make f continuous we require $f(0) = f(2\pi)$, or $2\pi a = \int_0^{2\pi} g\mathrm{d}\theta$. This is the same as we deduced in (b) above. By reversing the reasoning in (b), we can conclude from this that $H^1(S^1) = R^1$.

4.28 (a) We construct f as in the solution to exercise 4.24(c) above.

(b) Suppose M is simply connected, and let $\tilde{\alpha}$ be any closed one-form field sufficiently smooth on M. Then $\int_{\mathscr{C}} \tilde{\alpha}$ changes smoothly as the closed curve $\mathscr{C}$ is contracted. But $\mathscr{C}$ can always be made small enough to be entirely within a region in which the Poincaré lemma (§4.19) applies, in which $\int_{\mathscr{C}} \tilde{\alpha} = 0$. By continuity (i.e. by joining together a number of such small curves into a large one) it follows that $\int_{\mathscr{C}} \tilde{\alpha} = 0$ for any closed $\mathscr{C}$. Then by (a) above, $H^1(M) = 0$. The converse is similar. If $H^1(M) \neq 0$ there is a closed one-form $\tilde{\alpha}$ which is not exact. By (a) above it follows that $\int \tilde{\alpha} \neq 0$ for at least some closed curve $\mathscr{C}$ in M. If this curve could be smoothly deformed to a point then we would have $\int_{\mathscr{C}'} \tilde{\alpha} = 0$ for all sufficiently small contractions $\mathscr{C}'$ of $\mathscr{C}$. This would, as above, imply $\int_{\mathscr{C}} \tilde{\alpha} = 0$, a contradiction. So $\mathscr{C}$ cannot be shrunk to a point.

4.29 All and only linear combinations of $\{\bar{e}_1, \ldots, \bar{e}_m\}$ are annihilated by every one-form $\{\tilde{\omega}^{m+1}, \ldots, \tilde{\omega}^n\}$, so the complete ideal of these one-forms is the same as that of the original forms. Let $\tilde{\beta}$ be a q-form which annihilates every vector in X_P, and expand $\tilde{\beta}$ on the basis one-forms. Each term in $\tilde{\beta}$ is the wedge-product of q basis one-forms, and each term must contain at least one of $\{\tilde{\omega}^{m+1}, \ldots, \tilde{\omega}^n\}$. If not, that term would not annihilate every vector in X_P. This expansion of $\tilde{\beta}$ can be written as given in the exercise.

4.30 As in exercise 4.29, expand $\tilde{\gamma}$ on a one-form basis $\{\tilde{\alpha}_1, \ldots, \tilde{\alpha}_m, \tilde{\omega}^{m+1}, \ldots, \tilde{\omega}^n\}$. If $\tilde{\gamma}$ is in the ideal, each term has at least one of $\{\tilde{\alpha}_1, \ldots, \tilde{\alpha}_m\}$ and therefore satisfies (4.90). Conversely, if $\tilde{\gamma}$ is a q-form ($q \leqslant n - m$) and satisfies (4.90), construct a set of vectors $\{\bar{x}, \bar{y}_2, \ldots, \bar{y}_{m+q}\}$ in which $\bar{x}$ is in the annihilator X_P of the $\tilde{\alpha}_j$s, and the $\bar{y}_j$s are not. Let the $(m+q)$-form in (4.90) operate on this set. The only nontrivial terms occur when $\bar{x}$ is an argument of $\tilde{\gamma}$. If (4.90) is to vanish even with these terms, for arbitrary $\bar{y}_j$s, it must be that $\tilde{\gamma}(\bar{x}) = 0$ for any $\bar{x}$ in X_P. This means $\tilde{\gamma}$ is in the complete ideal. The remaining case, $q + m > n$, renders (4.90) an identity, but then $\tilde{\gamma}$ when expanded upon the above basis necessarily involves at least one $\tilde{\alpha}_i$ in each term, and so is in the ideal.

4.31 (a) Let $\tilde{\beta}$ be in the complete ideal of the set $\{\tilde{\alpha}_j\}$. Then since $\tilde{\beta} = \Sigma\, \tilde{\gamma}^j \wedge \tilde{\alpha}_j$ for some set $\{\tilde{\gamma}^j\}$, we have $\tilde{d}\tilde{\beta} = \Sigma\, (\tilde{d}\tilde{\gamma}^j \wedge \tilde{\alpha}_j + \tilde{\gamma}^j \wedge \tilde{d}\tilde{\alpha}_j)$. The first term is in the ideal; the second is also, since $\tilde{d}\tilde{\alpha}_j$ can be written as $\Sigma\, \tilde{\mu}^k \wedge \tilde{\alpha}_k$.

(b) Use (4.90) and the fact that a p-form for $p > n$ vanishes identically.

4.32 (b) Any curve satisfying $U = \text{const}$, $V = \text{const}$ has a tangent vector which annihilates $\tilde{d}U$ and $\tilde{d}V$, hence $\tilde{\alpha}$ and $\tilde{\beta}$, and therefore lies in $\mathscr{H}$.

(c) Use the test (4.90) to determine under what conditions $\tilde{d}\tilde{\alpha}$ and $\tilde{d}\tilde{\beta}$ are in the ideal. This involves considerable algebra, which is helped by the hint given in the problem. This makes, for instance, $\tilde{B} \wedge \tilde{d}\tilde{A} = 0$. The result is that we can write $\tilde{d}\tilde{\alpha} \wedge \tilde{\alpha} \wedge \tilde{\beta} = \tilde{d}f \wedge \tilde{d}g \wedge [-(\tilde{d}\tilde{C} + \tilde{A} \wedge \tilde{C} + \tilde{B} \wedge \tilde{F}) + f(\tilde{d}\tilde{A} + \tilde{B} \wedge \tilde{E}) + g(\tilde{d}\tilde{B} + \tilde{A} \wedge \tilde{B} + \tilde{B} \wedge \tilde{D})]$. This must vanish everywhere on the manifold. The term in square brackets is proportional to $\tilde{d}x \wedge \tilde{d}y$, which is independent of $\tilde{d}f \wedge \tilde{d}g$, so it must itself vanish. Since $\tilde{A}, \tilde{B}$, etc. are independent of f and g, this term vanishes if and only if the three terms in parentheses vanish separately. This gives the first three desired conditions. The remainder follow from $\tilde{d}\tilde{\beta} \wedge \tilde{\alpha} \wedge \tilde{\beta} = 0$.

(d) Factor $\tilde{d}x \wedge \tilde{d}y$ out of each term.

4.33 $\tilde{d}\tilde{\gamma} = \omega^2(\tilde{d}x \wedge \tilde{\alpha} + x\tilde{d}\tilde{\alpha}) + \tilde{d}y \wedge \tilde{\beta} + y\tilde{d}\tilde{\beta} = -\omega^2 y\tilde{d}x \wedge \tilde{d}t - \omega^2 x\tilde{d}y \wedge \tilde{d}t + \omega^2 x\tilde{d}y \wedge \tilde{d}t + \omega^2 y\tilde{d}x \wedge \tilde{d}t = 0$. Equation (4.97) is equally easy to show.

4.34 Consider the dot product $(\bar{\nabla}Y_{lm})\cdot({}^*\tilde{d}Y_{lm}) = g_{AB}(g^{AC}Y_{lm,C}) \times (\omega^{DB}Y_{lm,D}) = \omega^{DC}Y_{lm,C}Y_{lm,D}$ which vanishes by the antisymmetry of ω^{DC}. Since the metric is positive-definite, these vectors cannot be parallel unless they vanish, which happens only at isolated points if $l \neq 0$.

5.1 Take out $\tilde{d}S \wedge \tilde{d}T$ and multiply by P^2.

5.2 (a) $£_{\bar{U}}\tilde{\omega} = 0 = \tilde{d}[\tilde{\omega}(\bar{U})]$ since $\tilde{d}\tilde{\omega} = 0$. Since phase space satisfies the conditions of Poincaré's lemma (§4.19), there exists a function H such that $\tilde{\omega}(\bar{U}) = \tilde{d}H$, which leads to (5.16).
(b) Use $[£_{\bar{U}}, £_{\bar{V}}] = £_{[\bar{U},\bar{V}]}$ to show that if $\bar{U}$ and $\bar{V}$ are Hamiltonian, so is $[\bar{U}, \bar{V}]$. This is the bracket operation for their Lie algebra.

5.3 Antisymmetry of $\tilde{\omega}$.

5.4 Trivial. Use components.

5.5 As in exercise 5.2(a) above, $\tilde{\omega}(\bar{U}) = \tilde{d}H$.

5.6 (b) Just algebra.
(c) By (5.32), $\{f, \{g, h\}\} = \bar{X}_f\bar{X}_g(h)$; $\{g, \{h, f\}\} = -\{g, \{f, h\}\} = -\bar{X}_g\bar{X}_f(h)$; $\{h, \{f, g\} = X\{_{f,g}\}(h)$.

5.7 (b) $£_{\bar{U}}\tilde{\sigma} = 0$ because $£_{\bar{U}}\tilde{\omega} = 0$. But $£_{\bar{U}}\tilde{\sigma} = (\mathrm{div}_{\tilde{\sigma}}\bar{U})\tilde{\sigma}$.

5.8 Clearly $£_{\bar{X}_f}H = 0$, and since $\bar{X}_f$ has no momentum components $\partial f/\partial x^i = 0$ and $\partial f/\partial P_i = -U^i$. Thus $f = -U^iP_i$.

5.9 Use the transformation

$$(\Lambda^{i'}{}_j) = \begin{pmatrix} 1 & 0 & 0 & 0 \\ 0 & \cos\theta & -\sin\theta & 0 \\ 0 & \sin\theta & \cos\theta & 0 \\ 0 & 0 & 0 & 1 \end{pmatrix}.$$

5.10 (a) A three-form in a four-space has $C_3^4 = 4$ independent components.
(b) For example, $F_{[xy,z]} = 0 \Rightarrow F_{xy,z} + F_{zx,y} + F_{yz,x} = 0 = B_{z,z} + B_{y,y} + B_{x,x}$. This is (5.52c).

5.11 Matrix multiplication.

5.12 For example, $F^{t\nu}{}_{,\nu} = F^{tx}{}_{,x} + F^{ty}{}_{,y} + F^{tz}{}_{,z} = E_{x,x} + E_{y,y} + E_{z,z}$. This gives (5.52d).

5.13 (a) $({}^*\tilde{F})_{tx} = \frac{1}{2}(\omega_{yztx}F^{yz} + \omega_{zytx}F^{zy}) = F^{yz} = B_x$.
$({}^*\tilde{F})_{xy} = \frac{1}{2}(\omega_{tzxy}F^{tz} + \omega_{ztxy}F^{zt}) = F^{tz} = E_z$.
The whole matrix is the same as (5.53) with $B_i \to E_i, E_i \to -B_i$.
(b) Obvious from exercise 4.23.

5.14 (a) Prove by showing the first equation gives $\nabla \cdot \mathbf{B} = 4\pi\rho_m$, etc.
(b) $\tilde{d}\tilde{d}{}^*\tilde{F} = 0 \Rightarrow 0 = \tilde{d}{}^*\tilde{J} = \tilde{d}[\tilde{\omega}(\bar{J})] = (\mathrm{div}_{\tilde{\omega}}\bar{J})\tilde{\omega}$.

5.15 (a) Easily proved using components.
(b) Note that we had to restrict to $\mathscr{H}$ in order to integrate, since ${}^*\tilde{J}$ is a three-form.

(c) The restriction to $\mathscr{H}$ of $^*\tilde{J}$ is $J^t \tilde{d}x \wedge \tilde{d}y \wedge \tilde{d}z$. The restriction of $\tilde{F}$ to $\partial\mathscr{D}$ (a surface of const t and r) is $^*F_{\theta\phi}\tilde{d}\theta \wedge \tilde{d}\phi$. Now, $^*F_{\theta\phi} = \frac{1}{2}(\omega_{tr\theta\phi}F^{tr} + \omega_{rt\theta\phi}F^{rt}) = r^2 \sin\theta\, E_r$. (Recall equation (4.40) and exercise 4.21.) Therefore the integrals become, in conventional notation, $\int \rho\, d^3x = \oint E_r r^2 \sin\theta\, d\theta\, d\phi$.

5.16 (a) For example, consider the (t, x) component of (5.64): $F_{tx} = A_{x,t} - A_{t,x}$. Compare this with the usual definition, $E_x = \phi_{,x} + A_{x,t}$. Since $F_{tx} = -E_x$, the identifications follow. The remaining equations are consistent.

(b) $\phi \to \phi + f_{,t}; A^i \to A^i - \nabla_i f$.

(c) Static charge q at the origin: all components of **B** vanish and **E** $= qr^{-2}\mathbf{e}_r$. Only $(^*\tilde{F})_{\theta\phi}$ does not vanish, equalling (as in exercise 5.15) $q \sin\theta$. This gives $\alpha_{\phi,\theta} - \alpha_{\theta,\phi} = q \sin\theta$. Two possible solutions are $\{\alpha_\phi = -q\cos\theta, \alpha_\theta = \alpha_t = \alpha_r = 0\}$ and $\{\alpha_\theta = -q\phi \sin\theta, \alpha_\phi = \alpha_t = \alpha_r = 0\}$. These differ by a gauge transformation, and both render all other components of $\tilde{d}\tilde{\alpha}$ zero. But neither defines a well-behaved one-form. The first is undefined at the poles $\theta = 0$ and $\theta = \pi$; the second is multiple-valued.

5.17 Trivial.

5.18 $(\pounds_{\bar{U}}\bar{W})^i = [\bar{U}, \bar{W}]^i = U^j W^i{}_{,j} - W^j U^i{}_{,j} = U^t W^i{}_{,t} + U^\alpha W^i{}_{,\alpha} - W^\alpha U^i{}_{,\alpha}$, where sums on α run over only (x, y, z). Since $U^t = 1$ and $U^t{}_{,\alpha} = 0$ the result follows.

5.19 The results follow from $\Lambda^{t'}{}_x = \partial t'/\partial x = 0$, $\Lambda^t{}_{x'} = \partial t/\partial x' = 0$.

5.20 $dp = (\partial p/\partial\rho)\, d\rho + (\partial p/\partial S)\, dS$. This causes (5.77) to vanish because $dS \wedge dS = 0 = d\rho \wedge d\rho$.

5.21 Take the dual of (5.80), as in (5.82), and work out the components.

5.22 Use (3.37) in Cartesian coordinates to show that (5.85) and (5.86) reduce appropriately. Since these are tensor equations, their validity in one coordinate system assures their validity in all.

5.23 The isometry group $SO(3)$ clearly has elements which move any point of the sphere into any other: join them by a great circle and rotate about an axis perpendicular to that circle. So S^2 is homogeneous. The isotropy group of a point P is the set of all rotations which leave P fixed. This is obviously the subgroup $SO(2)$ of $SO(3)$, so S^2 is isotropic.

5.24 (a) Since V^i must vanish at P (the isotropy group leaves P fixed), its Taylor expansion is (5.96), for some matrix $A^i{}_j$. From (5.89) with $\Gamma^l{}_{ij} = 0$ in our coordinates (in which there is also no distinction between raised and lowered indices near P), we conclude that $A^i{}_j + A^j{}_i = 0$, which is (5.97).

(b) (5.98) is simple algebra.

(c) The isotropy group has the same Lie algebra as $SO(m)$, so the groups are identical at least in some neighborhood of the identity element. But a small neighborhood of P itself can be mapped 1–1 onto a neighborhood of the origin of R^m, and by (a) above the Killing fields can be mapped into one another to $O(x^2)$. Therefore their isotropy transformations can be put in 1–1 correspondence and the groups are identical.

(d) If $\mathbf{g}$ is not positive-definite, raising and lowering indices can involve sign changes, in our coordinates. Then (5.97) is properly $A_{ij} = -A_{ji}$, but $A^i{}_j \neq -A^j{}_i$. So the Lie algebra of the isotropy group does not involve antisymmetric matrices.

5.25 (a) The line $\theta = \text{const}$, $\phi = \text{const}$ is geometrically defined as an integral curve of $\bar{n}$. If the radial distance between spheres differed in different directions, the manifold would not be isotropic. Therefore g_{rr} is independent of θ and ϕ.

(b) Near P we can construct the coordinates of exercise 2.14 and transform to spherical polar coordinates via the standard flat-space transformation. These new coordinates are identical (as $r \to 0$) to those of (5.100), because the area of the spheres fixes r. Thus (5.102) is forced.

5.26 Algebra.

5.27 These correspond to $\zeta_{1m} = \text{const}$. It is easy to see that the norm of such a vector goes to zero as $r \to 0$, so it is in the isotropy group's algebra. The isotropy group $SO(3)$ is three-dimensional, and so is the set of all vectors generated by the three constants ζ_{1m}, so these vectors are the entire isotropy group.

5.28 Convert to Cartesian coordinates, or compute the norm of a vector with $V_m = 1$ and $\zeta_{1m} = 0$.

5.29 Clearly $f = 1 \Rightarrow S$ is E^3. $\partial/\partial x = \cos\phi \sin\theta\, \partial/\partial r + r^{-1} \cos\phi \cos\theta\, \partial/\partial\theta - r^{-1} \sin\phi \cos\theta\, \partial/\partial\phi$ is the vector generated by $V_1 = V_{-1} = (2\pi/3)^{1/2}$, $\zeta_{1m} = 0$.

5.30 From the second and third diagonal components of the matrices in (5.119) we conclude $r = \sin\chi/\sqrt{K}$ and $r = \sinh\chi/\sqrt{|K|}$ respectively. One needs to verify the first diagonal component. For example, in the case $K > 0$, $g_{\chi\chi} = g_{rr}(\partial r/\partial\chi)^2 = (1 - Kr^2)^{-1} \cos^2\chi/K = 1/K$, as required.

5.31 $w = r\cos\chi$, $x = r\sin\chi \sin\theta \sin\phi$, $y = r\sin\chi \sin\theta \cos\phi$, $z = r\sin\chi \cos\theta$. Then for example $g_{\theta\theta} = (\partial w/\partial\theta)^2 + (\partial x/\partial\theta)^2 + (\partial y/\partial\theta)^2 + (\partial z/\partial\theta)^2 = r^2 \sin^2\chi$. So with $r^2 = K^{-1}$ the identification is complete.

5.32 (a) The key point is that for $K < 0$, area/4π (radial distance)^2

$= \sinh^2\chi/\chi^2 > 1$. Consider a sphere in a submanifold of E^n. Because of E^n's positive metric, the distance from the centre of the sphere to the sphere along a curve in any submanifold of E^n is always larger than the 'true' radius of the sphere, so that for any spherically symmetric submanifold of E^n, area/4π (radial distance)$^2 \leqslant 1$. Therefore the open universe cannot be 'embedded' in any Euclidean space in a manner that preserves its metric. This is true even though the open universe has a positive-definite metric!

(b) Consider the hyperboloid $-t^2 + x^2 + y^2 + z^2 = K^{-1}(<0)$ in Minkowski space. Define the usual spherical coordinates by $x = r \sin\theta \times \cos\phi$, $y = r\sin\theta\sin\phi$, $z = r\cos\theta$, and then the submanifold is $t = (r^2 - K^{-1})^{1/2}$. The metric tensor has components $g_{rr} = -(\partial t/\partial r)^2 + (\partial x/\partial r)^2 + (\partial y/\partial r)^2 + (\partial z/\partial r)^2 = (1 - Kr^2)^{-1}$. Similarly $g_{\theta\theta} = r^2$, $g_{\phi\phi} = r^2\sin^2\theta$. So this hyperboloid is isometric to the open universe.

6.1 Just apply the definition of a tensor.

6.2 Algebra from (6.6), using $\bar{e}_{l'} = \Lambda^m{}_{l'}\bar{e}_m$.

6.3 Trivial.

6.4 In figure 6.1 the basis vector $\bar{e}_\theta$ is unchanged by transport in the θ-direction: $\nabla_{\bar{e}_\theta}\bar{e}_\theta = 0$. So from (6.6) we conclude $\Gamma^\phi{}_{\theta\theta} = \Gamma^\theta{}_{\theta\theta} = 0$. In figure 6.2 imagine that at D the vector is $\bar{e}_\phi$. Then its direction does not change as it is transported in the θ-direction, but its scale does, since $\bar{e}_\phi \to 0$ at the poles but the transported vector does not. In fact, $|\bar{e}_\phi| = \sin\theta$, so we conclude $\bar{e}_\phi(\theta + \delta\theta) - \bar{e}_\phi(\theta) = \bar{e}_\phi(\theta)\ [\delta(\sin\theta)/\sin\theta]$, or $\nabla_{\bar{e}_\theta}\bar{e}_\phi = \cot\theta\,\bar{e}_\phi$, or $\Gamma^\phi{}_{\phi\theta} = \cot\theta$, $\Gamma^\theta{}_{\phi\theta} = 0$. The corresponding derivatives in the $\bar{e}_\phi$ direction are harder to calculate because the curves $\theta = $ const are not great circles. Consider a point P at $\phi = 0$, $\theta = \theta_0$. In figure A.1 we have drawn a neighborhood of P the way it looks to someone standing there. The curve $\theta = \theta_0$ through P is not a straight line locally, but is tangent to a great circle at P which does

Fig. A.1

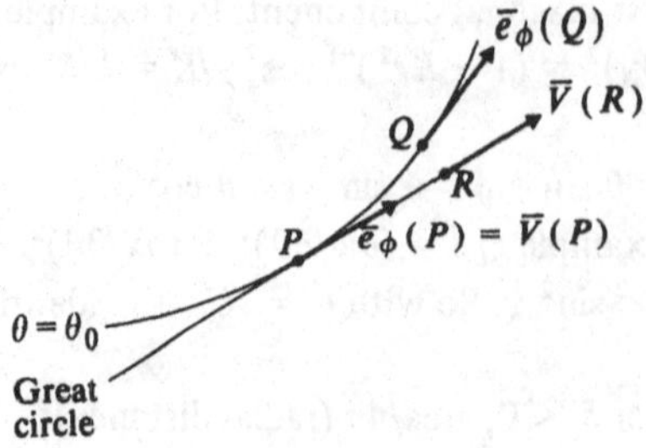

look straight. Consider the point Q at $\phi = \delta\phi \ll 1$ on the circle $\theta = \theta_0$. To calculate, say, $\nabla_{\bar{e}_\phi}\bar{e}_\phi$ at P, we need the difference $\bar{e}_\phi(Q) - \bar{e}_\phi(P)$, to first order in $\delta\phi$. Our coordinates are not Cartesian, so we cannot simply subtract components of vectors at different points. Instead, we construct a vector field $\bar{V}$ on the great circle by parallel-transporting $\bar{e}_\phi(P)$ along it, i.e. by keeping it tangent and of the same length. The point R with coordinate $\phi = \delta\phi$ is very near Q, their separation being $O(\delta\phi^2)$. Therefore to first order we can use $\bar{V}(R)$ as a reference and approximate $\bar{e}_\phi(Q) - \bar{e}_\phi(P) \approx \bar{e}_\phi(Q) - \bar{V}(R)$, which we *can* calculate simply by subtracting components. Now, in our coordinates $\bar{e}_\phi$ has components (0, 1) everywhere. So we must construct $\bar{V}$. The great circle is the intersection of the sphere $x^2 + y^2 + z^2 = 1$ with the plane $x = z \tan\theta_0$. In spherical coordinates this gives the equation of the great circle to be $\sin\theta = \sin\theta_0(1 - \cos^2\theta_0 \sin^2\phi)^{-1/2}$. Using ϕ as a parameter on it gives its tangent vector $(\mathrm{d}\theta/\mathrm{d}\phi, 1) = (\sin\theta_0 \cos\theta_0 \times \sin\phi/(1 - \cos^2\theta_0 \sin^2\phi), 1)$. At $\phi = 0$ (point P) this equals $\bar{e}_\phi(P)$, and at $\phi = \delta\phi$ (point R) it is $(\sin\theta_0 \cos\theta_0\, \delta\phi, 1)$ to first order. Again to first order, this has the same length as $\bar{e}_\phi(P)$, so this is in fact $\bar{V}(R)$. We therefore get $\nabla_{\bar{e}_\phi}\bar{e}_\phi = \lim_{\delta\phi\to 0}(\bar{e}_\phi(Q) - \bar{V}(R))/\delta\phi = (-\sin\theta_0 \times \cos\theta_0, 0)$. It follows that $\Gamma^\theta{}_{\phi\phi} = -\sin\theta\cos\theta$, $\Gamma^\phi{}_{\phi\phi} = 0$. A similar calculation for $\nabla_{\bar{e}_\theta}\bar{e}_\theta$ gives $\Gamma^\phi{}_{\theta\phi} = \cot\theta$, $\Gamma^\theta{}_{\theta\phi} = 0$.

6.5 $\langle\tilde{\omega}^j, \bar{e}_k\rangle = \delta^j{}_k$, so $\langle\nabla_i\tilde{\omega}^j, \bar{e}_k\rangle = -\langle\tilde{\omega}^j, \nabla_i\bar{e}_k\rangle = -\Gamma^j{}_{ki}$. Therefore $\nabla_i\tilde{\omega}^j$ is a one-form whose kth component is $-\Gamma^j{}_{ki}$, as in (6.8).

6.6 Use exercise 6.5 and follow the steps leading to equation (6.10).

6.7 As above.

6.8 In a coordinate basis, $[\bar{e}_i, \bar{e}_j] = 0$.

6.9 Need to show it is linear in its arguments. For example, $\mathbf{T}(\ ; f\bar{U}, \bar{V}) = \nabla_{f\bar{U}}\bar{V} - \nabla_{\bar{V}}(f\bar{U}) - [f\bar{U}, \bar{V}] = f(\nabla_{\bar{U}}\bar{V} - \nabla_{\bar{V}}\bar{U} - [\bar{U}, \bar{V}]) - \bar{U}\nabla_{\bar{V}}(f) + \bar{U}£_{\bar{V}}(f)$. The terms involving derivatives of f cancel, so $\mathbf{T}$ is indeed linear on that argument.

6.10 Similar to the proof of exercise 6.9.

6.11 One needs only to show it for scalars (where they are both the same) and vectors (which is the content of (6.13)). Since both derivative rules generalize to tensors of higher rank via (6.3), the result follows for all tensors.

6.12 Obvious.

6.13 Algebra.

6.14 (a) In a coordinate basis $[\bar{e}_i, \bar{e}_j] = 0$. $\nabla_i\nabla_j\bar{e}_k = \nabla_i\Gamma^l{}_{kj}\bar{e}_l = \Gamma^l{}_{kj,i}\bar{e}_l + \Gamma^l{}_{kj}\nabla_i\bar{e}_l = \Gamma^l{}_{kj,i}\bar{e}_l + \Gamma^l{}_{kj}\Gamma^m{}_{li}\bar{e}_m$. Antisymmetrizing on i and j and relabelling some indices gives the result.

(b) Obvious.

(c) (6.23a) is obvious from (6.19), but (6.23b) must be proved. In normal coordinates at P, $\Gamma^i{}_{jk}(P) = 0$ so that $R^l{}_{kij}(P) = \Gamma^l{}_{kj,i} - \Gamma^l{}_{ki,j}$. Then $3R^l{}_{[kij]} = \Gamma^l{}_{kj,i} - \Gamma^l{}_{ki,j} + \Gamma^l{}_{ik,j} - \Gamma^l{}_{jk,i} + \Gamma^l{}_{ji,k} - \Gamma^l{}_{ij,k} = 0$ because $\Gamma^i{}_{jk} = \Gamma^i{}_{kj}$.

(d) The four indices mean that we begin with n^4 components. Equation (6.23a) are $n^2 \cdot \frac{1}{2}n(n+1)$ separate relations, since l and k are free, while there are $\frac{1}{2}n(n+1)$ symmetric pairs (ij). (This is the same as the number of independent components of a symmetric $n \times n$ matrix.) Constraint (6.23b) is entirely independent of (6.23a) since it involves only $R^l{}_{k[ij]}$. There are $n(n-1)(n-2)/3!$ different antisymmetric triplets (kij) in this equation; with the free choice of l this gives the third term in (6.24).

6.15 In normal coordinates at P, $R^l{}_{kij;m} = R^l{}_{kij,m} = \Gamma^l{}_{kj,im} - \Gamma^l{}_{ki,jm}$; the first term is symmetric in (im), the second in (jm), so both vanish in (6.25). The Jacobi structure follows from (6.19).

6.16 (a) $\bar{e}_r = \cos\theta\, \bar{e}_x + \sin\theta\, \bar{e}_y$, $\bar{e}_\theta = -r\sin\theta\, \bar{e}_x + r\cos\theta\, \bar{e}_y \Rightarrow \nabla_{\bar{e}_\theta}\bar{e}_r = -\sin\theta\, \bar{e}_x + \cos\theta\, \bar{e}_y = r^{-1}\bar{e}_\theta \Rightarrow \Gamma^\theta{}_{r\theta} = r^{-1}$, $\Gamma^r{}_{r\theta} = 0$. Others are derived the same way.

(b) $V^r{}_{;r} = V^r{}_{,r}$. $V^r{}_{;\theta} = V^r{}_{,\theta} - rV^\theta$. $V^\theta{}_{;r} = V^\theta{}_{,r} + V^\theta/r$. $V^\theta{}_{;\theta} = V^\theta{}_{,\theta} + V^r/r$. $V^i{}_{;i} = V^r{}_{,r} + r^{-1}V^r + V^\theta{}_{,\theta} = r^{-1}(rV^r)_{,r} + V^\theta{}_{,\theta}$.

6.17 (a) $(\pounds_{\bar{V}}\tilde{\omega})_{i\ldots k} = \tilde{\omega}_{i\ldots k,l}V^l + \omega_{i\ldots k}V^l{}_{,l} = \omega_{i\ldots k;l}V^l + \omega_{i\ldots k}V^l{}_{;l}$. Then $V^l{}_{;l} = \text{div}_{\tilde{\omega}}\bar{V}$ for all $\bar{V}$ if and only if $\nabla\tilde{\omega} = 0$.

(b) $\omega_{i\ldots k;l} = \omega_{i\ldots k,l} - \Gamma^m{}_{ml}\omega_{i\ldots k} = (f_{,l} - f\Gamma^m{}_{ml})\epsilon_{i\ldots k}$.

6.18 (a) $\nabla_{\bar{V}}[\mathsf{g}(\bar{A},\bar{B})] = (\nabla_{\bar{V}}\mathsf{g})(\bar{A},\bar{B}) + \mathsf{g}(\nabla_{\bar{V}}\bar{A},\bar{B}) + \mathsf{g}(\bar{A},\nabla_{\bar{V}}\bar{B}) = (\nabla_{\bar{V}}\mathsf{g})(\bar{A},\bar{B})$. This vanishes for all $\bar{A}, \bar{B}, \bar{V}$ if and only if $\nabla\mathsf{g} = 0$.

(b) Equation (6.29) implies $g_{ij,k} = \Gamma^l{}_{ik}g_{lj} + \Gamma^l{}_{jk}g_{il}$. Adding the various gs together as on the right-hand side of (6.30) gives the result.

6.19 We need to show that (6.30) implies $\Gamma^j{}_{jk} = (\ln|g|^{1/2})_{,k}$. This is easy once we have proved that $g_{,k} = g^{ji}g_{ij,k}$. (This is true for the determinant of any matrix.) We begin with $g = \epsilon^{i\ldots k}g_{1i}\ldots g_{nk}$ $= g_{1i}(\epsilon^{ij\ldots k}g_{2j}\ldots g_{nk})$. From this we can see that if we *define* g^{i1} $= g^{-1}\epsilon^{ij\ldots k}g_{2j}\ldots g_{nk}$ then we have $1 = g_{1i}g^{i1}$. Moreover, $g_{2i}g^{i1} = 0$ by the antisymmetry of ϵ. We therefore have an explicit expression for the *inverse* of g_{ij}: $g^{ij} = ((n-1)!\,g)^{-1}\epsilon^{il\ldots m}\epsilon^{jk\ldots r}g_{lm}\ldots g_{rm}$. Now we use exercise 4.12(b), $g = \epsilon^{il\ldots m}\epsilon^{jk\ldots r}g_{ij}g_{lk}\ldots g_{mr}/n!$, to deduce $g_{,a} = n\epsilon^{il\ldots m}\epsilon^{jk\ldots r}g_{ij,a}g_{lk}\ldots g_{mr}/n! = g_{ij,a}g^{ji}$. The rest is easy.

6.20 Replace commas by semicolons in equation (3.37) and use $\nabla_i g_{jk} = 0$.

6.21 Algebra.

6.22 (b) One cannot simply subtract from equation (6.24) a new number representing the number of constraints in (6.33), because these new constraints may not all be independent of the previous ones. Instead, we begin anew and concentrate on pairs of indices. By (6.23a) there are $\frac{1}{2}n(n-1)$ independent pairs, so (6.33) implies R is a symmetric matrix in a space of $\frac{1}{2}n(n-1)$ dimensions, i.e. has $\frac{1}{2}[n(n-1)/2] \times [n(n-1)/2+1] = n(n-1)(n^2-n+2)/8$ independent components. Now, (6.23b) represents fewer independent constraints than before. Given all triples (kij) ($n(n-1)(n-2)/3!$ possible sets), does every choice of l give an independent constraint? No, because (6.23a) and (6.33) enable us to manipulate $3R_{l[kij]} = R_{lkij} + R_{ljki} + R_{lijk} = R_{klji} + R_{kilj} + R_{kjil} = 3R_{k[lji]}$. This means that we have new information only for every set of *four* indices, all different: $n(n-1)(n-2) \times (n-3)/4!$ constraints in all. The result is as given in the problem.

6.23 (a) From (6.33) and (6.23a).

6.24 A geodesic is defined by equation (6.16a). From this it is easy to see that $\nabla_{\bar{U}}\, \mathrm{g}(\bar{U}, \bar{U}) = 0$, so that if $\mathrm{g}(\bar{U}, \bar{U}) \neq 0$ small variations in the path will not change the sign of $\mathrm{g}(\mathrm{d}\bar{x}/\mathrm{d}\lambda, \mathrm{d}\bar{x}/\mathrm{d}\lambda)$. We will do the case of a space-like geodesic first. By the calculus of variations, $\delta \int (g_{ij}\dot{X}^i\dot{X}^j)^{1/2}\, \mathrm{d}\lambda$ when the path is changed by $\delta x^i(\lambda)$ is, to first order, $-\frac{1}{2}\int \delta x^i(\lambda)\,(-2\,\mathrm{d}(g_{ij}\dot{X}^j)/\mathrm{d}\lambda + g_{jk,i}\dot{X}^j\dot{X}^k)\,(g_{ij}\dot{X}^i\dot{X}^j)^{-1/2}\,\mathrm{d}\lambda = \int \delta x_i(\lambda)\,(\ddot{X}^i + \Gamma^i{}_{jk}\dot{X}^j\dot{X}^k)\,(g_{ij}\dot{X}^i\dot{X}^j)^{-1/2}\,\mathrm{d}\lambda$. (Here dots stand for $\mathrm{d}/\mathrm{d}\lambda$.) This implies a geodesic is an extremum. The proof for time-like geodesics is nearly identical. The case of a null geodesic is handled by breaking the integral into segments, in each of which the variation is time-like or space-like or null.

6.25 (a) $D_\mu\psi = \nabla_\mu\psi - \mathrm{i}A_\mu\psi$. $D_\mu(\mathrm{e}^{\mathrm{i}\phi(x)}\psi) = \nabla_\mu(\mathrm{e}^{\mathrm{i}\phi}\psi) - \mathrm{i}(A_\mu + \nabla_\mu\phi)\psi = \mathrm{e}^{\mathrm{i}\phi}(\nabla_\mu\psi - \mathrm{i}A_\mu\psi) = \mathrm{e}^{\mathrm{i}\phi}D_\mu\psi$.

(b) $D_\mu D_\nu\psi = \nabla_\mu\nabla_\nu\psi - \mathrm{i}A_\nu\nabla_\mu\psi - \mathrm{i}(\nabla_\mu A_\nu)\psi - \mathrm{i}A_\mu\nabla_\nu\psi - A_\mu A_\nu\psi$. Since $\nabla_\mu\nabla_\nu\psi = \nabla_\nu\nabla_\mu\psi$ we have $[D_\mu, D_\nu]\,\psi = -\mathrm{i}(\mathrm{d}\tilde{A})\psi$.

NOTATION

Symbols are listed with the page(s) on which they are defined or extended. For conventions on the placement of indices and the summation convention, see §2.21 and §2.26.

Types of tensor

$\bar{a}$, 13
$\tilde{b}$, 49, 116
F, 57
$\mathbf{r}$, 44
$\mathfrak{w}$, 129

Special tensors, etc.

g, g_{ij}, 65
g^{-1}, g^{ij}, 67
$\mathsf{R}(\ ,\ )$, $R^i{}_{jkl}$, 211
$\epsilon_{i\ldots j}$, $\epsilon^{i\ldots j}$, 128
δ_{ij}, 17
$\delta^{i\ldots j}_{k\ldots l}$, 130
$\Gamma^i{}_{jk}$, 205
$\Lambda^{i'}{}_{j}$, 61

Special spaces

E^n, 15
$GL(n, C)$, 100
$GL(n, R)$, 41
$L(n)$, 67
$O(n)$, 66
R, 2
R^n, 1
S^n, 26
$SO(n)$, 29, 98
$SU(n)$, 100
T_P, 34
$T^*{}_P$, 53
TM, 36
T^*M, 53
$U(n)$, 100

Operations

$[\ ,\ ]$, 11, 14, 102
${}^*\mathsf{T}$, ${}^*\tilde{B}$, 125, 126
$\times$, 38
$\otimes$, 59
$\int \tilde{\omega}$, 122
A^{T}, A^{-1}, 17
$\tilde{\mathrm{d}}$, 53, 134
$\det(A)$, 18
$\mathrm{div}_{\tilde{\omega}}$, 148, 149
$\exp(\ )$, 43
$f: M \to N$, 6
$f: x \mapsto y$, 6
$g \circ f$, 7
L_g, 92
$\pounds_{\bar{V}}$, 76
$\mathrm{tr}(A)$, 18
$V^j{}_{,i}$, 135
$V^j{}_{;i}$, 207
$\tilde{\alpha}|_W$, 120

$\tilde{\alpha} \wedge \tilde{\beta}$, 118
∇, $\nabla_{\bar{U}}$, ∇_i, 203–6
$\varpi(\bar{U})$, 50, 119
$\langle \varpi, \bar{U} \rangle$, 50

Miscellaneous
C^k, C^∞, 8
C^ω, 10
C^n_p, 117
$f(S)$, $f^{-1}(T)$, 6
$\binom{N}{N'}$, 57
$\partial(f_1, \ldots, f_n)/\partial(x_1, \ldots, x_n)$, 9
∂U, 144
1–1, 6

INDEX

accelerated observer, 71
adiabatic, 182, 183
adjoint transformation, 107
affine connection, 76, 201–222, 218, 221; linearity, 204; metric, 216; not a tensor field, 205; symmetric, 207, 208, 216; torsion, 207
affine parameter, 209, 218
analytic function, 9, 10
analytic manifold, 26
angular momentum operators, 85; as Killing vector fields, 89
annihilator, 154
annulling a form, 120, 153, 166
antiderivation, 134
antisymmetric tensor, 115
area, 113; tensor, 115
atlas, 25
automorphism, 107
axial eigenvalue, 90, 91
axial harmonics: scalar, 90; vector, 90
axial symmetry, 87, 89; and group theory, 91
axial vector, 126

basis, 14; Cartesian, 66; commutation coefficients, 211; coordinate, 34, 47, 56; dual, 55; for one-forms, 55; globally orthonormal, 69; handedness, 99, 115, 121, 132; Lorentz, 66; noncoordinate, 44, 211; orthonormal, 66, 68, 132, 184; transformation of, 60
basis transformation, 60; matrix of, 61
basis-invariance, 63
Betti number, 152
Bianchi identities, 212
big bang, 199
bijection, 7
boundary of a region, 144

canonical energy, 173
canonical momentum, 173, 221
canonical transformation, 168
Caratheodory's theorem, 165

Cartan, E., 113
chart, 24
Christoffel symbols, 205, 218; of two-sphere, 206; transformation law, 205
closed form, 138; locally exact, 138, 140; on a sphere, 150; sufficient condition for exactness, 142
closed ideal, 163
closed set of forms, 158
cofactor, 18
cohomology, 139, 142, 150; and simply connectedness, 152; classes of *n*-sphere, 151, 152
commutator, 44; of covariant derivatives, 210; of operators, 11; of vector fields, 44
compatibility: of connection and differential structure, 204; of connection and metric, 215, 216; of connection and volume form, 215, 216; of metric and volume form, 216
components: *see* under individual types of tensor
composition of two maps, 7
configuration space, 174
congruence, 43, 151
connection: affine, *see* affine connection; metric, 216
connection one-form, 220
conservation law, 89, 149, 158, 163, 171, 173, 182
conservative, 168
conserved quantities and Killing vector fields, 171
continuity: function, 7, 8; group, 12; map, 7; one-form, 52; space, 1, 2
continuum mechanics, 58
contraction, 50, 56, 59; of vector with form, 119
coordinate transformation, 31, 62, 255
coordinates, 23, 24; curvilinear, 70; normal, 210
Copernican principle, 189
cosmological principle, 189

cosmology, 161, 186–199; big bang, 199; closed, 198, 199; expanding, 199; flat, 198, 199; homogeneous, 187, 189; isotropic, 187, 189; open, 198, 199; standard model, 199
cotangent bundle, 53, 174, 175
covariant derivative, 184, 203; commutator of, 210; exponentiation of, 212; Jacobi identity for, 212; Leibniz rule for, 204; of a scalar field, 204; of a tensor field, 206; of a vector field, 203
cross-product, 125, 126; triple, 131
cross-section, 38, 53, 220
curl, 136, 176
curvature, 68, 214
curve, 30, 153; congruence, 43, 73, 151; geodesic, 208; parameter, 30; spacelike/timelike/null, 70; tangent vector, 32

density, scalar, 129; tensor, 129; weight, 129
diffeomorphism, 30, 73, 92, 188
differentiability class: of a function, 8; of a one-form, 53, 56
differential form, 68, 117; annulling, 120, 153, 166; closed, 138, 140, 142; degree of, 117; exact, 138, 140, 142, 163, 170; field, 120; independent components of, 117; integrable, 165; Lie derivative of, 142; restriction of, 120, 153, 188; sectioning, 120
differential forms: annihilator of, 154; closed ideal, 163; closed set, 158; commutation rule, 119; complete ideal, 154, 163; differential ideal, 155; surface-forming, 155
directional derivative, 33, 53
Dirac bra and ket, 51
Dirac delta function, 51
direct product, 59
directional derivation, 33, 53
discrete set, 2
distance function, 1, 3–5
distance on a manifold, 68, 70
distribution (of vector fields), 83
distribution (over functions), 51
divergence, 137, 176, 196; in spherical coordinates, 148; of a p-vector, 149; of a vector field, 147
divergence theorem, 147
domain of an operator, 11
dual, 186, 196; double, 128, 133; metric, 128; of a p-form, 125; of a p-vector, 125
duality of vectors and one-forms, 50

eigenvalue, 19, 65, 96, 99
eigenvector, 19
Einstein summation convention, 56, 62
electromagnetism, 175–181; as a gauge theory, 219–222; charge, 178; charge and topology, 179–80; current four-vector, 177, 178; Faraday tensor, 176, 221; gauge transformation, 180, 219, 221; magnetic monopoles, 178, 180; one-form potential, 180, 221, 219; plane waves, 181; polarization, 181; vector potential, 180
entropy, 163, 166, 167, 182, 183
equation of continuity, 149
equation of state, 163
equivalence class, 150
equivalence relation, 150
Ertel's theorem, 185
Euclidean space, 15, 65, 79, 121, 161, 176, 182, 197, 198, 214, 218
Euclidean vector algebra, 68, 125, 131, 132
Euclidean vector calculus, 70, 136, 137, 142, 147, 148, 182
exact form, 138, 163, 170; on a sphere, 149, 150
expansion, 186
exponential map, 167, 210, 215
exponentiation: of an operator, 43; of covariant derivative, 212
exterior derivative, 134; commutes with Lie derivative, 143; Leibniz rule for, 134

Faraday tensor, 176, 221
fiber bundle, 35, 36, 40, 53, 183, 220; base manifold, 36; cross-section, 38, 53, 220; fiber, 36, 40; global properties, 38; globally trivial, 38; locally trivial, 38, 40; principal, 42; projection, 37, 40; structure group, 40;
fixed-point theorem, 39, 151, 196
flat manifold, 172, 214
fluid: multicomponent, 164; perfect, 181; single-component, 163
fluid dynamics, 149
foliation, 81, 108, 188; leaf of, 81
frame bundle, 42
Frobenius' theorem: 81, 82, 153–155, 163, 165, 166
function, 5, 30: analytic, 9, 10; as a tensor, 58, 64; continuous, 7, 8; differentiable, 31
function space, 51
fundamental theorem of calculus, 134

Galilean spacetime, 182, 183, 218
gauge: curvature two-form, 221; theories, 219; transformation, 180, 219, 221
gauge-covariant derivative, 220
Gauss' theorem, 147, 148

geodesic curve, 208; affine parameter of, 209, 218; extremal length of, 218
geodesic deviation, 213, 214
geodesic equation, 208
geodesically complete manifold, 210
geometrical object, 62
$GL(n, C)$, 100
$GL(n, R)$, 41, 95; acting on R^n, 99; Lie algebra of, 97
gradient, 53; not naturally a vector, 54; of a vector field, 204; vector, 69, 71, 89
Grassmann algebra, 118, 125
group, 11–13; abstract, 105; homomorphism, 13; isomorphism, 12; isotropy, 188, 191, 192, 194, 197; Lie, *see* Lie group; Lorentz, 67, 192; permutation, 12; realization, 106; representation, 106; rotation, 29, 98, 99; translation, 12

Hamiltonian: equations, 167; function, 170, 171, 173, 174; vector field, 168, 170, 171
handedness, *see under* basis
harmonic oscillator, 153, 159
Hausdorff property, 3
Helmholtz circulation theorem, 185
Hermitian conjugate, 100
Hilbert space, 51, 108
homeomorphism, 40
homogeneous 187, 188, 191
homomorphism, 102,104
hypersurface, 79, 167, 178, 182, 183, 188; *see also* submanifold

ideal: closed, 163; complete, 154, 163; differential, 155
identity transformation, 17
image, 6
index notation, 73
index raising and lowering, 68
indices: antisymmetrized, 166; placement of, 56, 62
infinitely differentiable, 9
inner automorphism, 107
inner product, 15, 51, 71
integrability conditions, 138, 155
integration, 134; and orientation, 123; change of variables, 9; of forms, 121; of functions, 121
into (map), 7
invariance of a tensor field, 86; *see also* Killing vector field
inverse function theorem, 9, 35, 49
inverse image, 6
inverse map, 6
inversion, 99
isometry, 188, 190
isospin, 37
isotropic, 187, 189

Jacobi identity: for covariant derivatives, 212; for Lie derivatives, 78; for vector fields, 47; for Poisson brackets, 170
Jacobian, 9, 122, 129, 132
Jacobian matrix, 9, 35, 63

Killing vector field, 88, 108, 171, 188, 192, 195, 216; nonexistence for a general metric, 190
Killing's equation, 216
Klein–Gordon equation, 174, 219
Kronecker delta, 17

Lagrangian function, 167, 174
left-invariant vector field, 93–95
Leibniz rule: for covariant derivative, 204; for exterior derivative, 134; for Lie derivative, 78, 79
Levi–Civita symbol, 128; and determinants, 131; and p-delta symbol, 130; products of, 130
Lie algebra, 47, 92, 93, 95, 97, 101, 155, 170; Abelian, 93, 105; dimension of, 88; of invariant vector fields, 87; of isometry group, 190; of $SO(3)$, 100; of $SU(n)$, 101; structure constants, 93; subalgebra, 108
Lie bracket, 45, 75, 157; closure of a set of, 158; picture of, 46
Lie derivative, 68, 173, 182, 184; as a partial derivative, 78; commutes with exterior derivative, 143; components of, 79; of a differential form, 142; of a one-form, 79; of a scalar, 76; of a tensor, 79; of a vector field, 77;
Lie dragging, 73, 90, 209, 213; of a function, 73; of a one-form, 70, 79; of a region, 144; of a vector field, 75
Lie group, 12, 29, 87, 92, 188; abelian, 105; adjoint representation, 189; and invariance, 92; component of the identity, 97; covering group, 102; disconnected, 97; left and right translations, 92; Lie algebra of, 93; one-parameter subgroup of, 94, 95; simply connected, 102; tangent bundle, 94; transitive action, 188
Lie subalgebra, 192
linear combination, 14
linear independence, 14
linear transformation, 16; components of, 1
Liouville's theorem, 171
Lorentz frame, 70, 71
Lortentz group, 67, 192
Lorentz transformation, 67, 219

manifold, 23; analytic, 26; complex, 51; differentiable, 23; differential structure, 201; dimension of, 23; geodesically complete, 210; homogeneous, 188, 191; isotropic, 189; maximally symmetric, 191, 193; orientable, 41, 132, 175; Riemannian, 169, 201–222; simply connected, 102, 152; symplectic, 171
many-to-one, 6
map, 5; composition, 7; continuous, 7; dual, 125; exponential, 167, 210, 214; generated by a congruence, 73; into, 7; inverse, 6; many-to-one, 6; of p-forms to n-vectors, 125; one-to-one, 6; onto, 7; stereographic, 27
matrix, 50; anti-Hermitian, 100; block-diagonal form, 96; canonical form, 96, 97, 99; cofactors of, 18; determinant, 18, 131; diagonal, 65; inverse, 17, 19; nonsingular, 17; orthogonal, 65; singular, 17; trace, 19, 101; transpose, 17
Maxwell identities, 164, 169
Maxwell's equations, 175–181
metric, 36, 38, 51, 54, 76, 113, 169, 178, 184, 201, 215; Euclidean, 65; Minkowski, 66, 71; signature, 66, 69
metric connection, 216
metric dual, 128, 133
metric tensor, 64, 187, 214; as a map of vectors to one-forms, 67; canonical form, 66; indefinite, 66, 133; inverse, 67; negative-definite, 66; positive-definite, 66
metric tensor field, 68, 108; local flatness, 69; signature, 68
metric volume element, 132, 148, 160, 193
Minkowski metric, 66
Minkowski space, 70, 79, 179, 214, 217, 218, 219, 220
Möbius band, 39; not orientable, 121, 124; structure group, 42
multilinearity, 57

n-tuple, 1, 15, 23
n-vector, 125
neighborhood, 1; generalized, 5
Newtonian gravity, 187
norm, 14; Euclidean, 15
normal coordinates, 210
normal one-form, 147

$O(n)$, 66, 98; as a disconnected group, 98; dimension of, 99
one-form, 49; as a tensor, 58; basis, 55; components, 55; coordinate basis, 56; dual basis, 55; field, 52; normal, 147; picture of, 53
one-parameter subgroup, 94, 95; infinitesimal generator, 96
one-to-one, 6
onto (map), 7
open set in R^n, 2, 3
operator, 10; as a tensor, 58; domain of, 11; extension of, 11; multiplicative, 211
orientability, 41, 132, 175; internal, 121
orientation: external, 123; internal, 123
orthogonal group, *see* $O(n)$
outer product, 59, 64

p-delta symbol, 130; contraction of, 130
parallel transport, 202, 204; around a loop, 213
parallelism, 76, 201; global, 202, 214
parallelogram rule, 15
partial differential equations, 134, 137, 152; integrability conditions, 138, 155
permutation group, 12
phase space, 28, 168, 174; volume form, 171
Poincaré lemma, 140
Poisson bracket, 170
principle of equivalence, 218
principle of mediocrity, 189
principle of minimal coupling, 218
product space, 38
proper distance and time, 70
pseudo-norm, 15, 71

quantum mechanics, 51; commutation relations, 85
quotient space, 150

R, 2
R^n, 1
realization, 106; faithful, 106; group adjoint, 107; of $SO(3)$, 106; principal, 106; progressive, 106; retrograde, 106
relativity, 38, 188; general, 217, 218, 219; special, 66, 68, 70, 219
representation, 106; abstract, 110; adjoint, 107; irreducible, 109; of $SO(3)$, 106, 109
restriction of a form, 120, 153, 188
Ricci scalar, 217
Ricci tensor, 217
Riemann tensor, 211, 213; number of independent components, 212, 217
Riemannian geometry, 76, 184, 187
Riemannian manifold, 169, 201–222
right-invariant vector field, 94
rotation group, *see* $SO(3)$
rule of linearity, 16

scalar, 64; covariant derivative of, 204
sectioning a form, 120
shear, 186

signature, 66, 68, 69
similarity transformation, 19, 65
simply connected manifold, 102; and cohomology, 152
SO(2), 92
SO(3), 29, 98, 99, 100, 111, 160, 190, 192; double-valued representations of, 111; fundamental representation of, 111; global topology, 104; not simply connected, 105; realization of, 106; representation of, 106, 109
SO(*n*), 98, 189, 192
Sorkin's model for charge, 180
spherical harmonics, 108, 109; as eigenfunctions, 110; completeness of, 109, 110, 160; vector, 160, 161, 195
spherical symmetry, 92, 108, 192
spinor, 111
square intergrability, 10
Stoke's theorem, 144, 147, 149, 150, 159
stress tensor, 58
SU(2), 111; diffeomorphic to the three-sphere, 103; double covering of *SO*(3), 104; global topology, 104; Lie algebra of, 101
SU(*n*), 100, 101
subgroup, 12
submanifold, 79, 80, 90, 153, 157, 158, 188; one-forms of, 81; tangent vectors of, 80; *see also* hypersurface
symmetry: axial, 87, 89
symmetry group, Eucliedan, 66
symplectic: form, 171, 174; inner product, 171, 172, 174; manifold, 171

tangent bundle, 36, 37, 174; of a Lie group, 94; structure group, 41
tangent one-form, 53, 54
tangent space, 34
tangent vector, 32; space of, 33
Taylor expansion, 9, 167
tensor, 57; antisymmetric, 115; completely antisymmetric part of tensor, 116; completely antisymmetric tensor, 115; components of, 59; order, 68; symmetric, 64, 116; type of, 57
tensor equation, 64
tensor field, 57; components of, 59
tensor operation, 64
tensor product, 59
thermodynamics, 163–167; Caratheodory's theorem, 165; composite system, 164, 165; entropy, 166, 167, 182, 183; first law, 163; second law, 163, 166
topographical map, 53
topological space, 3
topology, 51; and charge 179, 180; global, 1, 28, 139, 140, 214; induced, 3, 5; local, 1; of *SO*(3) and *SU*(2), 104
torsion tensor, 207, 209
trace, 19, 101
transformation law: for basis one-forms, 61; for basis vectors, 61; for Christoffel symbols, 205; for one-form components, 61; for tensor components, 62; for vector components, 61
translation, 219
translation group, 12
transpose, 17

U(1)-bundle, 220
U(*n*), 100
unit matrix, 17
unitary group, 100
universe, *see* cosmology

valid tensor equation, 177
vector, 32–34; as a tensor, 58; components, 14, 32, 55; contravariant, 50, 62; covariant, 50, 62
vector field, 34, 42; components of, 34; covariant derivative, 203; gradient of, 204; Hamiltonian, 168, 170, 171; integral curves, 42
vector space, 13; as a manifold, 28, 172; complex, 16, 51; dimension, 14
vector subspace, 14
volume element, 113
volume form, 121, 171, 175, 177, 182, 201, 215, inverse, 126
vorticity, 182, 184; conservation of, 185

wedge product, 117; components of, 118; forms of arbitrary degree, 118
Weyl tensor, 217
Wheeler's model for charge, 179

zero-form, 117

9 780521 2988

Printed in the United States
By Bookmasters